Synthesis, Application and Biological Evaluation of Chemical Organic Compounds

Synthesis, Application and Biological Evaluation of Chemical Organic Compounds

Editors

Iliyan Ivanov
Stanimir Manolov

Basel • Beijing • Wuhan • Barcelona • Belgrade • Novi Sad • Cluj • Manchester

Editors

Iliyan Ivanov
Department of Organic
Chemistry
University of Plovdiv
Plovdiv
Bulgaria

Stanimir Manolov
Department of Organic
Chemistry
University of Plovdiv
Plovdiv
Bulgaria

Editorial Office
MDPI
St. Alban-Anlage 66
4052 Basel, Switzerland

This is a reprint of articles from the Special Issue published online in the open access journal *Processes* (ISSN 2227-9717) (available at: www.mdpi.com/journal/processes/special_issues/chemical_organic_compounds).

For citation purposes, cite each article independently as indicated on the article page online and as indicated below:

Lastname, A.A.; Lastname, B.B. Article Title. *Journal Name* **Year**, *Volume Number*, Page Range.

ISBN 978-3-0365-9067-7 (Hbk)
ISBN 978-3-0365-9066-0 (PDF)
doi.org/10.3390/books978-3-0365-9066-0

© 2023 by the authors. Articles in this book are Open Access and distributed under the Creative Commons Attribution (CC BY) license. The book as a whole is distributed by MDPI under the terms and conditions of the Creative Commons Attribution-NonCommercial-NoDerivs (CC BY-NC-ND) license.

Contents

About the Editors . . . . **vii**

Preface . . . . **ix**

Stanimir Manolov and Iliyan Ivanov
Special Issue: Synthesis, Application, and Biological Evaluation of Chemical Organic Compounds
Reprinted from: *Processes* **2023**, *11*, 2802, doi:10.3390/pr11092802 1

Stanimir Manolov, Dimitar Bojilov, Iliyan Ivanov, Gabriel Marc, Nadezhda Bataklieva and Smaranda Oniga et al.
Synthesis, Molecular Docking, Molecular Dynamics Studies, and In Vitro Biological Evaluation of New Biofunctional Ketoprofen Derivatives with Different N-Containing Heterocycles
Reprinted from: *Processes* **2023**, *11*, 1837, doi:10.3390/pr11061837 5

Chul Joong Kim, Bimal Kumar Ghimire, Seon Kang Choi, Chang Yeon Yu and Jae Geun Lee
Sustainable Bioactive Composite of *Glehnia littoralis* Extracts for Osteoblast Differentiation and Bone Formation
Reprinted from: *Processes* **2023**, *11*, 1491, doi:10.3390/pr11051491 23

Omar M. Khubiev, Victoria E. Esakova, Anton R. Egorov, Artsiom E. Bely, Roman A. Golubev and Maxim V. Tachaev et al.
Novel Non-Toxic Highly Antibacterial Chitosan/Fe(III)-Based Nanoparticles That Contain a Deferoxamine—Trojan Horse Ligands: Combined Synthetic and Biological Studies
Reprinted from: *Processes* **2023**, *11*, 870, doi:10.3390/pr11030870 44

A. F. M. Motiur Rahman, Ahmed H. Bakheit, Shofiur Rahman, Gamal A. E. Mostafa and Haitham Alrabiah
Procainamide Charge Transfer Complexes with Chloranilic Acid and 2,3-Dichloro-5,6-dicyano-1,4-benzoquinone: Experimental and Theoretical Study
Reprinted from: *Processes* **2023**, *11*, 711, doi:10.3390/pr11030711 57

Ming-Yu Chou, Chi-Pei Ou Yang, Wen-Ching Li, Yao-Ming Yang, Yu-Ju Huang and Ming-Fu Wang et al.
Evaluation of Antiaging Effect of Sheep Placenta Extract Using SAMP8 Mice
Reprinted from: *Processes* **2022**, *10*, 2242, doi:10.3390/pr10112242 76

Stanimir Manolov, Iliyan Ivanov, Dimitar Bojilov and Paraskev Nedialkov
Synthesis, In Silico, and In Vitro Biological Evaluation of New Furan Hybrid Molecules
Reprinted from: *Processes* **2022**, *10*, 1997, doi:10.3390/pr10101997 87

Mirza Nadeem Ahmad, Sohail Nadeem, Raya Soltane, Mohsin Javed, Shahid Iqbal and Zunaira Kanwal et al.
Synthesis, Characterization, and Antibacterial Potential of Poly(*o*-anisidine)/BaSO$_4$ Nanocomposites with Enhanced Electrical Conductivity
Reprinted from: *Processes* **2022**, *10*, 1878, doi:10.3390/pr10091878 100

Theodora-Venera Apostol, Mariana Carmen Chifiriuc, Laura-Ileana Socea, Constantin Draghici, Octavian Tudorel Olaru and George Mihai Nitulescu et al.
Synthesis, Characterization, and Biological Evaluation of Novel *N*-{4-[(4-Bromophenyl) sulfonyl]benzoyl}-L-valine Derivatives
Reprinted from: *Processes* **2022**, *10*, 1800, doi:10.3390/pr10091800 113

Eslam B. Elkaeed, Reda G. Yousef, Hazem Elkady, Ibraheem M. M. Gobaara, Aisha A. Alsfouk and Dalal Z. Husein et al.
The Assessment of Anticancer and VEGFR-2 Inhibitory Activities of a New 1*H*-Indole Derivative: In Silico and In Vitro Approaches
Reprinted from: *Processes* **2022**, *10*, 1391, doi:10.3390/pr10071391 . . . . **140**

Antonio Rosales Martínez and Ignacio Rodríguez-García
Hydrogen/Deuterium Exchange in Ambrox Could Improve the Long-Term Scent and Shelf Life of Perfumes
Reprinted from: *Processes* **2023**, *11*, 2358, doi:10.3390/pr11082358 . . . . **165**

About the Editors

Iliyan Ivanov

Prof. Iliyan Ivanov earned his M.Sc. in 1990 and his Ph.D. in 2003 from the University of Plovdiv, Bulgaria. His extensive research focuses on the practical application of the α-amidoalkylation reaction and the development of innovative approaches to synthesize N-heterocyclic compounds. Prof. Ivanov is best known for pioneering an alternative method for synthesizing isoquinoline analogues, a breakthrough that has since found successful applications in the creation of novel β-carbolines, quinazolinones, isochromans, and various N- and O-heterocyclic derivatives. With a rich academic background and a wealth of experience, Prof. Ivanov has authored over 70 publications in the field of heterocyclic compound synthesis.

Stanimir Manolov

Dr. Stanimir Manolov earned his B.Sc. in Computer Chemistry in 2008, followed by his M.Sc. in 2009, culminating in a Ph.D. in Organic Chemistry in 2015, all from the esteemed University of Plovdiv, Bulgaria. His academic journey has been marked by a fervent exploration of eco-friendly synthesis and the foundational principles of green chemistry, with a specific emphasis on pioneering innovative synthetic methods utilizing heterogeneous catalysts. Dr. Manolov's research extends to the realm of profen hybrids, a distinct class of compounds renowned for their potential applications in the pharmaceutical field. His work in this domain intricately weaves together precise in vitro biological assessments and cutting-edge in silico calculations to craft and refine compounds with augmented bioactivity. With an unwavering commitment to sustainable chemistry practices, Dr. Manolov actively contributes to the advancement of environmentally friendly processes for synthesizing biologically active compounds.

Preface

In the ever-evolving landscape of scientific inquiry, the synthesis, application, and biological evaluation of chemical organic compounds stand at the forefront of human innovation. The intricate dance of atoms and molecules, governed by the laws of chemistry, has led to breakthroughs that span industries, from pharmaceuticals to materials science and beyond. The profound impact of these compounds on our daily lives is undeniable, from the medications that heal us to the materials that build our world.

The journey embarked upon within the pages of this reprint is one of exploration, discovery, and application. It takes us from the fundamental principles of organic synthesis, where atoms are meticulously assembled into complex structures, to the practical applications of these compounds across diverse fields. It showcases how the synthesis of novel compounds has paved the way for innovations in drug development, materials science, and environmental science, among others. Moreover, it unravels the crucial process of biological evaluation, where the impact of these compounds on living organisms is meticulously studied to unlock their therapeutic potential.

As we embark on this journey, it is essential to acknowledge the collaborative spirit that drives scientific progress. The work presented here is a testament to the synergy between academia and industry, where researchers from various backgrounds come together to exchange ideas and push the boundaries of what is possible. It is also a tribute to the mentors, educators, and institutions that nurture the next generation of scientists, providing them with the tools and knowledge to continue this extraordinary voyage of discovery.

In closing, "Synthesis, Application, and Biological Evaluation of Chemical Organic Compounds" is a comprehensive resource for those immersed in the field and an accessible introduction for the curious mind. We hope that this reprint will inspire readers to appreciate the profound impact of chemical organic compounds on our world and, perhaps, ignite the spark of curiosity that leads to subsequent groundbreaking discoveries.

Welcome to the world of synthesis, application, and biological evaluation—a world where the boundaries of possibility are defined by the molecules we create, study, and harness for the betterment of humanity.

Iliyan Ivanov and Stanimir Manolov
Editors

Editorial

Special Issue: Synthesis, Application, and Biological Evaluation of Chemical Organic Compounds

Stanimir Manolov * and Iliyan Ivanov *

Department of Organic Chemistry, Faculty of Chemistry, University of Plovdiv, 24 Tsar Assen Str., 4000 Plovdiv, Bulgaria
* Correspondence: manolov@uni-plovdiv.net (S.M.); iiiliyan@abv.bg (I.I.)

Citation: Manolov, S.; Ivanov, I. Special Issue: Synthesis, Application, and Biological Evaluation of Chemical Organic Compounds. *Processes* 2023, 11, 2802. https://doi.org/10.3390/pr11092802

Received: 1 September 2023
Accepted: 19 September 2023
Published: 21 September 2023

Copyright: © 2023 by the authors. Licensee MDPI, Basel, Switzerland. This article is an open access article distributed under the terms and conditions of the Creative Commons Attribution (CC BY) license (https:// creativecommons.org/licenses/by/ 4.0/).

This Special Issue of *Processes*, entitled "Synthesis, Application, and Biological Evaluation of Chemical Organic Compounds", gathers the most recent work of leading researchers in a single forum. The contents include a broad range of synthesized or new organic compounds and nanostructures, the extraction of active components of plant and animal origin, the in silico analysis of such compounds, and biological assessment using various methodologies. In one of the presented studies, the authors [1] report the synthesis of chitosan/Fe(III)/deferoxamine nanoparticles. The developed nanoparticles have demonstrated exceptional antibacterial activity in vivo and in vitro, outperforming the conventional drugs ampicillin and gentamicin. Furthermore, the authors assert that the nanoparticles are risk-free regarding their potential toxicity. They discovered that adding iron ions to the chitosan matrix improves the ability of the resultant nanoparticles to damage the integrity of microbe membranes when compared to pure chitosan. The addition of deferoxamine to the produced nanoparticles significantly increases their ability to destroy the bacterial membrane. Ahmad and co-workers [2] reported the synthesis of poly(*o*-anisidine)/BaSO$_4$ nanocomposites via the oxidative polymerization of *o*-anisidine monomer with BaSO$_4$ filler for the composite materials' possible antibacterial capabilities. To attain optimal and controllable characteristics in the nanocomposites, the BaSO$_4$ filler ratio was varied between 1% and 10% with respect to the matrix. The FTIR measurements demonstrated a considerable interaction between POA and barium sulfate, as well as good UV-visible behavior of absorption. By altering the percentage load of the BaSO$_4$ filler, the conducting characteristics may be controlled. Furthermore, distinct bacterial strains, *Pseudomonas aeruginosa* and *Staphylococcus aureus*, were employed to assess the antibacterial activity of the POA/BaSO$_4$ nanocomposites. For *Staphylococcus aureus* and *Pseudomonas aeruginosa*, the highest inhibition zones of 0.8 and 0.9 mm were achieved using 7% and 10%, respectively.

In this Special Issue, a few articles have been published reporting the synthesis, full characterization, and in vitro and in silico biological assessment of novel compounds. A very interesting article was presented by Apostol and her colleagues [3]. They describe the synthesis and design of new compounds with an *L*-valine fragment and a 4-[(4-bromophenyl)sulfonyl]phenyl skeleton, which belong to *N*-acyl-α-amino acids, 4*H*-1,3-oxazol-5-ones, 2-acylamino ketones, and 1,3-oxazoles chemotypes. Antibacterial activity against bacterial and fungal strains, antioxidant activity utilizing DPPH, ABTS, and ferric-reducing power assays, and toxicity against the freshwater cladoceran Daphnia magna Strauss were all assessed. In addition, in silico investigations of the possible antibacterial action and toxicity were carried out by the authors. The findings regarding the antibacterial activity, antioxidant effect, and toxicity experiments, as well as in silico analysis, demonstrated the potential of *N*-{4-[(4-bromophenyl)sulfonyl]benzoyl}-*L*-valine and 2-{4-[(4-bromophenyl)sulfonyl]phenyl}-4-isopropyl-4*H*-1,3-oxazol-5-one for the development of new antimicrobial drugs to combat Gram-positive bacteria, particularly *Enterococcus faecium* biofilm-associated illnesses. Manolov and co-workers [4] reported the synthesis of a series of novel compounds between ketoprofen and nitrogen-containing heterocyclic compounds (piperidine, pyrrolidine, 1,2,3,4-tetrahydroquinoline, and 1,2,3,4-tetrahydroisoquinoline). The compounds were investigated for their anti-inflammatory and antioxidant properties in vitro. The hybrids' lipophilicity was assessed, both theoretically ($cLogP$) and empirically (R_M). The compounds' affinity for human serum albumin was estimated in silico using two software tools, and the stability of the predicted complexes was determined using molecular dynamics analysis. All new hybrids outperformed the reference compound, quercetin, in terms of HPSA activity. The in vitro results were validated by molecular docking. The maximum affinity for albumin is shown by hybrid molecule **3c** and compound **3d**. They are more potent anti-inflammatory agents than their forerunner, ketoprofen, as well as the regularly used ibuprofen. The same scientific group, from Plovdiv, Bulgaria [5], reported the synthesis of new hybrid molecules between divinylene oxide and the same nitrogen-containing heterocyclic compounds mentioned above. The anti-inflammatory, anti-arthritic, antioxidant, reducing, and chelating activities of the synthesized compounds were evaluated. When compared to the utilized ketoprofen (720.57 ± 19.78) standard, the less lipophilic molecules, H2 (60.1 ± 8.16) and H4 (62.23 ± 0.83), had an ATA that was approximately 12 times greater. The inhibition of albumin denaturation resulted in the newly produced hybrids being considered as promising anti-inflammatory medicines, as the expressed values were greater than the ketoprofen standard (126.58 ± 5.00), with the exception of H3 (150.99 ± 1.16). All reported compounds had high activity in terms of in vitro biological activities, making them excellent candidates for possible future medications. Elkaeed and colleagues investigated the anticancer and VEGFR-2 inhibitory properties of a novel 1H-indole derivative in vitro and in silico [6]. A novel 1H-indole derivative was designed to correspond to the reported properties of anti-VEGFR-2-approved drugs. The new compound's inhibitory potential was revealed by a molecular docking experiment that identified the pertinent binding sites. After that, six studies of MD simulation were carried out by the authors for 100 ns to verify the precise binding and optimal energy. Furthermore, MM-GBSA demonstrated flawless binding with a total exact energy of −40.38 Kcal/Mol. Using binding energy decomposition, the MM-GBSA tests identified the important amino acids in the protein–ligand interaction, revealing the diversity of interactions of compound **7** inside the VEGFR-2 enzyme. Because chemical **7** is novel, the authors' DFT experiments were used to optimize its molecular structure. In silico ADMET experiments revealed that compound **7** had a high drug-likeness value. It is interesting to note that compound **7** has a higher experimental in vitro prohibitory capacity than sorafenib, with an IC_{50} value of 25 nM. It is noteworthy that compound **10** showed potent inhibitory actions against two cancer cell lines (MCF-7 and HCT 116) with IC_{50} values of 12.93 and 11.52 µM, revealing good selectivity indices of 6.7 and 7.5, respectively.

Scientists from the Republic of Korea [7] set out to investigate and compare the coumarin-based compounds found in *G. littoralis* extracts, as well as the antioxidant and anti-osteoporotic properties of various *G. littoralis* extracts (leaf and stem, fruit, whole plant, and root extracts) on bone metabolism. In their research, they looked into how *G. littoralis* extract affected the growth and osteoblastic differentiation of MC3T3-E1 osteoblasts. The highest concentrations of scopoletin (53.0 mg/g) and umbelliferone (1.60 mg/g) were found in stem extracts when compared to the other samples. According to the findings presented here, ethanolic extracts of *G. littoralis* are an efficient osteoporosis preventative.

In their article [8], Rahman and colleagues describe the development of charge transfer (CT) complexes between organic and/or bioactive compounds, which is a crucial component in understanding molecule–receptor interactions. They created two novel CT complexes, procainamide–chloranilic acid and procainamide–2,3-dichloro-5,6-dicyano-1,4-benzoquinone, using the electron donor procainamide, electron acceptor chloranilic acid, and 2,3-dichloro-5,6-dicyano-1,4-benzoquinone. The stability of each complex was examined for the first time by utilizing spectroscopic properties such as the formation constant, molar extinction coefficient, oscillator intensity of the ionization potential, dipole moment, and standard free energy. Density functional theory (DFT) calculations were carried out, utilizing the ωB97XD/6-311++G(2d,p) level of theory to comprehend the complexes' non-covalent interactions. Both the DFT-computed interaction energies (ΔIEs) and the Gibbs free energies (ΔGs) matched, as observed experimentally. The DFT results strongly support the experimental findings.

Sheep placenta extract (SPE) is widely used in traditional medicine for its physiological effects, such as its wound-healing, antioxidant, and anti-inflammatory characteristics. However, the antiaging effects of SPE are uncertain. Chou and colleagues [9] studied the effect of SPE on aging using the senescence-accelerated mouse prone 8 (SAMP8) strain. After assessing age index characteristics, such as skin glossiness, spine lordosis, and kyphosis, it was discovered that SPE therapy reduced the aging index. Furthermore, they discovered that biochemical indicators such as lactic acid, glucose, ketone bodies, free fatty acids, tumor necrosis factor-alpha (TNF-α), and interleukin 6 (IL-6) did not change in the SPE-treated experimental group after 13 weeks. They discovered that lipid peroxidation (LPO) was reduced, while catalase and superoxide dismutase (SOD) activity was dramatically elevated in the brain tissues of SPE-treated male and female mice. SPE supplementation decreased the aging index and minimized the oxidative stress brought on by the aging process in mice without creating any harmful effects, suggesting the potential of SPE as a potent antiaging remedy.

Additionally, Rosales Martínez and Rodríguez-García shared their thoughts on how hydrogen/deuterium exchange in ambrox could lengthen the aroma of perfumes and extend their shelf lives [10]. Ambrox is a common ingredient in high-end perfumery since it is a natural marine compound with a wonderful ambergris-like scent. Improving the long-term aroma and shelf life of perfumes is a primary objective in the fragrance industry. To the best of the authors' knowledge, the exchange of hydrogen for deuterium to minimize the volatility of smell components has not yet been researched. In their opinion-type article, they share a new use of deuteration to synthesize deuterated ambrox in order to reduce volatility, improve long-term smell, and extend the shelf-life of perfumes.

The included articles demonstrate the importance and versatility of the search for new potential biologically active structures and their applications. All contributors show an impressive passion and professional abilities in the pursuit of new synthetic methods for producing new structural hybrid molecules, as well as those extracted from plant and animal sources. They use novel, proven, and dependable methods and technology to attain their goals. The new synthetic compounds and extracts described here have the potential to be used as new therapeutic formulations due to the variety of actions they exhibit, some of which are even more active than the standards with which they are compared.

As we reflect on the collective efforts and accomplishments showcased in this special issue, it is clear that the synthesis, application, and biological evaluation of chemical organic compounds represent a thriving and ever-evolving field of research. The potential for discovery remains boundless, and the pursuit of innovative solutions to global challenges continues to inspire scientists worldwide.

We extend our sincere gratitude to all the authors, reviewers, and contributors who have made this special issue possible. Your dedication to advancing the frontiers of knowledge is truly commendable. We hope that the research presented here serves as a source of inspiration and a catalyst for further exploration in the fascinating world of chemical organic compounds.

In the spirit of scientific inquiry and collaboration, we look forward to the continued advancement of this field and the remarkable discoveries that lie ahead. Thank you for joining us on this enlightening journey, and may our collective efforts continue to shape a brighter, more sustainable future.

We appreciate all of the contributors and the Special Issue's Editor-in-Chief, Mr Scott Pan, for their enthusiasm, as well as the editorial staff of *Processes*. Thank you all for the support.

Conflicts of Interest: The authors declare no conflict of interest.

References

1. Khubiev, O.M.; Esakova, V.E.; Egorov, A.R.; Bely, A.E.; Golubev, R.A.; Tachaev, M.V.; Kirichuk, A.A.; Lobanov, N.N.; Tskhovrebov, A.G.; Kritchenkov, A.S. Novel Non-Toxic Highly Antibacterial Chitosan/Fe(III)-Based Nanoparticles That Contain a Deferoxamine—Trojan Horse Ligands: Combined Synthetic and Biological Studies. *Processes* **2023**, *11*, 870. [CrossRef]
2. Ahmad, M.N.; Nadeem, S.; Soltane, R.; Javed, M.; Iqbal, S.; Kanwal, Z.; Farid, M.F.; Rabea, S.; Elkaeed, E.B.; Aljazzar, S.O.; et al. Synthesis, Characterization, and Antibacterial Potential of Poly(*o*-anisidine)/BaSO$_4$ Nanocomposites with Enhanced Electrical Conductivity. *Processes* **2022**, *10*, 1878. [CrossRef]
3. Apostol, T.-V.; Chifiriuc, M.C.; Socea, L.-I.; Draghici, C.; Olaru, O.T.; Nitulescu, G.M.; Visan, D.-C.; Marutescu, L.G.; Pahontu, E.M.; Saramet, G.; et al. Synthesis, Characterization, and Biological Evaluation of Novel N-{4-[(4-Bromophenyl)sulfonyl]benzoyl}-L-valine Derivatives. *Processes* **2022**, *10*, 1800. [CrossRef]
4. Manolov, S.; Bojilov, D.; Ivanov, I.; Marc, G.; Bataklieva, N.; Oniga, S.; Oniga, O.; Nedialkov, P. Synthesis, Molecular Docking, Molecular Dynamics Studies, and In Vitro Biological Evaluation of New Biofunctional Ketoprofen Derivatives with Different N-Containing Heterocycles. *Processes* **2023**, *11*, 1837. [CrossRef]
5. Manolov, S.; Ivanov, I.; Bojilov, D.; Nedialkov, P. Synthesis, In Silico, and In Vitro Biological Evaluation of New Furan Hybrid Molecules. *Processes* **2022**, *10*, 1997. [CrossRef]
6. Elkaeed, E.B.; Yousef, R.G.; Elkady, H.; Gobaara, I.M.M.; Alsfouk, A.A.; Husein, D.Z.; Ibrahim, I.M.; Metwaly, A.M.; Eissa, I.H. The Assessment of Anticancer and VEGFR-2 Inhibitory Activities of a New 1H-Indole Derivative: In Silico and In Vitro Approaches. *Processes* **2022**, *10*, 1391. [CrossRef]
7. Kim, C.J.; Ghimire, B.K.; Choi, S.K.; Yu, C.Y.; Lee, J.G. Sustainable Bioactive Composite of Glehnia littoralis Extracts for Osteoblast Differentiation and Bone Formation. *Processes* **2023**, *11*, 1491. [CrossRef]
8. Rahman, A.F.M.M.; Bakheit, A.H.; Rahman, S.; Mostafa, G.A.E.; Alrabiah, H. Procainamide Charge Transfer Complexes with Chloranilic Acid and 2,3-Dichloro-5,6-dicyano-1,4-benzoquinone: Experimental and Theoretical Study. *Processes* **2023**, *11*, 711. [CrossRef]
9. Chou, M.-Y.; Yang, C.-P.O.; Li, W.-C.; Yang, Y.-M.; Huang, Y.-J.; Wang, M.-F.; Lin, W.-T. Evaluation of Antiaging Effect of Sheep Placenta Extract Using SAMP8 Mice. *Processes* **2022**, *10*, 2242. [CrossRef]
10. Rosales Martínez, A.; Rodríguez-García, I. Hydrogen/Deuterium Exchange in Ambrox Could Improve the Long-Term Scent and Shelf Life of Perfumes. *Processes* **2023**, *11*, 2358. [CrossRef]

Disclaimer/Publisher's Note: The statements, opinions and data contained in all publications are solely those of the individual author(s) and contributor(s) and not of MDPI and/or the editor(s). MDPI and/or the editor(s) disclaim responsibility for any injury to people or property resulting from any ideas, methods, instructions or products referred to in the content.

Article

Synthesis, Molecular Docking, Molecular Dynamics Studies, and In Vitro Biological Evaluation of New Biofunctional Ketoprofen Derivatives with Different N-Containing Heterocycles

Stanimir Manolov [1], Dimitar Bojilov [1], Iliyan Ivanov [1,*], Gabriel Marc [2], Nadezhda Bataklieva [1], Smaranda Oniga [3], Ovidiu Oniga [2] and Paraskev Nedialkov [4]

1. Department of Organic Chemistry, Faculty of Chemistry, University of Plovdiv, 24 "Tsar Assen" Street, 4000 Plovdiv, Bulgaria; manolov@uni-plovdiv.net (S.M.); bozhilov@uni-plovdiv.net (D.B.); nadi_d@abv.bg (N.B.)
2. Department of Pharmaceutical Chemistry, Faculty of Pharmacy, "Iuliu Hațieganu" University of Medicine and Pharmacy, 41 Victor Babeș Street, 400010 Cluj-Napoca, Romania; marc.gabriel@umfcluj.ro (G.M.); ooniga@umfcluj.ro (O.O.)
3. Department of Therapeutic Chemistry, "Iuliu Hațieganu" University of Medicine and Pharmacy, 12 Ion Creangă Street, 400010 Cluj-Napoca, Romania; smaranda.oniga@umfcluj.ro
4. Department of Pharmacognosy, Faculty of Pharmacy, Medical University of Sofia, 2 Dunav Street, 1000 Sofia, Bulgaria; pnedialkov@pharmfac.mu-sofia.bg
* Correspondence: iiiliyan@abv.bg; Tel./Fax: +359-32-261-349

Citation: Manolov, S.; Bojilov, D.; Ivanov, I.; Marc, G.; Bataklieva, N.; Oniga, S.; Oniga, O.; Nedialkov, P. Synthesis, Molecular Docking, Molecular Dynamics Studies, and In Vitro Biological Evaluation of New Biofunctional Ketoprofen Derivatives with Different N-Containing Heterocycles. *Processes* 2023, 11, 1837. https://doi.org/10.3390/pr11061837

Academic Editors: Alexander Novikov and Andrea Temperini

Received: 25 May 2023
Revised: 15 June 2023
Accepted: 15 June 2023
Published: 17 June 2023

Copyright: © 2023 by the authors. Licensee MDPI, Basel, Switzerland. This article is an open access article distributed under the terms and conditions of the Creative Commons Attribution (CC BY) license (https:// creativecommons.org/licenses/by/ 4.0/).

Abstract: Herein, we report the synthesis of four new hybrid molecules between ketoprofen or 2-(3-benzoylphenyl)propanoic acid and N-containing heterocyclic compounds, such as piperidine, pyrrolidine, 1,2,3,4-tetrahydroquinoline, and 1,2,3,4-tetrahydroisoquinoline. The obtained hybrid compounds were fully characterized using ^{1}H- and ^{13}C-NMR, UV-Vis, and HRMS spectra. Detailed HRMS analysis is provided for all novel hybrid molecules. The compounds were assessed for their *in vitro* anti-inflammatory and antioxidant activity. The lipophilicity of the hybrids was determined, both theoretically (*cLogP*) and experimentally (R_M). The affinity of the compounds to the human serum albumin was assessed *in silico* by molecular docking study using two software, and the stability of the predicted complexes was evaluated by molecular dynamics study. All novel hybrids have shown very good *HPSA* activity, statistically close when compared to the reference—quercetin. The molecular docking confirmed the obtained *in vitro* results. Tetrahydroquinoline derivative **3c** and tetrahydroisoquinoline derivative **3d** have the highest affinity for albumin. They show stronger anti-inflammatory action than their predecessor, ketoprofen and the regularly used ibuprofen.

Keywords: ketoprofen; pyrrolidine; piperidine; 1,2,3,4-tetrahydroquinoline; 1,2,3,4-tetrahydroisoquinoline; hybrid molecules; *in vitro* biological activity; molecular docking; molecular dynamics

1. Introduction

N-heterocycles can be found in natural products and drug molecules and are indispensable components in the fields of organic synthesis, medicinal chemistry and materials science. The construction of these N-containing heterocycles by traditional methods usually requires the preparation of reactive intermediates. Through in recent decades, with the rapid advent of transition metal-catalyzed reactions, the synthesis of heterocycles from precursors with inert chemical bonds has become a challenge. Many have developed efficient methods for the preparation of N-heterocyclic compounds, such as aziridines, azetidines, indoles and quinolines and many others [1].

Pyrrolidine **1a** (Figure 1), also known as tetrahydropyrrole, is an organic compound with molecular formula $(CH_2)_4NH$. It is a cyclic secondary amine, also classified as a saturated heterocycle. Many modifications of pyrrolidine are found in natural and synthetic drugs and drug candidates [2].

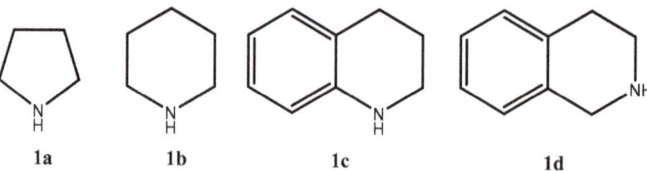

Figure 1. Structural formulas of pyrrolidine **1a**, piperidine **1b**, 1,2,3,4-tetrahydroquinoline **1c**, and 1,2,3,4-tetrahydroisoquinoline **1d**.

The pyrrolidine ring structure (Figure 1) is present in many natural alkaloids, such as nicotine and hygrin [3]. It is found in many medicines, such as procyclidine and bepridil. It also forms the basis for racetam compounds (e.g., piracetam and aniracetam). The amino acids proline and hydroxyproline are structurally derived from pyrrolidine [2]. Pyrrolidine is widely found in the literature as a fragment—part of a number of organic compounds possessing a number of biological activities, such as antiviral, anti-inflammatory, antibacterial, antidepressant, etc. [4].

Piperidine **1b** (Figure 1) is a key heterocyclic amine that is part of a number of pharmaceutical products and natural alkaloids, possessing a diverse range of biological activities [5]. Piperidine fragment **1b** is present in a vast array of natural alkaloids and approved pharmaceuticals [6]. For example, the alkaloid piperine gives a spicy taste in black paper [7]. This six-membered nitrogen-contained fully saturated ring is also found in stimulants (methylphenidate, pipradrol etc.), vasodilators (minoxidil), antipsychotics (droperidol, melperone, etc.), opioids (pethidine, fentanyl, etc.) and many others drugs.

The fully hydrogenated quinoline or 1,2,3,4-tetrahydroquinoline **1c** (Figure 1) is another example of a nitrogen-containing fragment that is widely distributed in nature and the same in pharmaceuticals. The structure of 1,2,3,4-tetrahydroquinoline **1c** is a very common structural motif and is found in numerous biologically active natural products and pharmacologically relevant therapeutic agents [8].

1,2,3,4-tetrahydroisoquinoline **1d** (Figure 1) is another cyclic secondary amine, widely distributed in nature, forming the isoquinoline alkaloids family. Many synthetic and natural molecules containing 1,2,3,4-tetrahydroisoquinoline skeleton have been reported to possess a wide range of pharmacological activities like antibacterial, anti-inflammatory, antifungal, antiviral, antimalarial, and anticancer, among others [9–14].

First synthesized back in 1967, ketoprofen continues to be widely used, and its interest continues to this day. It belongs to the group of nonsteroidal anti-inflammatory drugs (NSAIDs) belonging to the family of propionic derivatives of arylpropionic acid [15].

Ketoprofen (Figure 2) widespread use is mainly due to its antipyretic, analgesic and anti-inflammatory properties due to reversible inhibition of cyclooxygenase 1 and 2 (COX-1 and COX-2), which in turn reduces the production of pro-inflammatory prostaglandin precursors [16].

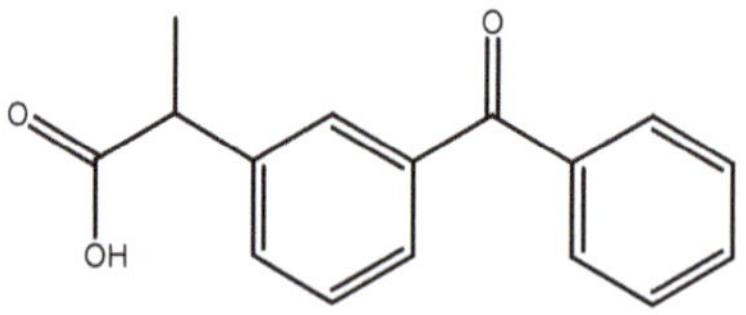

Figure 2. The structural formula of ketoprofen.

Because of the importance of these scaffolds in drug discovery and pharmaceutical chemistry, the development of new methodologies for the synthesis of new hybrids of

N-containing heterocyclic derivatives remains to be a very active field of research, as evidenced by the publication of more than 500 articles in the field in recent years.

Obtaining new hybrid molecules constructed from ketoprofen fragments attached to an N-containing heterocycle is particularly fascinating in order to examine its bio functionality.

2. Materials and Methods

2.1. General

The reagents were purchased from commercial suppliers (Sigma-Aldrich S.A. and Riedel-de Haën, Sofia, Bulgaria) and used as received. A Bruker NEO 400 (400/100 MHz ^{1}H/^{13}C) spectrometer was used for the recording of the NMR spectral data (BAS-IOCCP—Sofia, Bruker, Billerica, MA, USA). All compounds were analyzed in CDCl$_3$ at 400 MHz and 101 MHz for ^{1}H-NMR and ^{13}C-NMR, respectively. Chemical shifts were determined to tetramethylsilane (TMS) (δ = 0.00 ppm) as an internal standard; the coupling constants are given in Hz. Recorded NMR spectra were taken at room temperature (approx. 295 K). Absorbance was measured with a spectrophotometer Camspec M508, Leeds, UK. The high-resolution mass (HRMS) analysis was carried out on a Q Exactive Plus mass spectrometer with a heated electrospray ionization source (HESI-II) (Thermo Fisher Scientific, Inc., Bremen, Germany) coupled with an ultrahigh-performance liquid chromatography (UHPLC) system Dionex Ultimate 3000RSLC (Thermo Fisher Scientific, Inc.) consisting of 6-channel degasser SRD-3600, high-pressure gradient pump HPG-3400RS, autosampler WPS-3000TRS, column compartment TCC-3000RS, and narrow bore Hypersil GOLD™ C18 (2.1 × 50 mm, 1.9 µm) column. For the TLC analysis, precoated 0.2 mm Fluka silica gel 60 plates (Merck KGaA, Darmstadt, Germany) were used.

2.2. Synthesis

2.2.1. Synthesis of 2-(3-benzoylphenyl)propanoyl chloride 2

To ketoprofen (1 mmol, 0.254 g) dissolved in toluene (30 mL), an excess of thionyl chloride (1.2 mmol, 0.087 mL) was added. The reaction mixture was stirred under reflux for two hours. The excess of thionyl chloride and the toluene were removed under reduced pressure. The obtained 2-(3-benzoylphenyl)propanoyl chloride **2** was used without further purification.

2.2.2. Synthesis of Compounds 3a–d

To a solution of corresponding amines **1a–d** (1 mmol) in dichloromethane (30 mL), an equal amount of 2-(3-benzoylphenyl)propanoyl chloride **2** (1 mmol, 0.272 g) was added. After 10 min, triethylamine (1.2 mmol, 0.121 g) was added to the solution. After 30 min, the solution was washed with diluted hydrochloric acid, saturated solution of Na$_2$CO$_3$, and brine. The combined organic layers were dried over anhydrous Na$_2$SO$_4$, and the solvent was removed under reduced pressure. The new hybrid molecules were purified by filtration through short-column chromatography over neutral Al$_2$O$_3$.

3a *2-(3-benzoylphenyl)-1-(pyrrolidin-1-yl)propan-1-one.*

Light-yellow oil, Yield: 97% (0.299 g), R$_f$ = 0.45 (diethyl ether), ^{1}H NMR (400 MHz, CDCl$_3$) δ 7.74–7.69 (m, 2H), 7.65 (t, *J* = 1.8 Hz, 1H), 7.57 (ddd, *J* = 7.6, 1.7, 1.2 Hz, 1H), 7.54–7.48 (m, 2H), 7.43–7.33 (m, 3H), 3.76 (q, *J* = 6.9 Hz, 1H), 3.50–3.38 (m, 2H), 3.38–3.11 (m, 2H), 1.87–1.76 (m, 2H), 1.75–1.69 (m, 2H), 1.42 (d, *J* = 6.9 Hz, 3H) ^{13}C NMR (101 MHz, CDCl$_3$) δ 196.63 (Ph-$\underline{C}$O-Ph), 171.72 ($\underline{C}$=O), 142.09 (Ar), 137.92 (Ar), 137.53 (Ar), 132.54 (Ar), 131.49 (Ar), 130.08 (Ar), 129.26 (Ar), 129.03 (Ar), 128.75 (Ar), 128.30 (Ar), 46.41 ($\underline{C}$H$_2$-N), 46.11 ($\underline{C}$H$_2$-N), 44.73 ($\underline{C}$H), 26.10 ($\underline{C}$H$_2$), 24.17 ($\underline{C}$H$_2$), 20.12 ($\underline{C}$H$_3$). UV λ$_{max}$, MeOH: 276 (ε = 15,270) nm. HRMS Electrospray ionization (ESI) *m/z* calcd for [M+H]$^+$ C$_{20}$H$_{22}$NO$_2$$^+$ = 308.1645, found 308.1638 (mass error Δm = −2.27 ppm), calcd for [M+Na]$^+$ C$_{20}$H$_{21}$NO$_2$Na$^+$ = 330.1465, found 330.1458 (mass error Δm = −2.12 ppm).

3b *2-(3-benzoylphenyl)-1-(piperidin-1-yl)propan-1-one.*

Light yellow oil, Yield: 94% (0.302 g), R$_f$ = 0.85 (diethyl ether), ^{1}H NMR (400 MHz, CDCl$_3$) δ 7.73–7.69 (m, 2H), 7.62 (t, *J* = 1.8 Hz, 1H), 7.56 (dt, *J* = 7.6, 1.4 Hz, 1H), 7.54–7.49

(m, 1H), 7.46 (dt, J = 7.9, 1.5 Hz, 1H), 7.41 (dd, J = 8.2, 6.9 Hz, 2H), 7.36 (t, J = 7.6 Hz, 1H), 3.91 (q, J = 6.9 Hz, 1H), 3.50 (dddd, J = 105.8, 12.9, 6.5, 3.1 Hz, 2H), 3.27 (t, J = 5.5 Hz, 2H), 1.53–1.44 (m, 3H), 1.40 (d, J = 6.8 Hz, 3H), 1.33 (dtd, J = 17.1, 7.6, 4.6 Hz, 2H), 1.01 (dp, J = 12.1, 5.8 Hz, 1H). ^{13}C NMR (101 MHz, CDCl$_3$) δ 196.59 (Ph-$\underline{C}$O-Ph), 171.26 ($\underline{C}$=O), 142.74 (Ar), 138.08 (Ar), 137.54 (Ar), 132.54 (Ar), 131.23 (Ar), 130.05 (Ar), 128.96 (Ar), 128.79 (Ar), 128.59 (Ar), 128.31 (Ar), 46.65 ($\underline{C}$H$_2$-N), 43.19 ($\underline{C}$H$_2$-N), 42.82 ($\underline{C}$H), 26.13 (CH$_2$-$\underline{C}$H$_2$-CH$_2$), 25.52 (CH$_2$-$\underline{C}$H$_2$-CH$_2$), 24.49 (CH$_2$-$\underline{C}$H$_2$-CH$_2$), 20.64 ($\underline{C}$H$_3$). UV λ$_{max}$, MeOH: 276 (ε = 18,500) nm. HRMS Electrospray ionization (ESI) m/z calcd for [M+H]$^+$ C$_{21}$H$_{24}$NO$_2^+$ = 322.1802, found 322.1795 (mass error Δm = −2.17 ppm), calcd for [M+Na]$^+$ C$_{21}$H$_{23}$NO$_2$Na$^+$ = 344.1621, found 344.1614 (mass error Δm = −2.03 ppm).

3c *2-(3-benzoylphenyl)-1-(3,4-dihydroquinolin-1(2H)-yl)propan-1-one.*

Light-yellow oil, Yield: 95% (0.352 g), R$_f$ = 0.48 (petroleum/diethyl ether = 1/1), ^{1}H NMR (400 MHz, CDCl$_3$) δ 7.70–7.62 (m, 2H), 7.57–7.47 (m, 2H), 7.43–7.31 (m, 4H), 7.28 (t, J = 7.5 Hz, 1H), 7.16–6.90 (m, 4H), 4.33–4.21 (m, 1H), 3.85 (s, 1H), 3.50 (dt, J = 13.0, 6.7 Hz, 1H), 2.54–2.41 (m, 1H), 2.23 (s, 1H), 1.81 (dp, J = 13.2, 6.6 Hz, 1H), 1.71 (s, 1H), 1.44 (d, J = 6.9 Hz, 3H). ^{13}C NMR (101 MHz, CDCl$_3$) δ 196.61 (Ph-$\underline{C}$O-Ph), 171.52 ($\underline{C}$=O), 142.17 (Ar), 137.87 (Ar), 137.66 (Ar), 132.59 (Ar), 131.42 (Ar), 130.15 (Ar), 129.34 (Ar), 128.68 (Ar), 128.57 (Ar), 128.42 (Ar), 126.30 (Ar), 125.09 (Ar), 42.90 ($\underline{C}$H), 26.38 ($\underline{C}$H$_2$), 24.12 ($\underline{C}$H$_2$), 20.29 ($\underline{C}$H$_3$). UV λ$_{max}$, MeOH: 273 (ε = 6940) nm. HRMS Electrospray ionization (ESI) m/z calcd for [M+H]$^+$ C$_{25}$H$_{24}$NO$_2^+$ = 370.1802, found 370.1795 (mass error Δm = −1.89 ppm), calcd for [M+Na]$^+$ C$_{25}$H$_{23}$NO$_2$Na$^+$ = 392.1621, found 392.1613 (mass error Δm = −2.04 ppm).

3d *2-(3-benzoylphenyl)-1-(3,4-dihydroisoquinolin-2(1H)-yl)propan-1-one.*

Light-yellow oil, Yield: 95% (0.350 g), R$_f$ = 0.59 (diethyl ether), ^{1}H NMR (400 MHz, CDCl$_3$) δ 7.70–7.61 (m, 3H), 7.59–7.44 (m, 3H), 7.41–7.30 (m, 3H), 7.12–7.01 (m, 3H), 7.00–6.79 (m, 1H), 4.79–4.60 (m, 1H), 4.60–4.33 (m, 1H), 4.05–3.72 (m, 2H), 3.62–3.49 (m, 1H), 2.79–2.31 (m, 2H), 1.43 (dd, J = 6.9, 3.0 Hz, 3H). ^{13}C NMR (101 MHz, CDCl$_3$) δ 196.52 (Ph-$\underline{C}$O-Ph), 172.09 ($\underline{C}$=O), 142.32 (Ar), 138.17 (Ar), 137.43 (Ar), 133.92 (Ar), 133.41 (Ar), 132.58 (Ar), 131.27 (Ar), 131.25 (Ar), 130.04 (Ar), 129.00 (Ar), 128.90 (Ar), 128.77 (Ar), 128.31 (Ar), 126.69 (Ar), 126.55 (Ar), 126.28 (Ar), 47.39 ($\underline{C}$H$_2$), 44.70 ($\underline{C}$H$_2$), 43.39 ($\underline{C}$H), 29.19 ($\underline{C}$H$_2$), 20.65 ($\underline{C}$H$_3$). UV λ$_{max}$, MeOH: 235 (ε = 16,100) nm, 276 (ε = 10,400) nm. HRMS Electrospray ionization (ESI) m/z calcd for [M+H]$^+$ C$_{25}$H$_{24}$NO$_2^+$ = 370.1802, found 370.1795 (mass error Δm = −1.89 ppm), calcd for [M+Na]$^+$ C$_{25}$H$_{23}$NO$_2$Na$^+$ = 392.1621, found 392.1613 (mass error Δm = −2.04 ppm).

2.3. HRMS Analysis

Operating conditions for the HESI source used in a positive ionization mode were: +3.5 kV spray voltage, 320 °C capillary and probe heater temperature, sheath gas flow rate 36 a.u., auxiliary gas flow rate 11 a.u., spare gas flow rate 1 a.u. (a.u. refer to arbitrary values set by the Exactive Tune software) and S-Lens RF level 50.00. Nitrogen was used for sample nebulization and collision gas in the HCD cell. The aliquots of 1 μL of the solutions of the samples (ca. 20 μg mL^{-1}) were introduced into the mass spectrometer through the UHPLC system. Each chromatographic run was carried out isocratically with a mobile phase consisting of water-acetonitrile-methanol-acetic acid (25:50:25:0.2). The solvent flow rate was 300 μL min^{-1}. Full MS—ddMS^2 (Top5) was used as an MS experiment, where in full scan MS, the resolution, automatic gain control (AGC) target, maximum injection time (IT), and mass range were 70,000 (at m/z 200), 3 × 10^6, 100 ms, and m/z 100–500, respectively. The instrument parameters for ddMS^2 scans were as follows: the resolution was 17,500 (at m/z 200), AGC target was 1 × 10^5, maximum IT was 50 ms, loop count was 5, isolation window 2.0 m/z, stepped normalized collision energy (NCE) was set to 10, 20, 60. The data-depended (dd) settings were as follows: maximum ACG target was 5 × 10^4, dynamic exclusion was set to 1 s, preferred peptide match and switched on isotope exclusion were used. Xcalibur (Thermo Fisher Scientific, Waltham, MA, USA) ver. 4.0 was used for data acquisition and processing.

2.4. In Vitro Analysis

2.4.1. Hydrogen Peroxide Scavenging Activity (*HPSA*)

The Manolov et al. method was used to assess the hydrogen peroxide scavenging capability [17]. A 43 mM solution of H_2O_2 was prepared in a potassium phosphate buffer solution (0.2 M, pH 7.4). The analysis of the samples was carried out as follows: in test tubes, 0.6 mL H_2O_2 (43 mM), 1 mL sample/standard with different concentrations (20–1000 µg/mL), and 2.4 mL potassium phosphate buffer solution were mixed. The mixture was stirred and incubated in the dark for 10 min at 37 °C. Absorbance was measured at 230 nm with a spectrophotometer (Camspec M508, Leeds, UK) against a blank solution containing phosphate buffer and H_2O_2 without the sample. Ascorbic acid and quercetin were used as standards. The percentage *HPSA* of the samples was evaluated by comparing it with a blank sample and calculated using the following formula:

$$I, \%(HPSA) = \left[\frac{A_{blank} - (A_{TS} - A_{CS})}{A_{blank}}\right] \times 100$$

where A_{blank} is the absorbance of the blank sample, A_{CS} is the absorbance of the control sample, and A_{TS} is the absorbance of the test sample.

2.4.2. Inhibition of Albumin Denaturation (*IAD*)

In vitro, analysis of anti-inflammatory activity was assessed as inhibition of albumin denaturation (*IAD*). The analysis was performed according to Manolov et al. method [18] with minor modifications. The experiment was performed with human albumin. The solution of albumin (1%) was prepared in distilled water (pH 7.4). The tested compounds/standard were dissolved firstly in PBS, so the final concentration of the stock solution was 1000 µg/mL. Then a series of working solutions with different concentrations (20–500 µg/mL) in PBS were prepared. The reaction mixture was containing 2 mL test sample/standard of different concentrations and 1 mL albumin (1%). The mixture was incubated at 37 °C for 15 min and then heated at 70 °C for 15 min in a water bath. After cooling, the turbidity was measured at 660 nm with a spectrophotometer (Camspec M508, Leeds, UK). Ibuprofen and ketoprofen were used as standards. The experiment was performed three times. Percentage inhibition of albumin denaturation (*IAD*) was calculated against the control. The control sample is albumin with the same concentration dissolved in distilled water.

$$\%IAD = \left[\frac{A_{blank} - A_{sample}}{A_{blank}}\right] \times 100$$

2.4.3. Determination of Lipophilicity as cLlogP

The lipophilicity of the compounds was calculated using the software: ACD/ChemSketch/LogP Predictor v.14.08.

2.4.4. Molecular Docking

The molecular docking of the ligands was performed using AutoDock Vina 1.1.2 (ADV) and AutoDock 4.2 (AD) [19,20] against the human serum albumin (HSA), deposited with the entry code 7JWN in the Protein Data Bank [21]. Two software were chosen for molecular docking because although their names are similar, the principle on which they work is different. Because it is known that molecular docking studies may generate false positive results, the use of two software that works differently can be used for cross-validation of the results [22]. Supplementary, AD possesses an intrinsic tool that performs a rapid clustering analysis to confirm that the best binding pose is one of the most found poses from the total of poses generated.

For each compound, two ligand files were created, one for each R and S isomer using Avogadro 1.2.0 and following the previously reported protocol [23,24].

The preparation of the macromolecule as the target was performed according to the standard procedure previously reported by our group—removal of the co-crystallized molecules, addition of the polar hydrogen atoms and addition of charges [25].

The final preparation of the files of ligands and macromolecules was performed using AutoDockTools 1.5.6 [20].

Four main potential binding sites were targeted in the molecular docking study—Sudlow site I (subdomain IIA), Sudlow II (subdomain IIIA), site III and cleft. The reason for this choice is that, to the best of our knowledge, the other binding sites reported in the literature have a limited role in drug binding, and the sites we have chosen are the most important in drug binding to albumin [26–31].

The search space for each potential binding site was set as a cube, with sides equal to 20 for ADV and 54 for AD (spacing = 0.375). The cartesian coordinates of the center of the search space for each site were set as follows: for site Sudlow 1 x = 30.62, y = 25.50, z = 12.43, for site Sudlow 2 x = 5.95, y = 18.22, z = 21.06, for site 3 x = 30.15, y = 26.98, z = 37.99 and for cleft site x = 20.89, y = 21.74, z = 22.43.

In order to obtain results with higher reproducibility, ADV was requested to generate 20 poses for each ligand in each binding site, while AD was requested to generate 200 poses in order to perform the cluster analysis in the limit of 2 Å root mean square deviation of the coordinates of atoms.

The visualization of the results of the molecular docking study was performed using Chimera 1.10.2 [32].

2.4.5. Molecular Dynamics

To evaluate the stability in a time of the ligand–albumin complexes, some molecular dynamics simulations were performed with GROMACS 2023 [33] using the CHARMM36 force field [34] on a machine running Debian 11. The computer has an Intel Core 7700 K CPU and an NVIDIA RTX 3060 GPU. Their interoperability was based on CUDA 12. Lig-ands parametrization was made using the CgenFF server [34].

The protein and the ligands were placed in an orthorhombic box with a 1 nm gap at the sides and were filled using the TIP3P solvation water model [35]. The preparation of the systems and running of the simulations were made according to the previous works reported [36–38]. Briefly, the main parameters used for carrying out molecular dynamics studies will be presented. The constructed systems were neutralized by adding sodium ions. The energy of the systems was minimized for 5000 steps using the steepest descent method, and the convergence was reached until the maximum force <1000 KJ $mol^{-1} \cdot nm^{-1}$ to remove the steric clashes. The equilibration of the systems was made at NVT and NPT ensembles at 300 K for 100 ps (50,000 steps × 2 fs) each. The production of the simulations was run for 100 ns (50,000,000 × 2 fs) at 300 K and 1 bar with periodic boundary conditions on all axes.

Visualization of the evolution in time of the ligand–albumin complexes during the molecular dynamics simulation was made using VMD 1.9.4 [39].

3. Results and Discussion

3.1. Synthesis

In this article, we report the successful synthesis of four ketoprofen hybrid molecules with different N-containing compounds, such as pyrrolidine **1a**, piperidine **1b**, 1,2,3,4-tetrahydroquinoline **1c**, and 1,2,3,4-tetrahydroisoquinoline **1d** (Figure 1), as shown in Scheme 1. In Scheme 1, we report the synthesis of four novel ketoprofen hybrid compounds. Ning and co-workers report the synthesis and characterization of the fluorinated product of compound **3a** [40]. The second reported by us, molecule **3b**, partially consists of the molecule reported by a group of Turkish scientists [41]. There was no information available in chemical databases for compounds **3c** and **3d**.

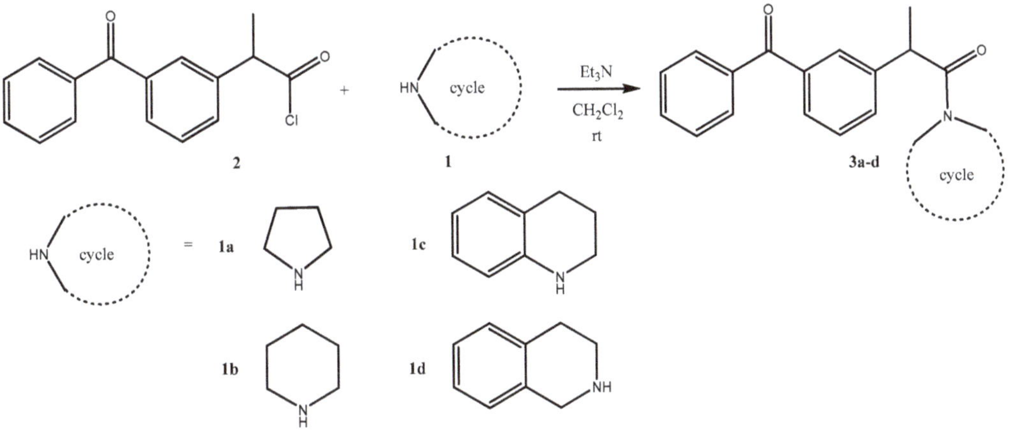

Scheme 1. Synthesis of hybrids **3a–d**.

3.2. Mass Analysis

Compounds **3a–d** have a common pharmacophore—ketoprofen. The only difference in these compounds is in the N-containing heterocyclic rings—pyrrolidine, piperidine, benzo[b]piperidine, and benzo[c]piperidine. We used mass spectrometry to probe their structure. The heterocyclic rings and ketoprofen are linked by a common structural fragment N-C(=O)-C. Under MS/MS conditions, three pathways of molecular ion fragmentation were established. The main fragmentation pathways of compounds **3a–d** include the cleavage of the N–C (path 1), C–C (path 2) bonds and the cleavage of the structural fragment C–C(=O)–C (path 3) connecting the two aromatic cores (Scheme 2).

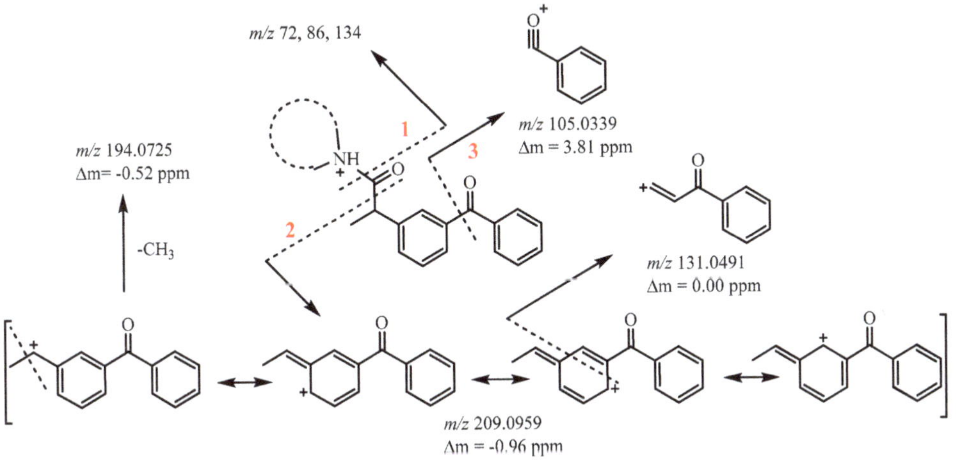

Scheme 2. General fragmentation scheme of ketoprofen derivatives **3a–d**.

Cleavage of the N-C bond (path 1) provides important information about the structure of the heterocyclic ring (m/z 72, 86, 134) (Scheme 2, Figures S14, S16, S18 and S20). Cleavage of the C-C bond (path 2) leads to a resonance-stable aromatic cation (m/z 209) characteristic of ketoprofen, established in our previous studies (Scheme 2) [42]. Under ESI-MS conditions, the same ion undergoes loss of the CH_3 radical to yield an ion with m/z 194. Furthermore, the fragment ion with m/z 131 results from the retrocyclization of the resonance cation m/z 209 (Scheme 2). Fragmentation carried out in route 3 gave an m/z 105 ion resulting from cleavage of the diphenyl ketone fragment (Scheme 2). Compounds **3c** and **3d** are isomers and have ion m/z 370 (Figures S17 and S19). In the fragmentation between the two isomers, a significant difference is observed, which is mainly expressed in the splitting of ion m/z 134 (Figures S18 and S20).

It is the product ion of both isomers, i.e., it corresponds to benzo[*b*]piperidine (1,2,3,4 tetrahydroquinoline nucleus) and benzo[*c*]piperidine (1,2,3,4 tetrahydroisoquinoline nucleus). In the MS/MS experiment, only benzo[*c*]piperidine was found to fragment to the characteristic ion with m/z 117 (Figure S20) [43]. Furthermore, a difference in the m/z 209 and m/z 105 ion intensities was observed in favor of benzo[*b*]piperidine, with the ratio between the two intensities being 2.17 and 1.45 for 209 and 105, respectively (Figures S18 and S20).

3.3. In Vitro Biological Assessment

All synthesized ketoprofen hybrids were tested for their *in vitro* inhibition of albumin denaturation (*IAD*) and hydrogen peroxide scavenging activity (*HPSA*). The obtained *in vitro* results are compared with the *in silico* predictions. The results of the study are presented in Table 1.

Table 1. *In vitro* results of the conducted biological activity.

Compounds	HPSA	IAD	$R_M \pm SD$	cLogP
	IC_{50}, µg/mL			
Ascorbic acid	24.84 ± 0.35	-	-	-
Quercetin	69.25 ± 1.82	-	-	-
Ibuprofen	-	81.50 ± 4.95	1.11 ± 0.010	3.72
Ketoprofen	-	126.58 ± 5.00	1.64 ± 0.006	3.59
3a	85.09 ± 0.24	167.02 ± 8.05	2.07 ± 0.010	4.10
3b	69.98 ± 0.50	130.28 ± 0.41	1.92 ± 0.005	4.61
3c	71.44 ± 0.27	77.18 ± 1.08	1.95 ± 0.008	5.91
3d	59.47 ± 0.36	73.59 ± 1.67	2.25 ± 0.010	5.09

3.3.1. Hydrogen Peroxide Scavenging Activity (*HPSA*)

Reactive oxygen species (ROS) are chemically reactive oxygen radicals and molecules [superoxide ($O_2^{\bullet-}$), hydroxyl ($^{\bullet}OH$), peroxyl ($ROO^{\bullet}$) and alkoxyl ($RO^{\bullet}$), HOCl, ozone (O_3), peroxynitrite ($ONOO^-$), singlet oxygen ($^1O_2$), and H_2O_2]. They are generated as a natural consequence of biological metabolism. Enzyme systems control ROS levels under physiological circumstances. They have been shown to cause harm to critical biological substances, such as phospholipids, proteins, and DNA. It has been established that the harm they inflict contributes to the development of a variety of illnesses (cancer, cardiovascular disease, atherosclerosis, and Alzheimer's disease) [44]. Even in a condition of physiological health, the harmful consequences of accumulating oxygen and its derivatives in the body contribute to a reduction in life expectancy [45].

The results for hydrogen peroxide scavenging activity (*HPSA*) and inhibition of albumin denaturation (*IAD*) are reported as IC_{50} values. Ascorbic acid, quercetin, ibuprofen and ketoprofen were utilized as standards. Because it is a function of R_f and is determined using thin-layer chromatography, R_M is dimensionless.

The current study focused on hydrogen peroxide scavenging. Hydrogen peroxide is a kind of oxidant that is constantly produced in living tissues as a result of a variety of metabolic activities. However, detoxification is critical in order to keep it from entering hazardous reactions like the Fenton reaction [46].

Ascorbic acid and quercetin were employed as controls. They are natural substances having antioxidant characteristics that have been proven. The values we obtained varied from 59.47 µg/mL to 85.09 µg/mL for the synthesized hybrids (Table 1).

The synthesized ketoprofen analogs had modest activity when compared to ascorbic acid (24.84 µg/mL), but when compared to quercetin, hybrid **3d** had greater activity, while hybrids **3b** and **3c** had activity similar to quercetin (Figure 3).

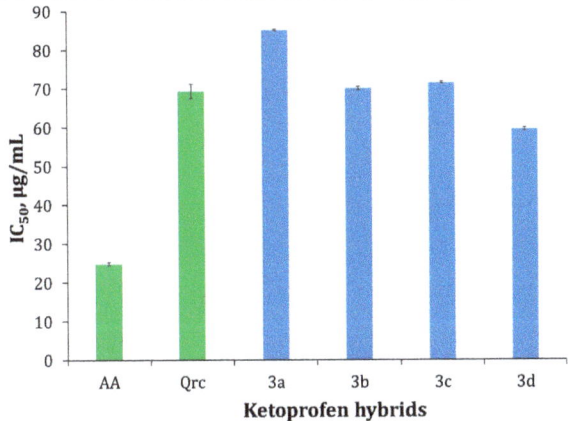

Figure 3. HPSA of the newly synthesized ketoprofen hybrid compounds. Ascorbic acid (AA) and quercetin (Qrc) were utilized as standards. HPSA results are given as IC_{50} (µg/mL).

Despite hydrogen peroxide's low level of reactivity, it can harm cells by creating hydroxyl radicals in them [47]. The most reactive radicals are hydroxy radicals, which are considered to be responsible for some tissue damage produced by inflammation. The superoxide anion radical ($O_2^{-\bullet}$) and H_2O_2 in living organisms are converted into $^\bullet OH$ and $^\bullet O_2$, which cause cell damage. The inflammatory process generates a superoxide anionic radical at the site of inflammation, which is coupled with the creation of other oxidizing species, such as $^\bullet OH$. It has been postulated that hydroxyl radical scavengers can function as protectors by lowering inflammation by reducing prostaglandin synthesis. As a result, removing H_2O_2 is critical in preventing the formation of $^\bullet OH$.

3.3.2. Inhibition of Albumin Denaturation (*IAD*)

Inflammation, according to contemporary thinking, is a beneficial process that occurs as a response to a disruption or sickness. An anti-inflammatory quality of a drug or therapy is the ability to prevent inflammation or swelling. Unlike opioids, which impact the central nervous system, anti-inflammatory medicines, which account for almost half of analgesics, reduce pain by lowering inflammation. Non-steroidal anti-inflammatory drugs (NSAIDs) and steroidal anti-inflammatory drugs (SAIDs) are two types of medications frequently used to treat inflammation. NSAIDs have a number of negative side effects, particularly stomach irritation that can result in gastric ulcers [48].

The essential purpose of derivatizing ketoprofen was to eliminate the irritating and unpleasant impact of the carboxyl group. Therefore, to alter the structure of ketoprofen, we employed a variety of *N*-containing heterocyclic compounds. The novel ketoprofen hybrids were examined for *IAD*. The ketoprofen hybids were compared to ibuprofen and ketoprofen, which are used to prevent inflammatory processes. The purpose of this study was to prevent albumin denaturation. This approach estimates the degree of denaturation resistance of the albumin molecule.

The study data showed that hybrids **3c** (77.18 µg/mL) and **3d** (73.59 µg/mL) possessed a higher degree of albumin protection against denaturation than the standards (Figure 4). The high activity of these compounds (**3c** and **3d**) is attributed to the presence of a 1,2,3,4-tetrahydroquinoline and 1,2,3,4-tetrahydroisoquinoline core, as well as the fact that they are less basic than **3a** and **3b**.

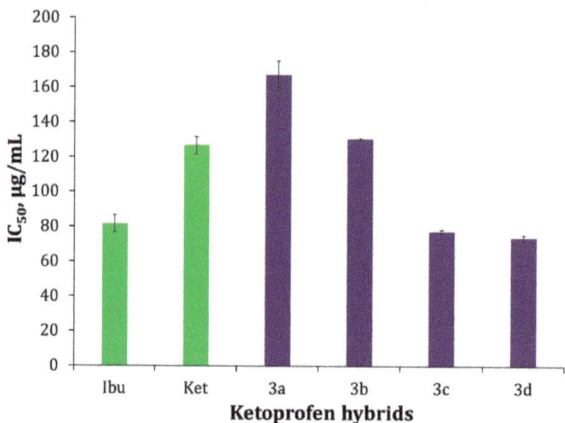

Figure 4. Inhibition of albumin denaturation (*IAD*) of newly created ketoprofen hybrid compounds. Ketoprofen (Ket) and ibuprofen (Ibu) used as benchmarks. Results for *IAD* are displayed as IC_{50} (μg/mL).

3.3.3. Experimental Determination of Lipophilicity (R_M)

Lipophilicity is a determining factor in the absorption of compounds, distribution in the body, passage through various membranes and biological barriers, metabolism, as well as excretion (these are the so-called ADME properties—absorption, distribution, metabolism, and excretion). Lipophilicity is an important factor that informs researchers to both predict and better understand the transport and importance of chemical molecules in physiological and ecological systems. It is of utmost importance for all "candidate" drugs, thanks to its extremely important role in the pharmaceutical and biotechnology industry. Lipophilicity can be measured experimentally or computed.

We used reversed-phase thin-layer chromatography to assess the lipophilicity of the resulting ketoprofen hybrids, a technique described by Hadjipavlou-Litina [49]. The results are shown in Table 1.

3.3.4. Molecular Docking

The results of the molecular docking of the enantiomers of compounds **3a–d** to the four sites of albumin given by AutoDock Vina are presented in Table 2, while the results of AutoDock are presented in Table 3.

Table 2. The results of the docking study were made using AutoDock Vina expressed as the binding affinity of the studied compounds to the four targeted sites of HSA expressed as a variation of Gibbs free energy (ΔG kcal/mol).

Compound	Isomer	Sudlow 1	Sudlow 2	Site 3	Cleft
3a	R	−8.5	−10.3	−11.5	−10.3
	S	−8.3	−10.5	−12.1	−10.0
3b	R	−8.8	−10.5	−11.7	−10.7
	S	−8.5	−11.1	−11.6	−10.2
3c	R	−10.3	−12.2	−12.9	−11.8
	S	−9.8	−12.5	−12.8	−11.2
3d	R	−10.0	−12.1	−12.4	−11.8
	S	−9.2	−12.0	−12.2	−11.6

Table 3. The results of the docking study were made using AutoDock expressed as the binding affinity of the studied compounds to the four targeted sites of HSA expressed as a variation of Gibbs free energy (ΔG kcal/moL) and the clustering analysis of the poses.

Compound	Isomer	Sudlow 1		Sudlow 2		Site 3		Cleft	
		ΔG	%C	ΔG	%C	ΔG	%C	ΔG	%C
3a	R	−8.30	20.5	−9.59	30.0	−10.80	49	−8.81	31.5
	S	−8.27	15.5	−9.74	14.5	−10.60	36	−8.79	26.5
3b	R	−8.81	16.5	−9.82	13.5	−11.47	116	−9.10	41.5
	S	−8.53	4.5	−10.16	24.0	−11.19	68	−9.12	18.5
3c	R	−9.80	18.5	−11.04	27.5	−12.65	104	−10.39	53.0
	S	−9.72	29.0	−11.41	40.5	−11.57	44	−9.95	44.5
3d	R	−9.81	7.0	−11.53	38.5	−12.80	56	−10.58	39.5
	S	−9.69	10.0	−11.44	36.0	−12.20	43	−10.46	49.0

%C: percent of conformations in the same 2Å RMSD cluster of atom coordinates.

The analysis of the results obtained after the docking of the ligands at the four albumin sites using AutoDock Vina shows a major difference in interactions, depending on the size of the heterocycle from the structure of the compounds and less influenced by the type of the isomer.

The compounds from the present series that have the highest affinity for albumin are tetrahydroquinoline derivative **3c** and tetrahydroisoquinoline derivative **3d**. Compounds **3a** and **3b** (pyrrolidine and piperidine, respectively) have, for the four studied sites, a lower affinity than compounds **3c** and **3d**.

Taking into account the affinity for the four sites, compounds exhibit the highest affinity for site 3. The affinity for the site Sudlow 2 is lower than for site 3, while the affinity for the cleft site is lower than for Sudlow 2. From the present series of compounds, they exhibit for Sudlow 1, the lowest affinity, compared to the other three sites.

The results of the docking study using AutoDock for cross-validation of the results given by AutoDock Vina share the same pattern, the affinity of the compounds for albumin being influenced by the size of the heterocycle from the structure of the compounds and less by the type of enantiomer. The most reproducible conformations of the compounds according to the RMSD of the atom coordinates are the ones for compounds **3c** and **3d**. For them, the percentage of conformations from their total conformations generated are found in the same cluster with the top binding conformation is higher than the percentage for **3a** and **3b**. Considering this observation, it can be concluded that compounds **3c** and **3d** have a more repetitive binding in the studied sites compared to compounds **3a** and **3b**, confirming this reproducibility through the repeatability of the conformations found in approximately the same area.

The depiction of the interaction between the ligands and albumin was presented for both enantiomers of a compound, chosen from the present series as the best binding pair of enantiomers on a specific site (Sudlow 2, site 3 and cleft). No depiction was made for any ligand in Sudlow 1 site because the molecular docking study performed on both software indicated that the respective site has a marginal role in the binding of the compounds from the present series.

The Sudlow 2 site, being mainly hydrophobic, comprised of Leu460, Val456, Leu457, Leu453, Leu387, Val433, Leu430 and Val426, easily fits the lipophilic moieties of compounds **3a–d**. Both enantiomers of compounds **3c** and **3d** are involved in a π-π stacking with Tyr411, while the amidic oxygen acts like a hydrogen bond acceptor from the phenol of Tyr411. The ketone of ligands can interact with the sidechain of Asn391 as a hydrogen bond acceptor and with the positively charged sidechain of Arg410 via an ion–dipole interaction.

Visual analysis of the binding poses of enantiomers of **3c** indicates that there are some differences in the binding mode of the two isomers (Figure 5), but mainly the difference between them is minor.

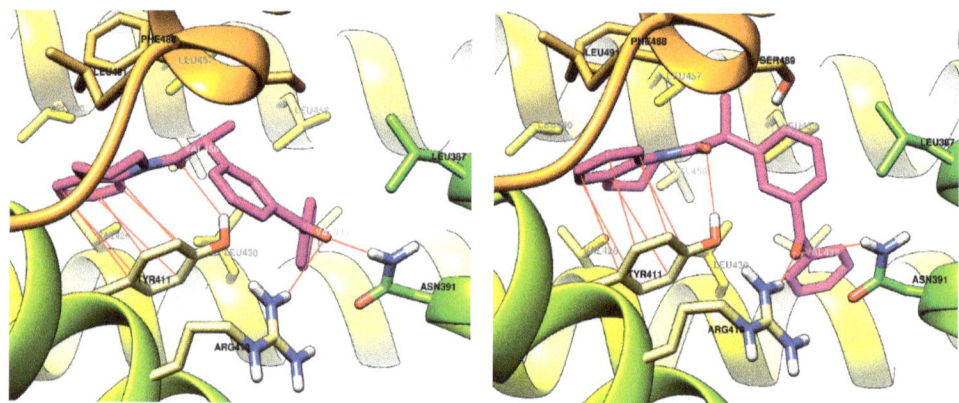

Figure 5. The best binding conformation of enantiomers of compound **3c R** (**left**) and **S** (**right**) in the Sudlow 2 site of albumin. Carbon atoms of **3c** are depicted in magenta.

The binding of **3d** enantiomers in site 3 of albumin is depicted in Figure 6. In both cases, the positively charged sidechain of Arg186 interacts with the benzene ring of tetrahydroisoquinoline fragment through a π-cation interaction. The pair Tyr161-Tyr138 are involved in a double π-π stacking with one of the benzenes of **3d**, for enantiomer **3dR** is expected to appear two supplementary interactions: one of the ketones with the peptide bridge Tyr138-Leu139 and a hydrogen bond between Tyr161 as a donor and the nitrogen atom of **3d** as acceptor.

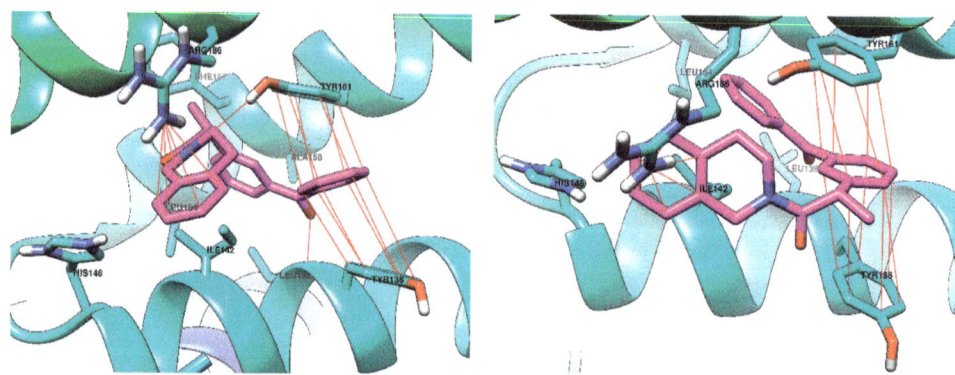

Figure 6. 3d in site 3. The best binding conformation of enantiomers of compound **3d R** (**left**) and S (**right**) in site 3 of albumin. Carbon atoms of **3d** are depicted in magenta.

The binding of **3d** enantiomers in the cleft of albumin is depicted in Figure 7. Both enantiomers are involved with the terminal benzene ring in a π-π stacking interaction with Tyr452. The ketone of **3d** is expected to interact with the sidechain of Asn429 via a hydrogen bond as an acceptor. The amidic oxygen of **3d** is expected to interact with the positively charged sidechain of Lys190 in the case of enantiomer R, while in the case of S enantiomer is expected to interact with the amide bridge between Val455-Val456. The tetrahydroisoquinoline fragment of **3d** of both enantiomers is predicted to interact with some hydrophobic residues, such as Leu463 or Pro421.

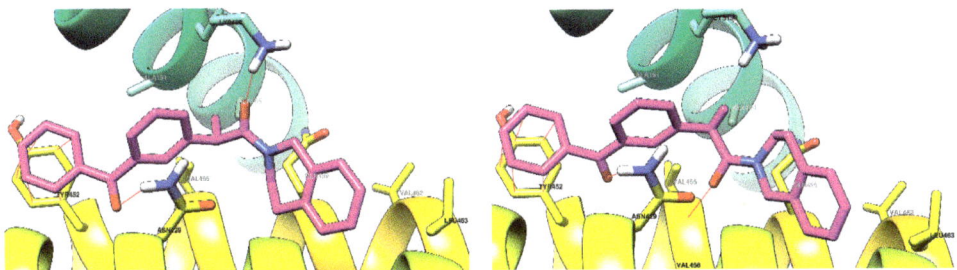

Figure 7. The best binding conformation of enantiomers of compound **3d R (left)** and **S (right)** in the cleft site of albumin. Carbon atoms of **3d** are depicted in magenta.

3.3.5. Molecular Dynamics

The stability of the albumin complexes with the best binding ligands in the molecular docking study performed with AutoDock vina with the variation of the Gibbs free energy in the first half of energies was evaluated during 100 ns of molecular dynamics simulation. According to the results of the molecular docking study, all the potential complexes of the studied compounds in site 3 of albumin were simulated, and the complexes of compounds **3c** and **3d** in the site Sudlow 2 and in the cleft site. None of the complexes were simulated with ligands docked into the Sudlow 1 site due to their low affinity to the respective site.

The stability of the protein–ligand simulated systems in the molecular dynamics study was expressed by calculating the average root-mean-square deviation (RMSD) of the backbone of the protein, the average root-mean-square deviation (RMSD) of the heavy atoms of ligands, the radius of gyration (RG) of the protein and the hydrogen bonds between the ligand and the protein.

An overview of the results of the molecular dynamics study is presented in Table 4 as the RMSD of the backbone of the protein, in Table 5 as the RMSD of the heavy atoms of ligands, Table 6 as RG of the protein, and in Table 7 the evolution of the hydrogen bonds between protein and ligands.

Table 4. The root mean square deviation of the backbone of the protein from the systems evaluated in the molecular dynamics study (nm).

System Evaluated	Sudlow 1	Sudlow 2	Site 3	Cleft
apo + 3a[R]	-	-	0.33	-
apo + 3a[S]	-	-	0.28	-
apo + 3b[R]	-	-	0.30	-
apo + 3b[S]	-	-	0.41	-
apo + 3c[R]	-	0.36	0.25	0.23
apo + 3c[S]	-	0.29	0.30	0.36
apo + 3d[R]	-	0.24	0.36	0.26
apo + 3d[S]	-	0.24	0.34	0.38
apo			0.38	

- not tested.

Table 5. The root mean square deviation of the heavy atoms of the ligands from the systems evaluated in the molecular dynamics study (nm).

System Evaluated	Sudlow 1	Sudlow 2	Site 3	Cleft
apo + 3a[R]	-	-	0.35	-
apo + 3a[S]	-	-	0.26	-
apo + 3b[R]	-	-	0.27	-

Table 5. Cont.

System Evaluated	Sudlow 1	Sudlow 2	Site 3	Cleft
apo + 3b[S]	-	-	1.34	-
apo + 3c[R]	-	0.24	0.32	0.41
apo + 3c[S]	-	0.25	0.23	0.20
apo + 3d[R]	-	0.47	0.59	0.27
apo + 3d[S]	-	0.26	0.62	0.36

- not tested.

Table 6. The radius of gyration of the protein from the systems evaluated in the molecular dynamics study (nm).

System Evaluated	Sudlow 1	Sudlow 2	Site 3	Cleft
apo + 3a[R]	-	-	2.85	-
apo + 3a[S]	-	-	2.78	-
apo + 3b[R]	-	-	2.84	-
apo + 3b[S]	-	-	2.89	-
apo + 3c[R]	-	2.75	2.78	2.77
apo + 3c[S]	-	2.80	2.82	2.77
apo + 3d[R]	-	2.79	2.80	2.81
apo + 3d[S]	-	2.79	2.77	2.78
apo		2.78		

- not tested.

Table 7. The average number of hydrogen bonds between the ligand and the protein in the systems evaluated in the molecular dynamics study (no/ns).

System Evaluated	Sudlow 1	Sudlow 2	Site 3	Cleft
apo + 3a[R]	-	-	0.01	-
apo + 3a[S]	-	-	0.12	-
apo + 3b[R]	-	-	0.12	-
apo + 3b[S]	-	-	0.21	-
apo + 3c[R]	-	0.17	0.46	0.05
apo + 3c[S]	-	0.02	0.19	0.02
apo + 3d[R]	-	0.00	0.06	0.02
apo + 3d[S]	-	0.33	0.37	0.03

- not tested.

It can be seen that compounds **3a** and **3b** (pyrrolidine and piperidine derivatives) gives complexes with albumin, which are less stable than those of compounds **3c** and **3d** (tetrahydroquinoline and tetrahydroisoquinoline hybrids). The increase of the nitrogen ring with a supplementary benzene ring leads to better stabilization of the albumin. On average, taking into account the data available for the complexes resulting from the binding of ligands into site 3, RMSD of the backbone of the protein is higher for complexes with compounds **3a** and **3b** than those with **3c** and **3d** (0.33 nm vs. 0.31 nm). The same trend is identified for the average RMSD of the ligands (0.56 nm vs. 0.44 nm) and the average RG of the protein (0.56 nm vs 0.44 nm). The highest stabilization of the backbone of the protein, expressed as the lowest RMSD of the backbone, was identified as **3cR** into site 3 and cleft, **3dR** into Sudlow 2 and cleft and **3dS** in Sudlow 2.

The changes in the position of the ligand expressed as RMSD of the heavy atoms of the ligand indicate that most of the predicted complexes are stable. Some exceptions were identified, such as compound **3bS** in site 3, **3cR** in cleft, **3dR** in Sudlow 2 and site 3 and **3dS** in site 3, which move significantly from their initial position.

Evaluating the hydrogen bonding between the ligand and the protein **3cR** and **3dS** are the ones that interact more via this type of bond than the other compounds.

Overall, the resumed data indicates that there is no obvious connection between the type of enantiomer of each compound and the parameters evaluated for the resulting complexes to express their stability. The stability of the complexes is influenced simultaneously by the type of nitrogen ring and the type of enantiomer.

Detailed information regarding the evolution of the stability of the complexes of compounds **3c** and **3d** in the Sudlow site 2 are presented in Table S1, of all **3a–d** compounds in site 3 of albumin in tables Tables S2–S4 for compounds **3c** and **3d** in the cleft site of albumin.

The data obtained after simulation of the complexes of compounds **3c** and **3d** into the Sudlow 2 site indicates that both compounds gave complexes with albumin with similar RMSD to the apo form of the protein. The complexes of **3d** (both enantiomers) lead to the best stabilization of the protein in terms of the RMSD of the protein backbone. Into the specified site, **3cR**, **3cS** and **3dS** have the lowest movement, compared to **3dR**, which has a significant change in position during the simulation.

3cR and **3dS** are supposed to have significantly more hydrogen bonds compared to the other enantiomer. Again, this observation confirms the previous observation that the interaction between each enantiomer and protein is influenced by the type of nitrogen ring and the type of enantiomer.

When docked into site 3 of albumin, compounds **3aS** and **3cR** gave the most stable complexes in the present series. Significant instability of the complexes given by the other compounds was identified as follows: **3bS** leaves site 3 after approximately 20 ns, and **3cR** suffers a significant change of position at approximately 35 ns, resulting in continuous changes in the coordinates of the atoms of the backbone of albumin, **3cS** gave a stable complex until 85 ns of simulation, while both enantiomers of **3d** won't reach a convergence, having a continuous movement into the site 3 of albumin.

The complex of **3dR** in site 3 of albumin is the most stable from the current series. A high degree of stability expressed in terms of RMSD of the protein backbone was identified too, but it didn't reach a convergence point during the simulation and the RMSD of the protein backbone was found to slowly and continuously increase until the simulation ended. Anyhow, both R enantiomers gave more stable complexes than S enantiomers when docked into the cleft of albumin. **3cS** moved less into the cleft than **3cR**, but the RMSD of the backbone was similar to the apo form and even decreased during simulation, compared to **3cS**. **3dS** exhibited a significant change in position at approximately 70 ns, affecting a little the RMSD of the backbone of the protein at that time, increasing the RMSD of the protein backbone from complex over the apo form.

4. Conclusions

In conclusion, we have obtained four novel hybrid compounds combining a ketoprofen skeleton and an N-containing hetero ring. The newly discovered molecules have been thoroughly characterized and were subjected to a comprehensive mass spectral analysis. According to the *in vitro* and *in silico* experiments, the hybrid compounds have considerable *HPSA* and *in vitro* anti-inflammatory action as measured by *IAD*. Despite their lipophilic character, compounds **3b-d** have *HPSA* values comparable to quercetin. To neutralize damaging radicals in the cell membrane, lipophilic antioxidants are required. *In vitro*, anti-inflammatory activity was evaluated by *IAD*, as well as by molecular docking and molecular dynamics. Ligand–albumin interactions were demonstrated for both enantiomers of compound **3a-d**, which were chosen from the current series as the best binding pair of enantiomers at a given site (Sudlow 2, site 3, and Cleft). The highest *in vitro* anti-inflammatory efficacy is shown by hybrids **3c** and **3d**, which stabilize the albumin macromolecule by forming ligand–albumin complexes with Sudlow 2, site 3, and cleft. This interaction is responsible for preventing albumin denaturation during inflammatory processes. The stability of the albumin macromolecule is due to the fact that hybrids **3c** and **3d** participate in π-π arrangement with Tyr411 (Sudlow 2), with the pair Tyr161-Tyr138 (site 3), and Tyr452 (cleft). Furthermore, H-bonds are generated with the amide, ketone, and oxygen with the polar amino acid residues implicated in the Sudlow 2, site 3, and cleft structures. The in-silico studies completely confirm our *in vitro* experimental results for anti-inflammatory effects. All of this demonstrates that the hybrid compounds we synthesized inherit ketoprofen's anti-inflammatory capabilities, making them excellent candidates for future medications.

Supplementary Materials: The following supporting information can be downloaded at: https://www.mdpi.com/article/10.3390/pr11061837/s1, Figure S1: ^{1}H-NMR spectrum of compound **3a**; Figure S2: ^{1}H-NMR spectrum of compound **3b**; Figure S3: ^{1}H-NMR spectrum of compound **3c**; Figure S4: ^{1}H-NMR spectrum of compound **3d**; Figure S5: ^{13}C-NMR spectrum of compound **3a**; Figure S6: ^{13}C-NMR spectrum of compound **3b**; Figure S7: ^{13}C-NMR spectrum of compound **3c**; Figure S8: ^{13}C-NMR spectrum of compound **3d**; Figure S9: UV spectrum of compound **3a**; Figure S10: UV spectrum of compound **3b**; Figure S11: UV spectrum of compound **3c**; Figure S12: UV spectrum of compound **3d**; Figure S13: ESI-HRMS of compound **3a**; Figure S14: Mass spectrum of **3a** obtained by positive ion ESI-MS/MS; Figure S15: ESI-HRMS of compound **3b**; Figure S16: Mass spectrum of **3b** obtained by positive ion ESI-MS/MS; Figure S17: ESI-HRMS of compound **3c**; Figure S18: Mass spectrum of **3c** obtained by positive ion ESI-MS/MS; Figure S19: ESI-HRMS of compound **3d**; Figure S20: Mass spectrum of **3d** obtained by positive ion ESI-MS/MS; Table S1: RMSD of protein backbone, RMSD of ligands and RG of protein in the molecular dynamics study when ligands **3c** and **3d** were docked into the Sudlow 2 site; Table S2: RMSD of protein backbone, RMSD of ligands and RG of protein in the molecular dynamics study when ligands **3a** and **3b** were docked into the site 3 of albumin; Table S3: RMSD of protein backbone, RMSD of ligands and RG of protein in the molecular dynamics study when ligands **3c** and **3d** were docked into the site 3 of albumin; Table S4: RMSD of protein backbone, RMSD of ligands and RG of protein in the molecular dynamics study when ligands **3c** and **3d** were docked into the cleft of albumin.

Author Contributions: Conceptualization, S.M. and D.B.; methodology, I.I.; software, G.M., S.M., S.O. and O.O.; validation, S.M., I.I. and D.B.; formal analysis, N.B., S.M., G.M., D.B., S.O. and P.N.; investigation, I.I.; resources, I.I. and O.O; data curation, S.M.; writing—original draft preparation, S.M., D.B. and G.M.; writing—review and editing, S.M. and D.B.; visualization, G.M.; supervision, I.I., S.O. and O.O; project administration, S.M.; funding acquisition, I.I. and O.O. All authors have read and agreed to the published version of the manuscript.

Funding: This research was funded by the Scientific Research Fund of the University of Plovdiv, grant number ФП23-ХФ-005.

Data Availability Statement: The data presented in this study are available in this article and supporting Supplementary Materials.

Conflicts of Interest: The authors declare no conflict of interest.

References

1. Qui, M.; Fu, X.; Fu, P.; Huang, J. Construction of aziridine, azetidine, indole and quinolone-like heterocycles via Pd-mediated C-H activation/annulation strategies. *Org. Biomol. Chem.* **2022**, *20*, 1339–1359. [CrossRef]
2. Kumar, P.; Satbhaiya, S. Chapter 6—proline and proline-derived otganocatalysts in the synthesis of heterocycles. In *Green Synthetic Approaches for Biologically Relevant Heterocycles. Volume 2: Green Catalytic Systems and Solvents Advances in Green and Sustainable Chemistry*, 2nd ed.; Elsevier: Amsterdam, The Netherlands, 2021; pp. 215–251. [CrossRef]
3. Carroll, I. Epibatidine analogs synthesized for characterization of nicotinic pharmacophores—A review. *Heterocycles* **2009**, *79*, 99–120. [CrossRef]
4. Amarouche, L.; Mehdid, M.; Brahimi, F.; Belkhadem, F.; Karmaoui, M.; Othman, A. Synthesis of some 2-substituted pyrrolidine alkaloid analogues: N-benzyl-2-(5-substituted 1,3,4-oxadiazolyl) pyrrolidine derivatives and pharmacological screening. *J. Saudi Chem. Soc.* **2022**, *26*, 101448. [CrossRef]
5. Singh, L.; Upadhyay, A.; Dixit, P.; Singh, A.; Yadav, D.; Chhavi, A.; Konar, S.; Srivastava, R.; Pandey, S.; Devkota, H.; et al. A Review of chemistry and pharmacology of piperidine alkaloids of pinus and related genera. *Curr. Pharm. Biotechnol.* **2022**, *23*, 1132–1141. [CrossRef]
6. Vitaku, E.; Smith, D.; Njardarson, J. Analysis of the structural diversity, substitution patterns, and frequency of nitrogen heterocycles among U.S. FDA approved pharmaceuticals. *J. Med. Chem.* **2014**, *57*, 10257–10274. [CrossRef] [PubMed]
7. Srinivasan, K. Black pepper and its pungent principle-piperine: A review of diverse physiological effects. *Crit. Rev. Food Sci. Nutr.* **2007**, *47*, 735–748. [CrossRef] [PubMed]

8. Sridharan, V.; Suryavanshi, P.; Menéndez, C. Advances in the chemistry of tetrahydroquinolines. *Chem. Rev.* **2011**, *111*, 7157–7259. [CrossRef]
9. Swidorski, J.; Liu, Z.; Yin, Z.; Wang, T.; Carini, D.; Rahematpura, S.; Zheng, M.; Johnson, K.; Zhang, S.; Lin, P.; et al. Inhibitors of HIV-1 attachment: The discovery and structure-activity relationships of tetrahydroisoquinolines as replacements for the piperazine benzamide in the 3-glyoxylyl 6-azaindole pharmacophore. *Bioorg. Med. Chem. Lett.* **2016**, *26*, 160–167. [CrossRef]
10. Liu, X.-H.; Zhu, J.; Zhou, A.; Song, B.-A.; Zhu, H.-L.; Bai, L.-S.; Bhadury, P.; Pan, C.-X. Synthesis, structure and antibacterial activity of new 2-(1-(2-(substituted-phenyl)-5-methyloxazol-4-yl)-3-(2-substituted-phenyl)-4,5-dihy-dro-1H-pyrazol-5-yl)-7-substituted-1,2,3,4-tetrahydroisoquinoline derivatives. *Bioorg. Med. Chem.* **2009**, *17*, 1207–1213. [CrossRef]
11. Kumar, A.; Katiyar, S.; Gupta, S.; Chauhan, P. Synthesis of new substituted triazino tetrahydroisoqiuinolines and β-carbolines as novel antileishmanial agents. *Eur. J. Med. Chem.* **2006**, *41*, 106–113. [CrossRef] [PubMed]
12. Zhu, J.; Lu, J.; Zhou, Y.; Li, Y.; Cheng, J.; Zheng, C. Design, synthesis, and antifungal activities *in vitro* of novel tetrahydroiso-quinoline compounds based on the structure of lanosterol 14α-demethylase (CYP51) of fungi. *Bioorg. Med. Chem. Lett.* **2006**, *16*, 5285–5289. [CrossRef]
13. Tiwari, R.; Singh, D.; Singh, J.; Chhillar, A.; Chandra, R.; Verma, A. Synthesis, antibacterial activity and QSAR studies of 1,2-disubstituted-6,7-dimethoxy-1,2,3,4-tetrahydroisoquinolines. *Eur. J. Med. Chem.* **2006**, *41*, 40–49. [CrossRef]
14. Faheem, R.; Kumar, B.; Sekhar, K.; Chander, S.; Kunjiappan, S.; Murugesan, S. Medicinal chemistry perspectives of 1,2,3,4-tetrahydroisoquinoline analogs—Biological activities and SAR studies. *RSC Adv.* **2021**, *11*, 12254–12287. [CrossRef] [PubMed]
15. Fischer, J.; Robin, G.C. *Analogue-Based Drug Discovery*; John Wiley & Sons: Hoboken, NJ, USA, 2006; p. 520. ISBN 9783527607495.
16. Jachak, S. Cyclooxygenase inhibitory natural products: Current status. *Curr. Med. Chem.* **2006**, *13*, 659–678. [CrossRef] [PubMed]
17. Manolov, S.; Ivanov, I.; Bojilov, D. Synthesis of New 1,2,3,4-Tetrahydroquinoline Hybrid of Ibuprofen and Its Biological Evaluation. *Molbank* **2022**, *2022*, M1350. [CrossRef]
18. Manolov, S.; Ivanov, I.; Bojilov, D. Microwave-assisted synthesis of 1,2,3,4-tetrahydroisoquinoline sulfonamide derivatives and their biological evaluation. *J. Serb. Chem. Soc.* **2021**, *86*, 139–151. [CrossRef]
19. Trott, O.; Olson, A. AutoDock Vina: Improving the speed and accuracy of docking with a new scoring function, efficient optimization, and multithreading. *J. Comput. Chem.* **2009**, *31*, 455–461. [CrossRef]
20. Morris, G.; Huey, R.; Lindstrom, W.; Sanner, M.; Belew, R.; Goodsell, D.; Olson, A. AutoDock4 and AutoDockTools4: Automated Docking with Selective Receptor Flexibility. *J. Comput. Chem.* **2009**, *30*, 2785–2791. [CrossRef]
21. Berman, H.M.; Westbrook, J.; Feng, Z.; Gilliland, G.; Bhat, T.N.; Weissig, H.; Shindyalov, I.N.; Bourne, P.E. The Protein Data Bank. *Nucleic Acids Res.* **2000**, *28*, 235–242. [CrossRef]
22. Makeneni, S.; Thieker, D.F.; Woods, R.J. Applying Pose Clustering and MD Simulations To Eliminate False Positives in Molecular Docking. *J. Chem. Inf. Model.* **2018**, *58*, 605–614. [CrossRef]
23. Hanwell, M.D.; Curtis, D.E.; Lonie, D.C.; Vandermeersch, T.; Zurek, E.; Hutchison, G.R. Avogadro: An advanced semantic chemical editor, visualization, and analysis platform. *J. Cheminform.* **2012**, *4*, 17. [CrossRef]
24. Stoica, C.I.; Marc, G.; Pîrnău, A.; Vlase, L.; Araniciu, C.; Oniga, S.; Palage, M.; Oniga, O. Thiazolyl-oxadiazole derivatives targeting lanosterol 14α-demethylase as potential antifungal agents: Design, synthesis and molecular docking studies. *Farmacia* **2016**, *64*, 390–397. Available online: https://farmaciajournal.com/wp-content/uploads/2016-03-art-11-Stoica_Cristina_Oniga_390-397.pdf (accessed on 18 March 2023).
25. Borlan, R.; Stoia, D.; Gaina, L.; Campu, A.; Marc, G.; Perde-Schrepler, M.; Silion, M.; Maniu, D.; Focsan, M.; Astilean, S. Fluorescent Phthalocyanine-Encapsulated Bovine Serum Albumin Nanoparticles: Their Deployment as Therapeutic Agents in the NIR Region. *Molecules* **2021**, *26*, 4679. [CrossRef] [PubMed]
26. Ascoli, G.; Domenici, E.; Bertucci, C. Drug binding to human serum albumin: Abridged review of results obtained with high-performance liquid chromatography and circular dichroism. *Chirality* **2006**, *18*, 667–679. [CrossRef] [PubMed]
27. Yang, F.; Zhang, Y.; Liang, H. Interactive Association of Drugs Binding to Human Serum Albumin. *Int. J. Mol. Sci.* **2014**, *15*, 3580–3595. [CrossRef]
28. Nishi, K.; Yamasaki, K.; Otagiri, M. Serum Albumin, Lipid and Drug Binding. *Subcell Biochem.* **2020**, *94*, 383–397. [CrossRef]
29. Czub, M.; Stewart, A.; Shabalin, I.; Minor, W. Organism-specific differences in the binding of ketoprofen to serum albumin. *IUCrJ* **2022**, *9*, 551–561. [CrossRef]
30. Zsila, F. Subdomain IB Is the Third Major Drug Binding Region of Human Serum Albumin: Toward the Three-Sites Model. *Mol. Pharm.* **2013**, *10*, 1668–1682. [CrossRef]
31. Mishra, V.; Heath, R. Structural and Biochemical Features of Human Serum Albumin Essential for Eukaryotic Cell Culture. *Int. J. Mol. Sci.* **2021**, *22*, 8411. [CrossRef]
32. Pettersen, E.; Goddard, T.; Huang, C.; Couch, G.; Greenblatt, D.; Meng, E.; Ferrin, T. UCSF Chimera—A visualization system for exploratory research and analysis. *J. Comput. Chem.* **2004**, *25*, 1605–1612. [CrossRef]
33. Abraham, M.J.; Murtola, T.; Schulz, R.; Páll, S.; Smith, J.C.; Hess, B.; Lindahl, E. GROMACS: High performance molecular simulations through multi-level parallelism from laptops to supercomputers. *SoftwareX* **2015**, *1–2*, 19–25. [CrossRef]
34. Vanommeslaeghe, K.; Hatcher, E.; Acharya, C.; Kundu, S.; Zhong, S.; Shim, J.; Darian, E.; Guvench, O.; Lopes, P.; Vorobyov, I.; et al. CHARMM general force field: A force field for drug-like molecules compatible with the CHARMM all-atom additive biological force fields. *J. Comput. Chem.* **2010**, *31*, 671–690. [CrossRef]

35. Jorgensen, W.; Chandrasekhar, J.; Madura, J.; Impey, R.; Klein, M. Comparison of simple potential functions for simulating liquid water. *J. Chem. Phys.* **1983**, *79*, 926–935. [CrossRef]
36. Lv, Z.; Wang, H.; Niu, X. Molecular dynamics simulations reveal insight into key structural elements of aaptamines as sortase inhibitors with free energy calculations. *Chem. Phys. Lett.* **2013**, *585*, 171–177. [CrossRef]
37. Jin, H.; Zhou, Z.; Wang, D.; Guan, S.; Han, W. Molecular Dynamics Simulations of Acylpeptide Hydrolase Bound to Chlorpyrifos-methyl Oxon and Dichlorvos. *Int. J. Mol. Sci.* **2015**, *16*, 6217–6234. [CrossRef]
38. Crișan, O.; Marc, G.; Nastasă, C.; Oniga, S.; Vlase, L.; Pîrnău, A.; Oniga, O. Synthesis and *in silico* approaches of new symmetric bis-thiazolidine-2,4-diones as Ras and Raf oncoproteins inhibitors. *Farmacia* **2023**, *71*, 254–263. Available online: https://farmaciajournal.com/wp-content/uploads/art-04-Crisan_Marc_Oniga_254-263.pdf (accessed on 22 March 2023).
39. Humphrey, W.; Dalke, A.; Schulten, K. VMD: Visual molecular dynamics. *J. Mol. Graph.* **1996**, *14*, 33–38. [CrossRef] [PubMed]
40. Ning, X.-Q.; Lou, S.-J.; Mao, Y.-J.; Xu, Z.-Y.; Xu, D.-Q. Nitrate-promoted selective C-H Fluorination of benzamides and benzenacetamides. *Org. Lett.* **2018**, *20*, 2445–2448. [CrossRef]
41. Uludağ, M.; Ergün, B.; Alkan, D.; Ercan, N.; Özkan, G.; Banoğlu, E. Stable ester and amide conjugates of some NSAIDs as analgesic and antiinflammatory compounds with improved biological activity. *Turk. J. Chem.* **2011**, *35*, 427–439. [CrossRef]
42. Manolov, S.; Ivanov, I.; Bojilov, D.; Nedialkov, P. Synthesis, *In Vitro* Anti-Inflammatory Activity, and HRMS Analysis of New Amphetamine Derivatives. *Molecules* **2023**, *28*, 151. [CrossRef]
43. Manolov, S.; Ivanov, I.; Bojilov, D.; Nedialkov, P. Synthesis, *in silico*, and *in vitro* biological evaluation of new furan hybrid molecules. *Processes* **2022**, *10*, 1997. [CrossRef]
44. Galano, A.; Macías-Ruvalcaba, N.; Campos, O.; Pedraza-Chaverri, J. Mechanism of the OH radical scavenging activity of nordihydroguaiaretic acid: A combined theoretical and experimental study. *J. Phys. Chem. B* **2010**, *114*, 6625–6635. [CrossRef]
45. Halliwel, B.; Gutterdge, J. *Free Radicals in Biology and Medicine*; Clarendon Press: Oxford, UK, 1985; p. 346.
46. Mansouri, A.; Makris, D.; Kefalas, P. Determination of hydrogen peroxide scavenging activity of cinnamic and benzoic acids employing a highly sensitive peroxyoxalate chemiluminescence-based assay: Structure-activity relationships. *J. Pharm. Biomed. Anal.* **2005**, *39*, 22–26. [CrossRef]
47. Ebrahimzadeh, M.; Nabavi, S.; Nabavi, S.; Bahramian, F.; Bekhradnia, A. Antioxidant and free radical scavenging activity of H. officinalis L. var. angustifolius, V. odorata, B. hyrcana and C. speciosum. *Pak. J. Pharm. Sci.* **2010**, *23*, 29–34.
48. Jayashree, V.; Bagyalakshmi, S.; Manjula Devi, K.; Richard Daniel, D. *In vitro* anti-inflammatory activity of 4-benzylpiperidine. *Asian J. Pharm. Clin. Res.* **2016**, *9*, 108–110. [CrossRef]
49. Pontiki, E.; Hadjipavlou-Litina, D. Synthesis and pharmacochemical evaluation of novel aryl-acetic acid inhibitors of lipoxygenase, antioxidants, and anti-inflammatory agents. *Bioorg. Med. Chem.* **2007**, *15*, 5819–5827. [CrossRef] [PubMed]

Disclaimer/Publisher's Note: The statements, opinions and data contained in all publications are solely those of the individual author(s) and contributor(s) and not of MDPI and/or the editor(s). MDPI and/or the editor(s) disclaim responsibility for any injury to people or property resulting from any ideas, methods, instructions or products referred to in the content.

Article

Sustainable Bioactive Composite of *Glehnia littoralis* Extracts for Osteoblast Differentiation and Bone Formation

Chul Joong Kim [1,†], Bimal Kumar Ghimire [2,†], Seon Kang Choi [3], Chang Yeon Yu [4] and Jae Geun Lee [1,*]

1. Research Institute of Biotechnology, Hwajin Bio Cosmetic, Chuncheon 24232, Republic of Korea
2. Department of Crop Science, College of Sanghuh Life Science, Konkuk University, Gwangjin, Seoul 05029, Republic of Korea
3. Department of Agricultural Life Sciences, Kangwon National University, Chuncheon 24341, Republic of Korea
4. Department of Bio-Resource Sciences, Kangwon National University, Chuncheon 24341, Republic of Korea
* Correspondence: leejeakun@hanmail.net; Tel.: +82-01091584996
† These authors contributed equally to this work.

Abstract: Different bone-related diseases are mostly caused by the disruption of bone formation and bone resorption, including osteoporosis. Traditional medicinal literature has reported the possible anti-osteoporotic properties of *Glehnia littoralis*. However, the chemical compounds in extracts that are responsible for bone metabolism are poorly understood. The present study aimed to explore and compare the coumarin-based compounds present in *G. littoralis* extracts, the antioxidant activities, and the anti-osteoporotic properties of different extracts of *G. littoralis* (leaf and stem, fruit, whole plant, and root extracts) on bone metabolism. This study analyzed *G. littoralis* extract effects on the proliferation and osteoblastic differentiation of MC3T3-E1 osteoblasts. Among the different tested samples, stem extracts had the highest scopoletin (53.0 mg/g), and umbelliferone (1.60 mg/g). The significantly ($p < 0.05$) highest amounts of imperatorin (31.9 mg/g) and phellopterin (2.3 mg/g), were observed in fruit and whole plant extracts, respectively. Furthermore, the results confirmed alkaline phosphatase activity, collagen synthesis, mineralization, osteocalcin content, and osterix and RUNX2 expression. *G. littoralis* extracts at concentrations greater than 20 µg/mL had particularly adverse effects on MC3T3-E1 cell viability and proliferation. Notably, cell proliferation was significantly elevated at lower *G. littoralis* concentrations. Comparatively, 0.5 µg/mL stem had a higher osteocalcin content. Of the four extract types, stem showed a higher collagen synthesis effect at concentrations of 0.5–5 µg/mL. Except for fruit extracts, *G. littoralis* extract treatment significantly elevated osterix gene expression. All *G. littoralis* extracts increased RUNX2 gene expression. The results described here indicate that *G. littoralis* ethanolic extracts can effectively prevent osteoporosis.

Keywords: *Glehnia littoralis*; osteoblast differentiation; collagen synthesis; alkaline phosphatase activity; osteoblast; osterix expression; RUNX2 expression

1. Introduction

Osteoporosis is an osteometabolic disorder caused by an imbalance in bone formation by osteoblasts and bone resorption by osteoclasts that make up bones [1,2], which leads to fragility, increased risk of fractures, and threatens mobility in the elderly [3]. According to the International Osteoporosis Foundation (IOF), approximately 200 million people worldwide suffer from osteoporosis [4,5]. Recent studies suggest that increases in inflammation-related cytokine secretion, the number of macrophages, and leukotriene B4, an inflammation-inducing factor [6–9], cause this disease. In women, osteoporosis is associated with a sharp decline in estrogen secretion in the postmenopausal stage, known as postmenopausal osteoporosis [10]. Synthetic therapeutic agents, such as parathyroid hormones, bisphosphonates, selective estrogen receptor modulators (SERMs), and hormone replacement therapy (HRT), are used to treat osteoporosis [11]. However, these drugs can have side effects, including hypercalcemia and osteosarcoma in postmenopausal women [12], esophageal gastric irritation, and cancer [13], and they can increase the chance of strokes, breast cancer, and coronary heart diseases [14]. Thus, it would be beneficial to discover natural anti-osteoporotic agents that minimize bone loss in postmenopausal women.

Osteoblast cells are critical in bone metabolism and are responsible for bone matrix synthesis and mineralization [15]. Cell culture is widely used to assess the activity of a substance in vitro, such as its osteoinductive potential. In particular, MC3T3 pre-osteoblasts have methodological advantages and facilities [16,17], meaning the differentiation of these cells into mature osteoblasts can be easily recognized by markers of osteoblastic metabolism, such as alkaline phosphatase (ALP), and by the degree of extracellular matrix (ECM) mineralization [18]. ALP is the most widely used biochemical marker for estimating osteoblastic activity [19]. This enzyme is associated with the ECM mineralization process and osteoblastic differentiation, as it is responsible for the maturation of the matrix, which will later be mineralized [18,20].

Current research focuses on natural materials and phytoestrogens for bone formation and resorption pathways related to bone metabolism. There is a constant search for alternatives that can help bone healing, in cases of injury, or favor rehabilitating the quality of the bone tissue. Several recent reports have indicated that phytoestrogen compounds in food and plants can effectively suppress the secretion of inflammation-inducing factors and inflammation-related cytokines associated with osteoporosis [21,22], and enhance or stimulate osteoblast activity [23]. Phytoestrogens act like estrogen and exist in the form of flavonoids, lignins, isoflavones, and coumestans, which share structural and functional similarities with synthetic estrogens [24,25]. The present study aimed to research therapeutic alternatives to treat osteoporosis based on drugs obtained from natural sources, mainly by observing their lower cost and incidence of adverse effects when compared to synthetic drugs.

Glehnia littoralis Fr. Schmidt et Miquel, a perennial marine herbaceous plant belonging to the Apiaceae family, is native to the sandy coastal area of eastern Asia, mainly Japan, Korea, China, Manchuria, Sakhalin, Okhotsk, the Kuril Islands, and North America from California to Alaska [26,27]. The leaves and flower buds are edible, their rhizomes and roots are used traditionally to treat lung diseases, tuberculosis, coughs, hemoptysis, and dyspnea [28,29], and they have diaphoretic, antipyretic, and analgesic effects [30,31]. Previous studies have reported the presence of phytochemicals such as phenolic acids, flavonoids, pyranocoumarins, and polysaccharides in *G. littoralis* extracts, which have various biological activities, including antitumor, antimicrobial, antioxidant, blood circulation-promoting and immunomodulatory properties [29]. Moreover, several previous studies have shown that these phytochemicals are effective in bone formation [32]. To our knowledge, no studies have reported the use of *G. littoralis* extracts as potential natural therapeutic agents to prevent osteoporosis.

Therefore, the main objectives of the present study were to identify and select the toxicity of different *G. littoralis* plant parts for use in the differentiation analysis of an MC3T3-E1 cell culture through a cell viability assay. In addition, this study investigated the proliferative potential of MC3T3-E1 cells treated with *G. littoralis* using a 3-(4,5-dimethylthiazol-2-yl)-2,5-diphenyltetrazolium bromide (MTT) assay. This study also compared the degree of osteoblastic differentiation between control cells and those treated with different *G. littoralis* concentrations by quantifying ALP. Finally, we evaluated the degree of matrix mineralization formed by MC3T3-E1 cells treated with different *G. littoralis* plant parts and concentrations to confirm its applicability as a natural material for osteoblast differentiation and bone formation.

2. Materials and Methods

2.1. Chemicals

All the chemicals and solvents used in the experiments were of analytical grade. All chemicals were obtained from Sigma-Aldrich (St. Louis, MO, USA).

2.2. Plant Material and Extract Preparation

The *G. littoralis* used in the experiment was grown and harvested from a field in Gangneung-si, Gangwon-do (Slonaemall, Gangneung, Republic of Korea) at 37°45'06" N latitude and 128°52'38" E longitude, in October 2020. The leaves, roots, fruits, and stems of the *G. littoralis* were washed with purified water and separated. The collected samples were dried at room temperature for 72 h. Approximately 2 g of the finely ground samples and placed in a conical flask containing 40 mL of 80% (v/v) ethanol. The mixture was filtered through filter paper to remove debris, and the solvents were evaporated at 41 °C in a rotary vacuum evaporator (Eyela, SB-1300, Shanghai Eyela Co., Ltd., Shanghai, China) and then lyophilized using a freeze dryer (PVTFD 300R) (IlShinBioBase, Yangju, Republic of Korea). The collected extracts were mixed with 80% methanol and stored in a refrigerator at 4 °C until further analysis.

2.3. Determination of Total Phenolic Content (TPC)

The total phenolic content of different samples was measured by following the Folin-Ciocalteu procedure, the method Singleton and Rossi [33] described previously. An aliquot of 100 µL of the plant extract (at a concentration of 1 mg mL^{-1}) and 500 µL of the Folin-Ciocalteu reagent (1:3 v/v) were mixed with 500 µL distilled water in a test tube and shaken for 5 min at room temperature. Then, 500 µL sodium carbonate (10%) was added to the solution, and the mixture was left to rest for 1 h. Deionized distilled water served as a blank. Then, the absorbance of the obtained mixture was taken using a ultraviolet-visible (UV-Vis) spectrophotometer (Jasco V530 UV-VIS spectrophotometer, Tokyo, Japan) at 725 nm against the blank. The results are expressed as mg of the gallic acid equivalent (GAE) per g of the dry weight (DW). A calibration curve was prepared using 20–500 mg/L gallic acid (R^2 = 0.9980).

2.4. Determination of Total Flavonoid Content (TFC)

Total flavonoid content was measured following the method described by Moreno et al. [34]. An aliquot of 500 µL of plant extract (at a concentration of 1 mg mL^{-1}) was mixed with 100 µL KCH$_3$COO (1 M) and 100 µL of 10% Al(NO$_3$)$_3$ in a 10 mL test tube and homogenized manually, and incubated for 50 min at room temperature. Then, the absorbance was measured using a spectrophotometer (Jasco V530 UV-VIS spectrophotometer, Tokyo, Japan) at 415 nm against the blank. The experiment was carried out in triplicate, and the results are expressed as the mean ± standard deviation in mg of the quercetin equivalent (QE) per g of the dry sample. A calibration curve was prepared using 20–500 mg/L quercetin (R^2 = 0.9970).

2.5. LC/UVD Quantitative Analysis of Coumarin-Based Compounds

Quantitative analysis of coumarin-based compounds was performed by using LC/UVD with various solvents. An UltiMate 3000 HPLC system (Thermo Fisher Scientific Lin., San Jose, CA, USA) coupled with a UV detector (Thermo Fisher Scientific, San Jose, CA, USA) was applied for quantitative analysis of coumarin-based compounds. The separation of each compound was achieved by a column (5 µm, 250 mm × 4.6 mm, Bischoff Analysentechnik und-geräte GmbH., Leonberg, Garmany). The mobile phase comprised water (A) and acetonitrile (B). The gradient elution conditions were: 0–5 min, 80% A; 5–60 min, 80–0% A; 60–70 min, 0% A; 70–71 min, 0–80% A; 71–80 min, 80% A. The temperature of the column oven was maintained at 30 °C. A flow rate of 1.0 mL/min was used and the injection volume was 10 µL. The chromatograms of the compounds were acquired at 203 nm. Standard chemical compounds such as scopoletin, umbelliferone, phellopterin, and imperatorin were obtained from Sigma-Aldrich (St. Louis, MO, USA). Stock standard chemicals (1000 ng/mL) were prepared in 100% methanol. The stock solutions were maintained at 4 °C and used to construct calibration curves after appropriate dilution. Amounts of different quantified compounds were calculated as mean values from HPLC analyses based on the calibration curves of the corresponding standard compounds. The regression equation of the standard compounds were as follows.

$$\text{Scopoletin, } y = 0.482x + 0.077, R^2 = 0.999$$

$$\text{Umbelliferone, } y = 1.005x + 0.027, R^2 = 0.999$$

$$\text{Imperatorin, } y = 0.996x + 0.017, R^2 = 0.999$$

$$\text{Phellopterin, } y = 0.993x + 0.010, R^2 = 0.999$$

2.6. Antioxidant Activity

2.6.1. Evaluating the 2,2-Diphenyl-1-picrylhydrazyl (DPPH) Radical Scavenging Assay

The free radical scavenging capacity of *G. littoralis* samples was measured using a DPPH radical scavenging assay, following the method Chung et al. [35] described previously. In triplicate, an aliquot of 200 µL of different sample extracts was mixed with 4.5 mL DPPH (0.004% in methanol). The reaction mixture was homogenized manually and incubated at room temperature (25 °C) for 40 min. The mixture was shaken and kept in dark conditions for 45 min. Then, the absorbance value was taken using a spectrophotometer (Jasco V530 UV-VIS spectrophotometer, Tokyo, Japan) at 517 nm. A blank was prepared by replacing DPPH with 80% methanol in the reaction medium. BHT was used as the positive control.

The free radical scavenging ability of the sample was measured from the following equation:

$$\text{DPPH scavenging activity (\%)} = (\text{Abs}_{control} - \text{Abs}_{sample})/\text{Abs}_{control} \times 100\%$$

where $\text{Abs}_{control}$ represent the absorbance of the mixture + methanol, and Abs_{sample} represent the absorbance of the mixture + plant extract. The antioxidant activity was expressed as the capacity to scavenge or reduce the DPPH radical by 50%; that is, the amount of antioxidant compounds required to scavenge or reduce the initial concentration of DPPH by 50%.

2.6.2. Evaluation of the 2,2′-Azino-bis(3-ethylbenzothiazoline-6-sulfonic Acid (ABTS) Assay

The antioxidant activity was determined using the ABTS radical method described elsewhere [35]. Briefly, the ABTS solution used in the experiments was formed by mixing 7.4 mM ABTS and 2.6 mM potassium persulphate (1:1, *v/v*). The reaction solution was incubated for 12 h at room temperature. The solution was then diluted with 80% methanol until a solution with an absorbance of 0.70 ± 0.01 was achieved. Then, 1 mL diluted ABTS was mixed with 100 mL of the sample. Then, the absorbance was taken using a spectrophotometer (Jasco V530 UV-VIS spectrophotometer, Tokyo, Japan) at 734 nm. Trolox was used as a positive control. The standard curve was prepared from various concentration of trolox (500 µM, 600 µM, 700 µM, 800 µM, 900 µM, and 1000 µM).

The ABTS radical scavenging ability of the samples was measured from the following equation:

$$\text{ABTS scavenging activity} = (\text{Abs}_{control} - \text{Abs}_{sample})/\text{Abs}_{control} \times 100$$

where $\text{Abs}_{control}$ represent the absorbance of the ABTS solution + methanol, and Abs_{sample} represent the absorbance of the ABTS solution + test sample.

2.7. Cell Culture

Osteoblast MC3T3-E1 cells derived from mouse bones were purchased from the American Type Culture Collection (ATCC, Manassas, VA, USA). Osteoblast MC3T3-E1 cells were prepared in α-minimum essential medium (α-MEM, Gibco, Grand Island, NE, USA) with 10% fetal bovine serum (FBS; Gibco, Grand Island, NE, USA), 100 U/mL penicillin (CO_2 incubator (MCO-230AIC-PK, Panasonic, Kadoma, Japan) using a culture medium supplemented with Gibco (Grand Island, NE, USA)), and 100 U/mL streptomycin (Gibco, Grand Island, NE, USA) at 37 °C, 5% CO_2, and 95% air. The cells were washed with phosphate-buffered saline (PBS; pH 7.4; Gibco, Grand Island, NE, USA) and 0.25% trypsin-2.65 mM EDTA (Gibco, Grand Island, NE, USA), and sub-cultured by changing the culture medium every two days.

2.8. Cell Viability

Cell viability was analyzed according to the protocol described by Denizot and Land [36] by dispensing osteoblast MC3T3-E1 cells in a 96-well plate at 2×10^3 cells/well and subsequently pre-incubating them in a CO_2 incubator for 24 h. After incubation, the MC3T3-E1 cells were treated with different concentrations of plant sample for 72 h. The 3-(4,5-dimethylthiazol-2-yl)-2,5-diphenyltetrazolium bromide (MTT solution) at $5~\text{mg} \cdot \text{mL}^{-1}$ was added to each well and maintained at 37 °C for 4 h. Then, the resultant formazan crystals were dissolved in dimethyl sulfoxide (DMSO). The absorbance of the mixture was taken using a microplate reader (Thermo Fisher Scientific Instrument Co., Ltd., Shanghai, China) at 570 nm.

2.9. Cytotoxicity

The cytotoxicity of different plant parts was evaluated according to a previously described protocol [37,38]. Osteoblast MC3T3-E1 cells were dispensed in a 96-well plate at a density of 2×10^3 cells/mL, then pre-incubated in a CO_2 incubator for 24 h. After incubation, the cells were treated with various concentrations of samples (ranging from 0 to 200 µg/mL) and incubated for 48 h. The cell culture solution (10 µL) was added to a 96-well plate containing 40 µL PBS. After placing the LDH reagent in each well, the mixture was placed in a dark place for 45 min at 25 °C. After the reaction was terminated by mixing 50 µL of stop solution, the cytotoxicity was evaluated by quantifying the plasma membrane damage. LDH is a stable enzyme present in all cell types and is rapidly released into the culture medium when the plasma membrane is damaged. Cell membrane integrity was evaluated by measuring the LDH leakage levels from the cells using the LDH Cytotoxicity Assay Kit (BioVision, Inc., Milpitas, CA, USA). The absorbance value of the samples was measured using a microplate reader (Thermo Fisher Scientific Instrument Co., Ltd., Shanghai, China) at 492 nm.

2.10. ALP Activity

MC3T3-E1 cells were seeded as described above, and the ALP activity of the media was measured by following the method described by Liu et al. [39]. Initially, MC3T3-E1 cells were dispensed in a 48-well plate at 5×10^4 cells/well and pre-incubated in a CO_2 incubator for 24 h. Then, the MC3T3-E1 cells were treated with different sample concentrations (0–20 µg/mL). Subsequently, the culture medium was replaced and supplemented with 100 µL chemiluminescent substrate for alkaline phosphatase (CSPD; Roche, Basel, Switzerland), added to 20 µL total cell lysate, and reacted for 30 min. The amount of protein in the total cell lysate was measured using CSPD (Life Technologies, Carlsbad, CA, USA) according to the Protein Assay Kit (Bio-Rad, Hercules, CA, USA), and the ALP activity value was expressed as the fold change per µg of total protein.

2.11. Collagen Synthesis Rate

Collagen content was quantified by a Sirius Red-based colorimetric assay as described by Park et al. [40]. The MC3T3-E1 cells were cultured in an osteogenic medium containing

10 mM β-glycerophosphate, 5 nM dexamethasone, 50 g/mL ascorbic acid, and *G. littoralis* (0 μg/mL and 20 μg/mL). After seven days, the cells were cultured in a medium containing bovine serum albumin (BSA) (3%) for two days. The cultured cells were gently washed twice with PBS, followed by fixation with Bouin's fluid (8.3% formaldehyde and 4.8% acetic acid in saturated aqueous picric acid, $(O_2N)_3C_6H_2OH$) for 1 h. After fixation, the fixative fluids were removed, and the cultured cells were washed with tap water for 10 min. The cultured cells were air-dried and stained with Sirius red dye (0.1% saturated picric acid) for 1 h on a shaker. The stained cells were washed with PBS and observed under a microscope. For quantitative analysis, the stained cell layer was washed extensively with 0.01 N HCl to remove the non-bound dye. The stained cells were dissolved in 0.2 mL of 0.1 N and shaken for 30 min. Then, the absorbance was measured at 540 nm.

2.12. Mineralization Content

Calcium content was measured in osteoblasts using the method described by Zakłos-Szyda et al. [41]. Osteoblast MC3T3-E1 cells were aliquoted in a 24-well plate at 2×10^4 cells/well in a CO_2 incubator at 37 °C. The cells were treated with plant samples at a concentration of 0–20 μg/mL and placed in an incubator for seven and 14 days to induce mineral deposition. The culture medium was changed at three-day intervals. For harvesting, the MC3T3-E1 cells were washed with PBS twice and fixed with 4% paraformaldehyde for 2 h. The cells were then stained with 40 mM Alizarin Red S (pH 4.5). The stained cells were then rinsed four times with distilled water and observed under an inverted microscope. Calcified nodules appearing bright red were confirmed, and 100 mM cetylpyridinium chloride (Sigma) solution was added to each well to elute the stain. The eluted stain (100 μL) was added to a 96-well microplate. The absorbance of solubilized calcium-bound Alizarin Red S was taken at 570 nm using a spectrophotometer (Eppendorf AG 22331; Hamburg, Germany). Calcium deposition was expressed as the molar equivalent of calcium.

2.13. Osteocalcin Content

Osteocalcin concentration was determined by the test method of the OCN ELISA kit (Takara Bio, Kusatsu, Japan) by following the process described by Bukhari et al. [42]. Initially, 100 μL of the cell culture solution was added to a 96-well plate coated with antibodies and incubated for 1 h at 37 °C. After removing unbound material with washing buffer (50 mM Tris, 200 mM NaCl, and 0.2% Tween 20), horseradish peroxidase (HRP)-conjugated streptavidin was added to each well and incubated at 37 °C for 1 h to bind to the antibodies. After washing the wells five times, tetramethylbenzidine (TMB) solution was added to each well and incubated at room temperature for 20 min in the dark. HRP catalyzed the conversion of a chromogenic substrate (TMB) to a colored solution, with a color intensity proportional to the amount of protein present in the sample. After adding the stop solution, the absorbance of the sample in each well was measured at 450 nm. Results are presented as the percentage change in activity compared to the untreated control. The concentration of osteocalcin in the serum was measured using the same method as that used for cultured cell samples.

2.14. mRNA Expression Rate

Total RNA was isolated from cells at specific times using TRIzol reagent (Invitrogen, Waltham, MA, USA) with DNase treated with RNase-free DNase (35 U/mL) for 50 min at 37 °C according to the method described by Liu et al. [43] and Matsubara et al. [44]. The total RNA present in each sample was quantified by measuring the absorbance at 260 nm using a spectrophotometer (Eppendorf AG 22331; Hamburg, Germany). Gene expression was measured by adding cDNA to a PCR mixture containing EXPRESS SYBR Green qPCR Supermix (Bio Prince, Seongsu, Seoul, Republic of Korea). Real-time PCR was performed using the Rotor-Gene Q (Qiagen, Düsseldorf, Germany). The reaction was carried out at 95 °C for 20 s, 60 °C for 20 s, and 72 °C for 25 s for 40 cycles of amplification. Relative mRNA expression of specific genes was standardized with glyceraldehyde 3-phosphate dehydrogenase (GAPDH), and sequencing was performed using PCR primer sequences (Table 1).

Table 1. Forward and reverse primers sequences used in this study.

Primer Name		Sequence
Osterix	Forward	5′-AGCGACCACTTGAGCAAACAT-3′
	Reverse	5′-GCGGCTGATTGGCTTCTTCT-3′
RUNX2	Forward	5′-CGGCCCTCCCTGAACTCT-3′
	Reverse	5′-TGCCTGCCTRGGGATCTGTA-3′

2.15. Statistical Processing

The results were presented as the mean ± standard deviations of three trials independent (n=3). The data was compared by one-way ANOVA using the SPSS program (Statistical Package for Social Science, Version 24), followed by Duncan's multiple range test, considering significant differences at P values < 0.05.

3. Results

3.1. The Total Phenolic Contents (TPCs) and Total Flavonoid Contents (TFCs)

The total phenolic contents (TPCs) and total flavonoid contents (TFCs) of stems, leaves, roots, and whole plant extracts of *G. littoralis* extracts were calculated from the linear regression equation of the gallic acid standard calibration curve. The TPC of different plant parts ranged from 4.27 ± 0.03 to 23.29 ± 0.43 mgGAE/g dry sample. As Figure 1 shows, the leaf and stem (GLSE) and root (GLRE) extracts had the highest and lowest TPCs, respectively. The total flavonoid contents (TFCs) of stems, leaves, roots, and whole plant extracts of *G. littoralis* extracts were determined using colorimetric analysis. The flavonoid content of different *G. littoralis* extract parts ranged from 0.28 ± 0.04 to 4.93 ± 0.30 mgQE/g dry sample. The results showed that fruit (GLFE) extracts contained a higher TFC, followed by GLFE, whole plant (GLAE), and GLRE extracts.

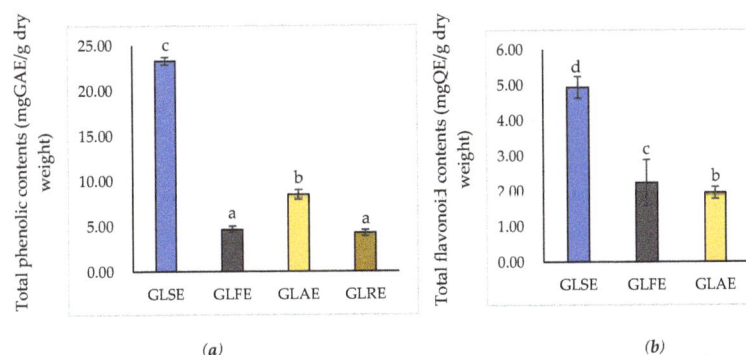

Figure 1. (a) Total phenolic contents (TPC) and (b) total flavonoid contents (TFC) in *G. littoralis* extracts. The mean values followed by the same letter are not significantly different based on the DMRT ($p < 0.05$). GLSE: *G. littoralis* leaf, stem extracts, GLFE: *G. littoralis* fruit extracts, GLAE: *G. littoralis* all extracts, GLRE: *G. littoralis* root extracts.

3.2. LC/UVD Quantitative Analysis of Coumarin-Based Compounds

LC/UVD quantitative analysis of four coumarin components varied widely with the leaf, stem, fruit, whole plant, and root extracts of *G. littoralis* (Table 2). Among the different tested samples, GLSE extracts had the highest scopoletin (53.0 mg/g), and umbelliferone (1.6 mg/g). The lowest amount of scopoletin was observed in GLRE (8.5 mg/g), and of umbelliferone, in GLFE (0.8 mg/g). The significantly ($p < 0.05$) highest amount of imperatorin (31.9 mg/g) and phellopterin (2.3 mg/g) were observed in GLFE and GLAE extracts, respectively (Table 2). It was confirmed that the total content of four types of coumarin was higher in the order of GLSE, GLRE, GLAE, and GLFE (Figure 2).

Table 2. LC/UVD quantitative analysis compounds of extracts from *G. littoralis* (Unit: mg/g, dry weight).

Sample [1]	Scopoletin	Umbelliferone	Imperatorin	Phellopterin	Total
GLSE	53.0 ± 0.2 [d]	1.6 ± 0.0 [d]	2.0 ± 0.0 [a]	N.D. [2]	56.6 ± 0.2 [d]
GLFE	17.7 ± 0.0 [b]	0.8 ± 0.0 [a]	8.9 ± 0.1 [b]	1.1 ± 0.4 [b]	28.5 ± 0.5 [a]
GLAE	24.5 ± 1.1 [c]	1.0 ± 0.0 [b]	15.1 ± 0.6 [c]	0.6 ± 0.1 [a]	41.1 ± 1.8 [b]
GLRE	8.5 ± 0.0 [a]	1.4 ± 0.1 [c]	31.9 ± 0.1 [d]	2.3 ± 0.0 [c]	44.2 ± 0.1 [c]

[1] GLSE: *G. littoralis* leaf, stem extracts, GLFE: *G. littoralis* fruit extracts, GLAE: *G. littoralis* plant extracts, GLRE: *G. littoralis* root extracts. [2] N.D.: not detected. Each value is means ± SD of three replicate tests. The mean values followed by the same letter are not significantly different based on the DMRT ($p < 0.05$).

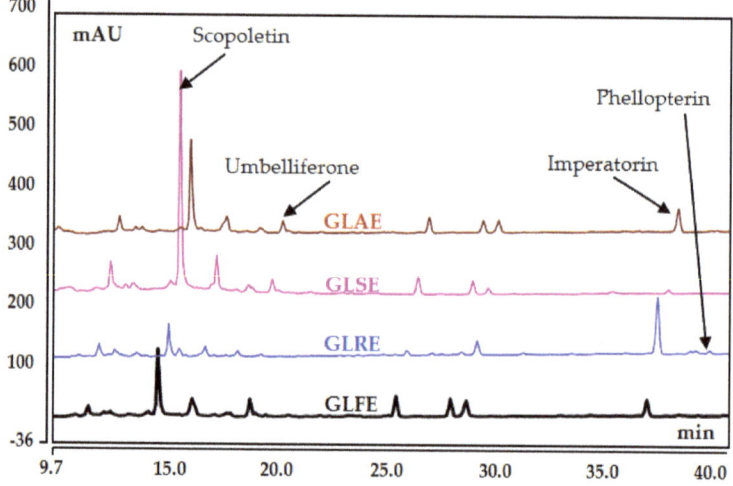

Figure 2. Representative chromatograms of compounds of extracts from *G. littoralis*.

3.3. Antioxidant Activity

Figure 3 presents the radical scavenging potential of the GLSE, GLFE, GLAE, and GLRE extracts. The results showed that the antioxidant activity of different plant parts varied significantly and appeared in a concentration-dependent manner. Likewise, GLAE extracts had higher antioxidant properties, as represented by the lower half-maximal inhibitory concentration (IC$_{50}$) (718.40 ± 14.025 µg mL^{-1}). Furthermore, all other plant part extracts, except GLRE extracts, possessed a better radical scavenging potential, indicating the presence of higher amounts of antioxidant compounds. The results showed that the ABTS$^+$ radical scavenging potential of different *G. littoralis* extract parts varied significantly ($p < 0.05$), with GLAE extracts exhibiting higher ABTS radical scavenging activity (16,348.41 ± 1315.26 µg mL^{-1}). Comparatively, the GLRE extract showed the lowest antioxidant activity (Figure 3). Two different antioxidant assays were carried out, and these antioxidant assays were positively and significantly correlated with TPC and TFC of *G. littoralis* (Table S1).

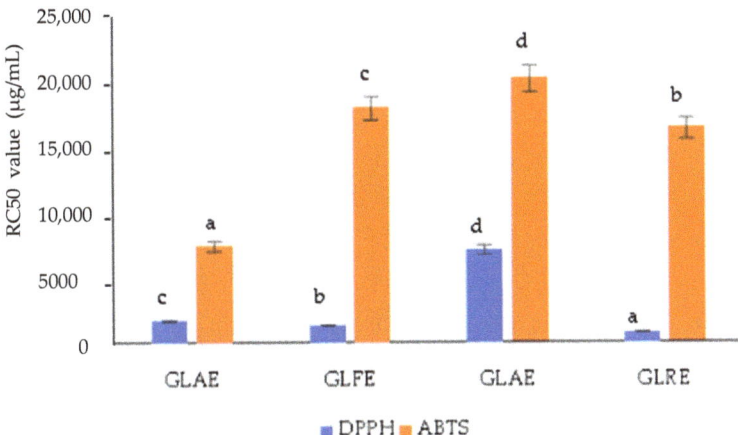

Figure 3. Antioxidant activity (DPPH and ABTS radical assay) in *G. littoralis* extracts. Mean values followed by the same letter are not significantly different based on the DMRT ($p < 0.05$). GLSE: *G. littoralis* leaf, stem extracts, GLFE: *G. littoralis* fruit extracts, GLAE: *G. littoralis* all extracts, GLRE: *G. littoralis* root extracts.

3.4. Cell Viability Test

We used osteoblastic precursor cell lines (MC3T3-E1 cells) derived from *Mus musculus* to induce the expression of osteoblast markers and investigate the effect of *G. littoralis* on bone metabolism. In this study, we attempted to determine the effect of *G. littoralis* extract on in vitro osteogenic induction using MC3T3-E1 cells. We performed a pilot study in which *G. littoralis* extract concentrations varied from 0.5 to 200 μg/mL to determine the optimal concentration. Cell viability studies were performed for different *G. littoralis* parts, including GLSE, GLFE, GLAE, and GLRE extracts. Initially, the purified *G. littoralis* ethanol extracts were re-suspended in 70% ethanol, and various plant extract concentrations were added to the MC3T3-E1 cell culture. The viability of cells varied considerably among the different plant parts used. The viable cell number increased in up to 5 μg/mL of *G. littoralis* treatment. Then, the higher concentrations showed a significant decrease in the cell population in a concentration-dependent manner (Figure 4), indicating *G. littoralis*'s cellular toxicity properties. Lower concentrations of the plant extracts (between 0 and 2 μg/mL) did not significantly affect the cell viability. Among the different extraction samples, GLSE at a concentration of 0.5–10 μg/mL showed 88.5 ± 1.0% to 96.6 ± 3.9% cellular viability, while these values varied in GLFE (90.7 ± 1.2% to 94.5 ± 0.9%), GLAE (85.8 ± 1.6% to 95.4 ± 1.2%), and GLRE (80.0 ± 0.7% to 97.7 ± 0.4%) at the same concentrations (Figure 4). In addition, the lactate dehydrogenase (LDH) cytotoxicity assay proved that plant extracts beyond 5 μg/mL were toxic to the cell lines (Figure 5). We used 5 μg/mL of *G. littoralis* extract in the subsequent experiments to avoid cytotoxicity and promote MC3T3-E1 cell growth.

3.5. ALP Activity

We evaluated ALP activity to assess how *G. littoralis* extracts affect osteogenic induction (Figure 6). All the sample extracts (GLSE, GLAE, GLRE, and GLFE) significantly increased ALP activity in MC3T3-E1 cells. Comparatively, higher ALP activities were observed at sample concentrations of 0.5 μg/mL. Increasing the treated sample concentration resulted in a decrease in ALP activity. Among the treated samples, GLSE at a concentration of 0.5–20 μg/mL showed higher ALP activity ranging from 158.4 ± 7.9% to 125.9 ± 11.5%, respectively, while the lowest ALP activity was observed in GLFE at a concentration of 0.5–20 μg/mL ranging from 127.4 ± 2.9% to 84.1 ± 3.7%, respectively. The results indicate that *G. littoralis* extract enhanced the ALP activity required for osteoblast formation and ECM mineralization in MC3T3-E1 cells.

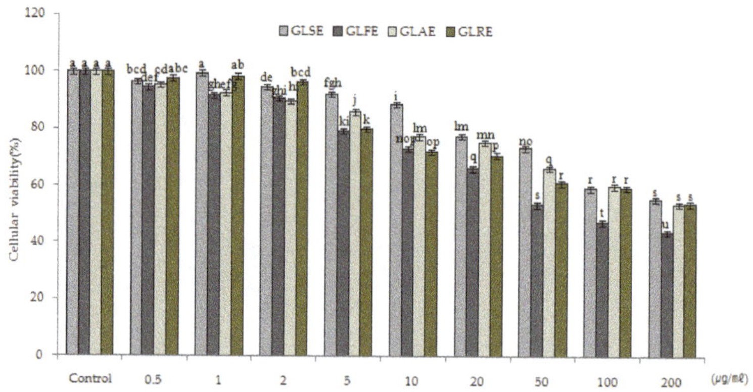

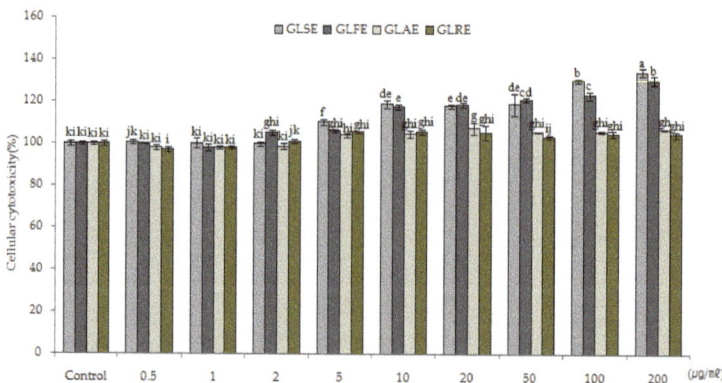

Figure 4. Cell viability of extracts from each part of *G. littoralis* in osteoblastic MC3T3-E1 cell line. Each value is the mean ± standard deviation of nine replicate tests. The mean values followed by the same letter are not significantly different based on the DMRT ($p < 0.05$). GLSE: *G. littoralis* leaf, stem extracts, GLFE: *G. littoralis* fruit extracts, GLAE: *G. littoralis* all extracts, GLRE: *G. littoralis* root extracts.

Figure 5. Cell cytotoxicity of extracts from each part of *G. littoralis* in osteoblastic MC3T3-E1 cell line. Each value is the mean ± standard deviation of nine replicate tests. Mean values followed by the same letter are not significantly different based on the DMRT ($p < 0.05$). GLSE: *G. littoralis* leaf, stem extracts, GLFE: *G. littoralis* fruit extracts, GLAE: *G. littoralis* all extracts, GLRE: *G. littoralis* root extracts.

3.6. Collagen Synthesis Rate

GLSE, at a concentration of 0.5–5 µg/mL, showed a higher collagen synthesis effect than the other extract types (Figure 7). The results indicate that the GLFE methanolic extract had phytochemicals that inhibited the collagen synthesis effect. As Figure 1 shows, GLFE extracts showed lower collagen synthesis effects at concentrations ranging from 1 to 20 µg/mL, which decreased in a concentration-dependent manner. GLFE extract concentrations higher than 2 µg/mL did not affect collagen synthesis. However, no significant difference was observed between GLAE and GLRE extracts in collagen synthesis at any of the concentrations used in the experiment.

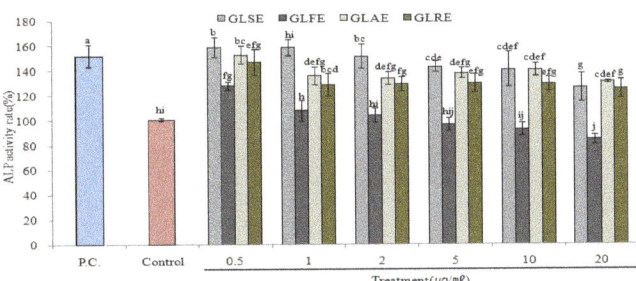

Figure 6. ALP activity of extracts from each part of *G. littoralis* in osteoblastic MC3T3-E1 cell line. Each value is the mean ± standard deviation of nine replicate tests. Mean values followed by the same letter are not significantly different based on the DMRT ($p < 0.05$). P.C.: Positive control (ascorbic acid (50 μg/mL), β-glycerophosphate (100 mM). GLSE: *G. littoralis* leaf, stem extracts, GLFE: *G. littoralis* fruit extracts, GLAE: *G. littoralis* all extracts, GLRE: *G. littoralis* root extracts.

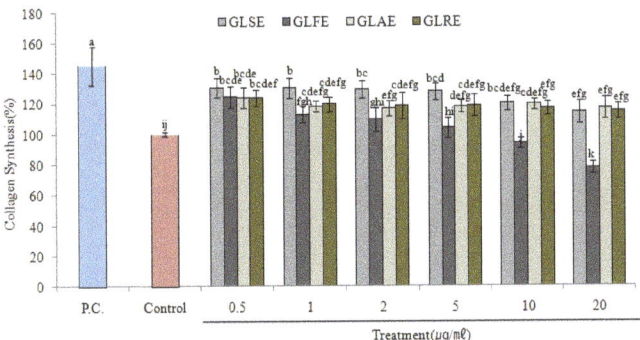

Figure 7. Collagen synthesis effects of extracts from each part of *G. littoralis* in osteoblastic MC3T3-E1 cell line. Each value is mean ± standard deviation of nine replicate tests. Mean values followed by the same letter are not significantly different based on the DMRT ($p < 0.05$). P.C.: Positive control (ascorbic acid (50 μg/mL), β-glycerophosphate (100 mM). GLSE: *G. littoralis* leaf, stem extracts, GLFE: *G. littoralis* fruit extracts, GLAE: *G. littoralis* all extracts, GLRE: *G. littoralis* root extracts.

3.7. GL Extracts Enhanced Osteoblast Mineralization

After seven days of treatment with *G. littoralis* extracts, we measured the mineralization content in MC3T3-E1 cells. Among the different samples, GLAE showed a higher mineral content (131.9 ± 4.7%), followed by GLSE (130.8 ± 2.9%), GLRE (127.8 ± 3.4%), and GLFE (119.9 ± 5.6%), at a concentration of 0.5 μg/mL. *G. littoralis* extracts at a concentration of 0.5 μg/mL induced higher mineralization levels in all the groups compared to 20 μg/mL treatments (Figure 8). GLFE showed lower mineralization and decreased in a concentration-dependent manner. GLAE showed higher mineralization at all concentrations used (0.5–20 μg/mL), indicating that the whole plant extracts contain more bioactive compounds responsible for osteoblast formation.

3.8. Osteocalcin Content

We investigated the effects of *G. littoralis* extracts on the degree of osteocalcin production during the late stage of osteoblast differentiation. As Figure 9 shows, *G. littoralis* extract significantly increased osteocalcin production in MC3T3-E1 cells. This is the first report describing the inhibitory osteoporotic properties of *G. littoralis* using the MC3T3-E1 in vitro system. Osteocalcin content levels varied with different concentrations of the tested plant parts. Except in the case of GLAE, an increase in the sample concentration resulted in reduced osteocalcin content. Comparatively, 0.5 µg/mL GLSE resulted in a higher osteocalcin content, indicating that the phytochemical responsible for producing the protein in osteoblast cells is more present in this extract. However, increasing the GLFE concentration resulted in a decrease in osteocalcin content in a concentration-dependent manner. Comparatively, 20 µg/mL GLFE displayed lower osteocalcin production than the negative control, indicating that a higher GLFE concentration is cytotoxic and inhibits osteoblast formation.

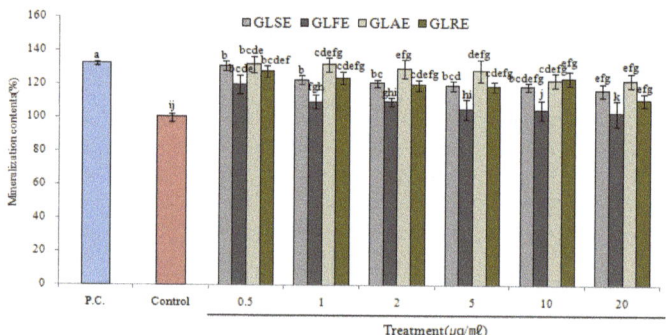

Figure 8. Mineralization contents of extracts from each part of *G. littoralis* in osteoblastic MC3T3-E1 cell line. Each value is the mean ± standard deviation of nine replicate tests. Mean values followed by the same letter are not significantly different based on the DMRT ($p < 0.05$). P.C.: Positive control (ascorbic acid (50 µg/mL), β-glycerophosphate (100 mM). GLSE: *G. littoralis* leaf, stem extracts, GLFE: *G. littoralis* fruit extracts, GLAE: *G. littoralis* all extracts, GLRE: *G. littoralis* root extracts.

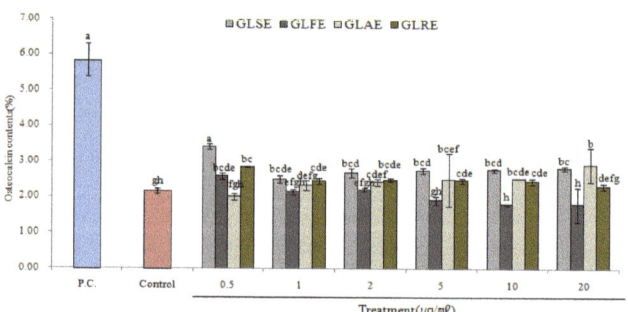

Figure 9. Osteocalcin contents of extracts from each part of *G. littoralis* in osteoblastic MC3T3-E1 cell line. Each value is mean ± standard deviation of nine replicate tests. The mean values followed by the same letter are not significantly different based on the DMRT ($p < 0.05$). P.C.: Positive control (ascorbic acid (50 µg/mL), β-glycerophosphate (100 mM). GLSE: *G. littoralis* leaf, stem extracts, GLFE: *G. littoralis* fruit extracts, GLAE: *G. littoralis* all extracts, GLRE: *G. littoralis* root extracts.

3.9. mRNA Expression Rate

We confirmed how *G. littoralis* extracts affect osteoblast differentiation by analyzing the expression patterns of prominent osteoblast marker genes Table 3. After seven days of treatment with different *G. littoralis* extracts, a real-time polymerase chain reaction (PCR) was performed to investigate the effect on osterix and RUNX2 mRNA expression. As the results show, the expression patterns of both osteoblastic genes changed when treated with the *G. littoralis* extracts. The findings showed that, except in the case of GLFE extracts, *G. littoralis* extract treatment significantly elevated osterix gene expression, and RUNX2 gene expression was increased by all *G. littoralis* extracts. Comparatively, RUNX2 expression significantly increased after treatment with lower *G. littoralis* extract concentrations (ranging from 0.5 to 2 µg/mL).

Table 3. Effect of *G. littoralis* extracts on Osterix and RUNX2 mRNA expression in osteoblastic MC3T3-E1 cells.

Sample	Concentration (µg/mL)	mRNA Expression Rate (Fold) Osterix [1]	RUNX2
P.C.	AA (50 µg/mL) + BGP (100 mM)	1.820 ± 0.072 [c]	2.531 ± 0.050 [a]
Control	-	1.000 ± 0.000 [ki]	1.000 ± 0.000 [k]
GLSE	0.5	1.510 ± 0.046 [ef]	2.327 ± 0.023 [ab]
	1	1.530 ± 0.041 [de]	2.220 ± 0.090 [b]
	2	1.620 ± 0.033 [d]	2.190 ± 0.250 [b]
	5	1.820 ± 0.053 [c]	1.940 ± 0.150 [c]
	10	1.520 ± 0.021 [ef]	1.533 ± 0.029 [fg]
	20	1.430 ± 0.003 [f]	1.410 ± 0.017 [gh]
GLFE	0.5	0.990 ± 0.020 [ki]	1.920 ± 0.080 [cd]
	1	0.970 ± 0.070 [lm]	1.830 ± 0.070 [cde]
	2	0.960 ± 0.010 [lm]	1.800 ± 0.090 [cde]
	5	0.937 ± 0.055 [lm]	1.700 ± 0.120 [def]
	10	0.900 ± 0.070 [lm]	1.310 ± 0.040 [ghij]
	20	0.880 ± 0.050 [m]	1.110 ± 0.050 [jk]
GLAE	0.5	1.280 ± 0.020 [gh]	1.280 ± 1.113 [jk]
	1	1.250 ± 0.070 [gh]	1.250 ± 1.517 [fg]
	2	1.240 ± 0.080 [gh]	1.240 ± 1.660 [ef]
	5	1.200 ± 0.080 [hi]	1.200 ± 1.353 [ghi]
	10	1.130 ± 0.060 [ij]	1.130 ± 1.190 [hijk]
	20	1.080 ± 0.020 [jk]	1.080 ± 1.073 [jk]
GLRE	0.5	1.080 ± 0.050 [jk]	1.150 ± 0.030 [jk]
	1	1.320 ± 0.030 [g]	1.060 ± 0.040 [k]
	2	1.430 ± 0.080 [f]	1.230 ± 0.120 [hij]
	5	1.850 ± 0.020 [c]	1.190 ± 0.150 [hij]
	10	2.020 ± 0.120 [b]	1.080 ± 0.070 [jk]
	20	2.220 ± 0.050 [a]	1.010 ± 0.030 [k]

[1] Each value is mean ± standard deviation of nine replicate tests. Mean values within a column followed by the same letter are not significantly different based on the DMRT ($p < 0.05$). P.C.: Positive control (ascorbic acid (50 µg/mL), β-glycerophosphate (100 mM), AA: Ascorbic acid, BGP: β-glycerophosphate. GLSE: *G. littoralis* leaf, stem extracts, GLFE: *G. littoralis* fruit extracts, GLAE: *G. littoralis* all extracts, GLRE: *G. littoralis* root extracts.

4. Discussion

Plants contain various antioxidants; these antioxidants play an important role in protecting plants from oxidative stress and signals, when ingested, and they can act as natural antioxidants to help prevent disease [45]. Among these compounds, the phenolic compounds present in the plants are mostly responsible for the antioxidant properties [46]. In the present study, the DPPH and ABTS radical scavenging assays presented wide variations in antioxidant activity values for different plant parts. We studied the relationship between the different parameters by obtaining Pearson's correlation coefficients. The antioxidant potentials estimated by both assays were somewhat different and showed a high correlation between them ($p < 0.05$, r = 0.894). The variation in antioxidant potential could be due to the various antioxidant compound responses to the different radicals present in each assay. Excessive reactive oxygen species (ROS) production can cause lipid and protein oxidation, damage DNA integrity, and simultaneously cause tissue damage [47]. Several previous studies have shown the involvement of ROS in bone remodeling by enhancing osteoclastic bone resorption and decreasing osteoblast cell formation [48,49]. In the present study, all *G. littoralis* extracts effectively scavenged the DPPH and ABTS radicals, indicating that the *G. littoralis* extracts had potential antioxidant activity and protected the MC3T3-E1 cells from degeneration and death. Moreover, Pearson's correlation analysis revealed a significant correlation between DPPH and TPC and TFC in *G. littoralis* extracts. A similar correlation was also observed between ABTS and TFC. Phenolic compounds, such as coumarins and their derivatives, have been reported as dominant *G. littoralis* phytochemical components [48,49]. In *G. littoralis* extracts, phenolic compounds, including caffeic acid, vanillic acid, ferulic acid, chlorogenic acid, rutin, quercetin, kaempferol, and coumarins and their derivatives, have been identified [50–52]. In the present study, coumarin-based flavonoids such as scopoletin, umbelliferone, imperatorin, phellopterin have been detected in *G. littoralis* extracts. These compounds are mostly responsible for the antioxidant properties. For instance, scopoletin has been involved in considerable antioxidant activities by scavenging ROS, especially hydrogen peroxide (H_2O_2) scavenging activity, ferrous ion (Fe^{2+}) chelating activity, and activity against superoxide anion radicals ($O_2^{\bullet-}$), and OH-radicals [53]. Moreover, Um et al. [53] isolated scopoletin and umbelliferone from *Glehnia littoralis* and demonstrated a reactive oxygen species (ROS) scavenging ability of about 60% or more compared to a control. Imperatorin (IMP) has been reported in several plants with antioxidant properties [54–57]. Methanolic extracts containing umbelliferone have been shown to exhibit an efficient pro-oxidant activity [58] and inhibit lipid peroxidation [59]. Others observed that the treatment of umbelliferone has been shown to inhibit the intracellular ROS production in irradiated lymphocytes and effectively restore the mitochondrial membrane and inhibited gamma radiation-induced DNA damage [60].

Numerous studies observed that oxidative stress enhances the differentiation and function of osteoclasts [61]. ROS-induced oxidative stress has been shown to involve the suppression of bone formation and the stimulation of osteoclast resorption [62]. Present results indicate that phenolic compounds such as scopoletin, umbelliferone, imperatorin, and phellopterin present in the *G. littoralis* extracts inhibited ROS formation to suppress the excessive bone breakdown by osteoclasts. Moreover, several studies have reported the anti-osteoporotic properties of coumarins by suppressing the interaction of advanced glycation end-products (AGE) and their receptors [63]. Treatment using scopoletin prevented bone loss in diabetic mice by increasing bone turnover of bone-degrading osteoclasts and bone-forming osteoblasts. It has been shown that treatment with imperatorin in rats promoted osteogenesis and suppressed the osteoclast differentiation [64]. The authors found that the imperatorin activates AKT that leads to the inactivation of GSK3β that causes the activation of β-catenin and accumulation of β-catenin in the nucleus [65–68]. It was believed that the activation of β-catenin plays an important role in the suppression of osteoblast differentiation [69]. Thus, it can be inferred that imperatorin could induce osteogenesis via the AKT/GSK3β/β-catenin pathway [64], indicating that imperatorin present in the GL extracts could be responsible for bone growth and inhibition of resorption. Furthermore, umbelliferone prevented trabecular bone matrix degradation and osteoclast formation in bone tissue [70]. The authors reported that umbelliferone is closely associated with the dysfunction of osteoclasts attributed to defects in osteoclast survival and differentiation [65]. In addition, Li et al. [71] reported that phellopterin inhibits Ca^{2+} influx induced by the stimulation of voltage-gated and receptor-dependent calcium channels [72,73]. Therefore, in the present study, these compounds, together with the other polyphenols present in the *G. littoralis* extracts, strongly favored MC3T3-E1 cell differentiation. Furthermore, it can be suggested that *G. littoralis* effectively contributes to the prevention of oxidative damage to bone tissues via antioxidant action and its phytochemicals.

ALP, a typical protein product, is associated with osteoblast growth and differentiation and is expressed and increased during the active matrix maturation phase immediately after the cell proliferation period [18,20,74]. Although the exact ALP mechanism of action is poorly understood, it is believed that these enzymes are responsible for bone mineralization [75]. Therefore, it is important to examine the effect of *G. littoralis* extracts on MS3T3-E1 cell ALP activity. In the present study, all *G. littoralis* extracts effectively accelerated ALP activity in a dose-dependent manner. Moreover, some *G. littoralis* extracts showed higher ALP activity than the positive control, indicating that the different phytochemicals present in the extracts may be necessary for osteoblast differentiation. We hypothesize that

G. littoralis extracts are associated with osteoblast proliferation and differentiation from a newly synthesized protein component.

In this study, maximum ALP activity was observed at the lowest *G. littoralis* extract concentration (0.5 µg/mL), which was confirmed in GLSE and GLAE extracts. Similar to our findings, the aqueous extracts of rooibos promoted ALP activity and mineralization [76]. Moreover, there is abundant evidence that dietary phytochemicals have osteoprotective effects. Caffeic acid regulates bone remodeling by inhibiting osteoclastogenesis, bone resorption, and osteoblast apoptosis [77]. Chlorogenic acid extracted from *Cortex Eucommiae* inhibited a decrease in bone mineral density [78]. Quercetin inhibits osteoblast apoptosis, osteoclastogenesis, and oxidative stress [79]. Jang et al. [80] reported similar results in *A. rugosa*, stating that some of the phenolic compounds present in *A. rugosa* effectively suppressed osteoclasts [81,82]. Flavonoids, such as orientin, quercetin, and luteolin, have shown blastogenic effects by increasing ALP activity and mineralization in rooibos [83,84]. Treating osteoblast cells with various phenolic compounds increases ALP synthesis and decreases the expression of antigens involved in osteoblast immune functions, which may improve bone mineral density [85]. Flavonoids, such as icariin and naringin, were found to regenerate bone tissues by increasing ALP activity and osteopontin content [86–89]. Another study observed increases in osteoblast proliferation, and several other reports have provided convincing data about phytochemicals and their association with osteoclast formation in vitro [90]. Although this study did not determine the exact composition of phenolic compounds in *G. littoralis*, the phenolic compounds from *G. littoralis* could be crucial in modulating the bone formation process through the osteoblast formation process and ALP production.

It has been reported that osteoblasts produce biochemical markers, such as type I collagen, ALP, and osteopontin, which are important components for matrix maturation and mineralization [74]. In the present study, phenotypic markers, such as collagen and osteocalcin, mainly associated with the later stage of osteoblast differentiation and were elevated in the MS3T3-E1 cells treated with *G. littoralis* extract. This indicated that *G. littoralis* extracts were vital in osteoblast differentiation. GLSE showed greater collagen synthesis and upregulated osteoblastic MC3T3-E1 cell proliferation and differentiation by enhancing ALP activity and mineralization compared to the other extract types. There is increasing interest in both in vitro and in vivo research that phenolic compounds may favorably improve osteoporosis. Sparse experimental data show that phenolic acids may have in vitro estrogenic activity. Bioactive compounds, such as β-estradiol, reportedly significantly increase osteoblastic cell proliferation, DNA and protein content, and ALP activity [91]. Phenolic acids may act on osteoblasts by binding to their estrogen receptors, found in osteoblastic cells [92].

Bone mineralization refers to the deposition of calcium and minerals in cells. It acts as a reservoir for calcium and phosphorus in the bone, maintains bone elasticity and flexibility, and provides compactness and mechanical resistance to the bone [93]. During the postmenopausal period, estrogen deficiency causes a decrease in the absorption of micronutrients in the body [94]. In the present study, different GL extracts showed various degrees of elevation in calcium levels, possibly due to affecting calcium absorption and contributing to matrix deposition during osteogenesis [95]. In this test, the maximum mineralization content was observed at the lowest concentration (0.5 µg/mL) of GLSE, GLFE, and GLRE extracts. Our results are consistent with those reported by Yun et al. [96], who observed an increase in calcium deposition in osteoblast MC3T3-E1 cells treated with lower concentrations of chrysanthemum extract. Prak et al. [97] reported similar results in 10 µg/mL of seaweed extracts. Osteocalcin is a non-collagenous protein in the bone secreted into osteoblasts and used as a biochemical marker for bone formation [98]. Osteocalcin is associated with changes in bone turnover rate in bone metabolism and is reflected in the rate of bone formation. The osteocalcin carboxyl group is removed and released into the circulation due to pH acidification of the bone when osteoclasts resorbed it [99].

Several transcription factors are involved in osteoblast differentiation. Most importantly, RUNX2 and osterix are genes that differentiate mesenchymal stem cells into immature osteoblasts and are osteoblast-specific transcription factors required for osteoblast differentiation and bone formation [100]. RUNX2, the earliest identifiable marker, is known as "a master gene" for osteoblast differentiation and is associated with ALP and osterix upregulation [101,102]. It has been argued that RUNX2 triggers osteocalcin expression by binding to the cis-acting elements of the osteocalcin promoter region of osteogenic genes to initiate the expression of ALP, osteopontin, bone sialoprotein (BSP), and osteocalcin [103,104]. Osterix is an osteoblast-specific transcription factor containing a zinc finger. It maintains strong expression in mesenchymal cells and the periosteum and is expressed in cells, such as chondrocytes and the bone matrix [105].

In the present study, we determined the gene expression patterns of osteoblast differentiation markers to understand how *G. littoralis* extracts induce mineralization. The results showed that *G. littoralis* extract treatments significantly elevated osterix and RUNX2 gene expression and enhanced the production of proteins involved in osteoblast production, such as type 1 collagen and osteocalcin. RUNX2 gene upregulation by the cells and their ALP activity and mineralization have also been reported in *Davallia formosana* extracts [106]. Previous studies have shown that phenolic compounds of different plant species extracts can induce the proliferative capacity and maturation of osteoblastic cells by improving ALP activity and increasing calcium ion deposition in the ECM [107,108]. It has been reported that changes in osteoblastic cell activity by phenolic compounds occur through the modulation of different transcription factors, such as osterix, osteocalcin, and bone morphogenic proteins (BMPs), by activating the genes involved [108]. The phenolic compounds of various plant species induce osteoblast cell differentiation through the expression of osterix and RUNX2 markers, which are associated with bone maturation [109–113]. In another study, daidzein, present in soy, acted as a phytoestrogen via osteoblast proliferation and differentiation by activating the BMP/Smad signaling pathway [114]. In the present study, all the GL extracts showed higher expression levels of mRNA expression rate of RUNX2 than control. Furthermore, it has been shown that imperatorin promotes the maturation and differentiation of osteoblast by increasing the expression of RUNX2; thus, it is closely associated with early stage osteogenic differentiation [115–117]. In the present study, all *G. littoralis* extracts increased RUNX2 gene expression. The results described here indicate that *G. littoralis* ethanolic extracts can effectively prevent osteoporosis. These results indicate that the phenolic compounds in GL extracts may synergistically induce osteoblastic cell proliferation to a greater extent than a single compound. Moreover, the results further suggest that phytochemicals other than phenolic compounds may be present in the *G. littoralis* extracts, causing the osteoblastic proliferation of MC3T3-E1 cells. Because the extracts of GLAE with stronger antioxidant activity show stronger anticancer activities, it is implied that the contents of flavonoids in GL are responsible not only for its antioxidant activities but also effectively prevent osteoporosis.

5. Conclusions

This study is the first to demonstrate that *G. littoralis* extracts can enhance osteoblast cell formation. The data produced at the molecular level suggest that *G. littoralis* extracts effectively induced osteoblast cell ALP production and mineralization. However, further research is required to establish the detailed mechanism involved in *G. littoralis*'s anti-osteoporotic potential by identifying the active components present in *G. littoralis* extracts. This study suggests that *G. littoralis* extracts have phytoestrogenic properties that may enable the development of therapeutic agents to prevent osteoporosis.

Supplementary Materials: The following supporting information can be downloaded at: https://www.mdpi.com/article/10.3390/pr11051491/s1. Table S1. Pearson's correlation coefficient among antioxidant activities, total phenolic contents, and total flavonoid contents.

Author Contributions: C.J.K. and B.K.G. contributed by doing experiments and writing the manuscript. C.Y.Y. supervised the experiments. S.K.C. and J.G.L. contributed by analyzing phenolic compounds and editing the manuscript. All authors have read and agreed to the published version of the manuscript.

Funding: This work was supported by funding from Hwajin Bio Cosmetic, Chuncheon 24232, Korea. Also, this work was supported by funding from the KU research professor program.

Informed Consent Statement: Not applicable.

Data Availability Statement: Not applicable.

Conflicts of Interest: The authors declare no conflict of interest.

References

1. Rachner, T.D.; Khosla, S.; Hofbauer, L.C. Osteoporosis: Now and the future. *Lancet* **2011**, *377*, 1276–1287. [CrossRef]
2. Rodan, G.A.; Martin, T.J. Therapeutic approaches to bone diseases. *Science* **2000**, *289*, 1508–1514. [CrossRef]
3. Kung, A. Management of osteoporosis in Hong Kong. *Clin. Calcium* **2004**, *14*, 108–111.
4. Kanis, J.A. *WHO Technical Report*; University of Sheffield: Sheffield, UK, 2007; p. 66.
5. IOF (International Osteoporosis Foundation). Facts and Statistics. International Osteoporosis Foundation Website. Available online: http://www.iofbonehealth.org/factsstatistics (accessed on 18 January 2014).
6. Udagawa, N.; Takahashi, N.; Jimi, E.; Matsuzaki, K.; Tsurukai, T.; Itoh, K.; Nakagawa, N.; Yasuda, H.; Goto, M.; Tsuda, E.; et al. Osteoblasts/stromal cells stimulate osteoclast activation through expression of osteoclast differentiation factor/RANKL but not macrophage colony-stimulating factor. *Bone* **1999**, *25*, 517–523. [CrossRef]
7. Suda, T.; Ueno, Y.; Fujii, K.; Shinki, T. Vitamin D and bone. *J. Cell. Biochem.* **2003**, *88*, 259–266. [CrossRef]
8. Meghji, S.; Sandy, J.R.; Scutt, A.M.; Harvey, W.; Harris, M. Stimulation of bone resorption by lipoxygenase metabolites of arachidonic acid. *Prostaglandins* **1988**, *36*, 139–149. [CrossRef]
9. Garcia, C.; Boyce, B.F.; Gilles, J.; Dallas, M.; Qiao, M.; Mundy, G.R.; Bonewald, L.F. Leukotriene B4 stimulates osteoclastic bone resorption both in vitro and in vivo. *J. Bone Miner. Res.* **1996**, *11*, 1619–1627. [CrossRef] [PubMed]
10. Gruber, H.E.; Ivey, J.L.; Baylink, D.L.; Mathews, M.; Nelp, W.B.; Sisom, B.; Chestnut, C.H. Long-term calcitonin therapy in post-menopausal osteoporosis. *Metabolism* **1984**, *33*, 295–303. [CrossRef] [PubMed]
11. Tasadduq, R.; Gordon, J.; AL-Ghanim, K.A.; Lian, J.B.; Wijnen, A.J.V.; Stein, J.L.; Stein, G.S.; Shakoori, A.R. Ethanol Extract of *Cissus quadrangularis* Enhances Osteoblast Differentiation and Mineralization of Murine Pre-Osteoblastic MC3T3-E1 Cells. *J. Cell. Physiol.* **2017**, *232*, 540–547. [CrossRef] [PubMed]
12. Hamadeh, I.S.; Ngwa, B.A.; Gong, Y. Drug induced osteonecrosis of the jaw. *Cancer Treat. Rev.* **2015**, *41*, 455–464. [CrossRef]
13. Abrahamsen, B. Adverse effects of bisphosphonates. *Calcif. Tissue Int.* **2010**, *86*, 421–435. [CrossRef] [PubMed]
14. Rossouw, J.E.; Anderson, G.L.; Prentice, R.L.; LaCroix, A.Z.; Kooperberg, C.; Stefanick, M.L.; Jackson, R.D.; Beresford, S.A.; Howard, B.V.; Johnson, K.C.; et al. Writing group for the Women's Health Initiative Investigators. Risks and benefits of estrogen plus progestin in healthy postmenopausal women: Principal results from the women's health Initiative randomized controlled trial. *J. Am. Med. Assoc.* **2002**, *288*, 321–333.
15. Adluri, R.S.; Zhan, L.; Bagchi, M.; Maulik, N.; Maulik, G. Comparative effects of a novel plant-based calcium supplement with two common calcium salts on proliferation and mineralization in human osteoblast cells. *Mol. Cell. Biochem.* **2010**, *340*, 73–80. [CrossRef]
16. Kodama, H.; Amagai, Y.; Sudo, H.; Kasai, S.; Yamamoto, S. Establishment of a clonal osteogenic cell line from newborn mouse calvaria. *Jpn. J. Oral Biol.* **1981**, *23*, 899–901. [CrossRef]
17. Quarles, L.D.; Yohay, D.L.; Lever, L.W.; Caton, R.; Wenstrup, R.J. Distinct proliferative and differentiated stages of murine MC3T3-E1 cells in culture: An in vitro model of osteoblast development. *J. Bone Miner. Res.* **1992**, *7*, 683–692. [CrossRef] [PubMed]
18. Golub, E.E.; Boesze-Battaglia, K. The role of alkaline phosphatase in mineralization. *Curr. Opin. Orthop.* **2007**, *18*, 444–448. [CrossRef]
19. Jeong, J.C.; Lee, J.W.; Yoon, C.H.; Lee, Y.C.; Chung, K.H.; Kim, M.G.; Cheorl-Ho Kim, C.H. Stimulative effects of *Drynariae Rhizoma* extracts on the proliferation and differentiation of osteoblastic MC3T3-E1 Cells. *J. Ethnopharmacol.* **2005**, *96*, 489–495. [CrossRef]
20. Lian, J.B.; Stein, G.S. Concepts of osteoblast growth and differentiation: Basis for modulation of bone cell development and tissue formation. *Crit. Rev. Oral Biol. Med.* **1992**, *3*, 269–305. [CrossRef]
21. Ammon, H.P.T.; Mack, T.; Singh, G.B.; Safayhi, H. Inhibition of Leukotriene B4 Formation in Rat Peritoneal Neutrophils by an Ethanolic Extract of the Gum Resin Exudate of *Boswellia serrata*. *Planta Med.* **1991**, *57*, 203–207. [CrossRef]

22. Jun, A.Y.; Kim, H.J.; Park, K.K.; Son, K.H.; Lee, D.H.; Woo, M.H.; Kim, Y.S.; Lee, S.K.; Chung, W.Y. Extract of *Magnoliae Flos* inhibits ovariectomy-induced osteoporosis by blocking osteoclastogenesis and reducing osteoclast-mediated bone resorption. *Fitoterapia* **2012**, *83*, 1523–1531. [CrossRef]
23. Mohan, S.; Kutilek, S.; Zhang, C.; Shen, H.G.; Kodama, Y.; Srivastava, A.K.; Wergedal, J.E.; Beamer, W.G.; Baylink, D.J. Comparison of bone formation responses to parathyroid hormone (1–34), (1–31), and (2–34) in mice. *Bone* **2000**, *27*, 471–478. [CrossRef]
24. Anderson, J.; Garner, S. Phytoestrogens and bone Bailieres. *Clin. Endocrinol. Metab.* **1998**, *12*, 543–557.
25. Fitzpatrick, L. Selective estrogen receptor modulators and phytoestrogens: New therapies for the postmenopausal women. *Mayo Clin. Proc.* **1999**, *74*, 601–607. [CrossRef] [PubMed]
26. Jing, Y.; Li, J.; Zhang, Y.; Zhang, R.; Zheng, Y.; Hu, B.; Wu, L.; Zhang, D. Structural characterization and biological activities of a novel polysaccharide from *Glehnia littoralis* and its application in preparation of nano-silver. *Int. J. Biol. Macromol.* **2021**, *183*, 1317–1326. [CrossRef] [PubMed]
27. Rozema, J.; Bijwaard, P.; Prast, G.; Broekman, R. Ecophysiological adaptation of coastal halophytes from foredunes and salt marshes. *Vegetatio* **1985**, *62*, 499–521. [CrossRef]
28. Cassileth, B.R.; Rizvi, N.; Deng, G.; Yeung, K.S.; Vickers, A.; Guillen, S.; Woo, D.; Coleton, M.; Kris, M.G. Safety and pharmacokinetic trial of docetaxel plus an Astragalus-based herbal formula for non-small cell lung cancer patients. *Cancer Chemother. Pharmacol.* **2009**, *65*, 67–71. [CrossRef]
29. Ng, T.B.; Liu, F.; Wang, H.X. The antioxidant effects of aqueous and organic extracts of *Panax quinquefolium*, *Panax notoginseng*, *Codonopsis pilosula*, *Pseudostellaria heterophylla*, and *Glehnia littoralis*. *J. Ethnopharmacol.* **2004**, *93*, 285–288. [CrossRef]
30. Chiang Su New Medicinal College (de.). *Dictionary of Chinese Crude Drug'*; Shanghai Scientific Technologic Publisher: Shanghai, China, 1977; p. 644.
31. Masuda, T.; Takasugi, M.; Anetai, M. Psoralen and other linear furanocoumarins as phytoalexins in *Glehnia littoralis*. *Phytochemistry* **1998**, *47*, 13–16. [CrossRef]
32. Hwang, Y.H.; Ha, H.; Kim, R.; Cho, C.W.; Song, Y.R.; Hong, H.D.; Kim, T. Anti-Osteoporotic Effects of Polysaccharides Isolated from Persimmon Leaves via Osteoclastogenesis Inhibition. *Nutrients* **2018**, *10*, 901. [CrossRef]
33. Singleton, V.L.; Rossi, J.A., Jr. Colorimetry of total phenolics with phosphomolybdic-phosphotungstic acid reagents. *Am. J. Enol. Vitic.* **1965**, *16*, 144–158. [CrossRef]
34. Moreno, M.I.; Isla, M.I.; Sampietro, A.R.; Vattuone, M.A. Comparison of the free radical-scavenging activity of propolis from several regions of Argentina. *J. Ethnopharmacol.* **2000**, *71*, 109–114. [CrossRef] [PubMed]
35. Chung, I.-M.; Chelliah, R.; Oh, D.-H.; Kim, S.-H.; Yu, C.Y.; Ghimire, B.K. *Tupistra nutans* wall. root extract, rich in phenolics, inhibits microbial growth and α-glucosidase activity, while demonstrating strong antioxidant potential. *Braz. J. Bot.* **2019**, *42*, 383–397. [CrossRef]
36. Denizot, F.; Lang, R. Rapid colorimetric assay for cell growth and survival: Modifications to the tetrazolium dye procedure giving improved sensitivity and reliability. *J. Immunol. Methods* **1986**, *89*, 271–277. [CrossRef] [PubMed]
37. Kim, B.M.; Kim, G.T.; Kim, E.J.; Lim, E.G.; Kim, S.Y.; Kim, Y.M. Extract from *Artemisia annua* Linné induces apoptosis through the mitochondrial signaling pathway in HepG2 cells. *J. Korean Soc. Food Sci. Nut.* **2016**, *45*, 1708–1716. [CrossRef]
38. Sewing, S.; Boess, F.; Moisan, A.; Bertinetti-Lapatki, C.; Minz, T.; Hedtjaern, M.; Tessier, Y.; Schuler, F.; Singer, T.; Roth, A.B. Establishment of a predictive *in vitro* assay for assessment of the hepatotoxic potential of oligonucleotide drugs. *PLoS ONE* **2016**, *11*, 159431. [CrossRef] [PubMed]
39. Liu, X.W.; Ma, B.; Zi, Y.; Xiang, L.B.; Han, T.Y. Effects of rutin on osteoblast MC3T3-E1 differentiation, ALP activity and Runx2 protein expression. *Eur. J. Histochem.* **2021**, *65*, 3195. [CrossRef]
40. Park, E.K.; Jin, H.S.; Cho, D.Y.; Kim, J.H.; Kim, M.C.; Choi, C.W.; Jin, Y.L.; Lee, J.W.; Park, J.H.; Chung, Y.S.; et al. The effect of *Lycii radicis* cortex extract on bone formation in vitro and in vivo. *Molecules* **2014**, *19*, 19594–19609. [CrossRef]
41. Zakłos-Szyda, M.; Nowak, A.; Pietrzyk, N.; Podsedek, A. *Viburnum opulus* L. juice phenolic compounds influence osteogenic differentiation in human osteosarcoma saos-2 cells. *Int. J. Mol. Sci.* **2020**, *21*, 4909. [CrossRef]
42. Bukhari, S.N.A.; Hussain, F.; Thu, H.E.; Hussain, Z. Synergistic effects of combined therapy of curcumin and *Fructus Ligustri Lucidi* for treatment of osteoporosis: Cellular and molecular evidence of enhanced bone formation. *J. Integr. Med.* **2019**, *17*, 38–45. [CrossRef]
43. Techaniyom, P.; Tanurat, P.; Sirivisoot, S. Osteoblast differentiation and gene expression analysis on anodized titanium samples coated with graphene oxide. *Applied Surface Science* **2020**, *526*, 146646. [CrossRef]
44. Matsubara, T.; Kida, K.; Yamaguchi, A.; Hata, K.; Ichida, F.; Meguro, H.; Aburatani, H.; Nishimura, R.; Yoneda, T. BMP2 regulates Osterix through Msx2 and Runx2 during osteoblast differentiation. *J. Biol. Chem.* **2008**, *283*, 29119–29144. [CrossRef] [PubMed]
45. Kasote, D.M.; Katyare, S.S.; Hegde, M.V.; Bae, H. Significance of Antioxidant Potential of Plants and its Relevance to Therapeutic Applications. *Int. J. Biol. Sci.* **2015**, *11*, 982–991. [CrossRef] [PubMed]
46. Womeni, H.M.; Djikeng, F.T.; Tiencheu, B.; Linder, M. Antioxidant potential of methanolic extracts and powders of some Cameroonian spices during accelerated storage of soybean oil. *Adv. Biol. Chem.* **2013**, *3*, 304–313. [CrossRef]
47. Naka, K.; Muraguchi, T.; Hoshii, T.; Hirao, A. Regulation of reactive oxygen species and genomic stability in hematopoietic stem cells. *Antioxid. Redox Signal* **2008**, *10*, 1883–1894. [CrossRef] [PubMed]
48. Bai, X.C.; Lu, D.; Liu, A.L.; Zhang, Z.M.; Li, X.M.; Zou, Z.P.; Zeng, W.S.; Cheng, B.L.; Luo, S.Q. Reactive oxygen species stimulates receptor activator of NF-kappaB ligand expression in osteoblast. *J. Biol. Chem.* **2005**, *280*, 17497–17506. [CrossRef]

49. Lee, N.K.; Choi, Y.G.; Baik, J.Y.; Han, S.Y.; Jeong, D.W.; Bae, Y.S.; Kim, N.; Lee, S.Y. A crucial role for reactive oxygen species in RANKL-induced osteoclast differentiation. *Blood* **2005**, *106*, 852–859. [CrossRef]
50. Kitajima, J.; Okamura, C.; Ishikawa, T.; Tanaka, Y. Coumarin glycosides of *Glehnia lifforalis* root and rhizoma. *Chem. Pharm. Bull.* **1998**, *46*, 1404–1407. [CrossRef]
51. Lee, J.W.; Lee, C.; Jin, Q.; Yeon, E.T.; Lee, D.; Kim, S.Y.; Han, S.B.; Hong, J.T.; Lee, M.K.; Hwang, B.Y. Pyranocoumarins from *Glehnia littoralis* inhibit the LPS-induced NO production in macrophage RAW 264.7 cells. *Bioorganic Med. Chem. Lett.* **2014**, *24*, 2717–2719. [CrossRef]
52. Malik, A.; Kushnoor, A.; Saini, V.; Singhal, S.; Kumar, S.; Yadav, Y.C. In vitro antioxidant properties of Scopolet in. *J. Chem. Pharm. Res.* **2011**, *3*, 659–665.
53. Um, Y.R.; Lee, J.I.; Lee, J.L.; Kim, H.J.; Yea, S.S.; Seo, Y.W. Chemical constituents of the halophyte *Glehnia littoralis*. *J. Korean Chem. Soc.* **2010**, *54*, 701–706. [CrossRef]
54. Nasser, M.I.; Zhu, S.; Hu, H.; Huang, H.; Guo, M.; Zhu, P. Effects of imperatorin in the cardiovascular system and cancer. *Biomed. Pharmacother.* **2019**, *120*, 109401. [CrossRef] [PubMed]
55. Bertina, R.; Chena, Z.; Martínez-Vázquez, M.; García-Argáez, A.; Froldi, G. Vasodilation and radical-scavenging activity of imperatorin and selected coumarinic and flavonoid compounds from genus Casimiroa. *Phytomedicine* **2014**, *21*, 586–594. [CrossRef] [PubMed]
56. Shen, D.Y.; Chao, C.H.; Chan, H.H.; Huang, G.J.; Hwang, T.L.; Lai, C.Y.; Lee, K.H.; Thang, T.D.; Wu, T.S. Bioactive constituents of *Clausena lansium* and a method for discrimination of aldose enantiomers. *Phytochemistry* **2012**, *82*, 110–117. [CrossRef] [PubMed]
57. Adebajo, A.C.; Iwalewa, E.O.; Obuotor, E.M.; Ibikunle, G.F.; Omisore, N.O.; Adewunmi, C.O.; Obaparusi, O.O.; Klaes, M.; Adetogun, G.E.; Schmidt, T.J.; et al. Pharmacological properties of the extract and some isolated compounds of Clausena lansium stem bark: Anti-trichomonal, antidiabetic, antiinflammatory, hepatoprotective and antioxidant effects. *J. Ethnopharmacol.* **2009**, *122*, 10–19. [CrossRef] [PubMed]
58. Kassim, N.K.; Rahmanil, M.; Ismail, A.; Sukari, M.A.; Ee, G.C.L.; Nasir, N.M.; Awang, K. Antioxidant activity-guided separation of coumarins and lignan from *Melicope glabra* (Rutaceae). *Food Chem.* **2013**, *139*, 87–92. [CrossRef]
59. Singh, R.; Singh, B.; Singh, S.; Kumar, N.; Kumar, S.; Arora, S. Umbelliferone-An antioxidant isolated from *Acacia nilotica* (L.) Willd. Ex. Del. *Food Chem.* **2010**, *120*, 825–830. [CrossRef]
60. Kanimozhi, G.; Prasad, N.R.; Ramachandran, S.; Pugalendi, K.V. Umbelliferone modulates gamma-radiation induced reactive oxygen species generation and subsequent oxidative damage in human blood lymphocytes. *Eur. J. Pharmacol.* **2011**, *672*, 20–29. [CrossRef]
61. Luyen, B.T.L.; Tai, B.H.; Thao, N.P.; Lee, S.H.; Jang, H.D.; Lee, Y.M.; Kim, Y.H. Evaluation of the Anti-osteoporosis and Antioxidant Activities of Phenolic Compounds from *Euphorbia maculate*. *J. Korean Soc. Appl. Biol. Chem.* **2014**, *57*, 573–579. [CrossRef]
62. Zhang, J.K.; Yang, L.; Meng, G.L.; Yuan, Z.; Fan, J.; Li, D.; Chen, J.Z.; Shi, T.Y.; Hu, H.M.; Wei, B.Y.; et al. Protection by salidroside against bone loss via inhibition of oxidative stress and bone-resorbing mediators. *PLoS ONE* **2013**, *8*, e57251. [CrossRef]
63. Lee, E.J.; Kang, Y.H. Coumarin Boosts Optimal Bone Remodeling Through Blocking AGE-RAGE Interaction in Diabetic Osteoblasts and Osteoclasts | Current Developments in Nutrition | Oxford Academic. *Curr. Dev. Nutr.* **2020**, *4*, 395.
64. Yan, D.Y.; Tang, J.; Chen, L.; Wang, B.; Weng, S.; Xie, Z.; Wu, Z.Y.; Shen, Z.; Bai, B.; Yang, L. Imperatorin promotes osteogenesis and suppresses osteoclast by activating AKT/GSK3 β/β-catenin pathways. *J. Cell. Mol. Med.* **2020**, *24*, 2330–2341. [CrossRef] [PubMed]
65. Lin, F.X.; Zheng, G.Z.; Chang, B.O.; Chen, R.C.; Zhang, Q.H.; Xie, P.; Li, X.D. Connexin 43 modulates osteogenic differentiation of bone marrow stromal cells through GSK-3beta/Beta-catenin signaling pathways. *Cell. Physiol. Biochem.* **2018**, *47*, 161–175. [CrossRef]
66. Galli, C.; Piemontese, M.; Lumetti, S.; Manfredi, E.; Macaluso, G.M.; Passeri, G. GSK3b-inhibitor lithium chloride enhances activation of Wnt canonical signaling and osteoblast differentiation on hydrophilic titanium surfaces. *Clin. Oral Implant. Res.* **2013**, *24*, 921–927. [CrossRef]
67. Wang, J.; Guan, X.; Guo, F.; Zhou, J.; Chang, A.; Sun, B.; Cai, Y.; Ma, Z.; Dai, C.; Li, X.; et al. miR-30e reciprocally regulates the differentiation of adipocytes and osteoblasts by directly targeting low-density lipoprotein receptor-related protein 6. *Cell Death Dis.* **2013**, *4*, e845. [CrossRef] [PubMed]
68. Li, J.P.; Zhuang, H.T.; Xin, M.Y.; Zhou, Y.L. MiR-214 inhibits human mesenchymal stem cells differentiating into osteoblasts through targeting β-catenin. *Eur. Rev. Med. Pharmacol. Sci.* **2017**, *21*, 4777–4783.
69. Wei, W.; Zeve, D.; Suh, J.M.; Wang, X.; Du, Y.; Zerwekh, J.E.; Dechow, P.C.; Graff, J.M.; Wan, Y. Biphasic and dosage-dependent regulation of osteoclastogenesis by β-catenin. *Mol. Cell. Biol.* **2011**, *31*, 4706–4719. [CrossRef]
70. Kwak, S.C.; Baek, J.M.; Lee, C.H.; Yoon, K.H.; Lee, M.S.; Kim, J.Y. Umbelliferone Prevents Lipopolysaccharide-Induced Bone Loss and Suppresses RANKL-Induced Osteoclastogenesis by Attenuating Akt-c-Fos-NFATc1 Signaling. *Int. J. Biol. Sci.* **2019**, *15*, 2427–2437. [CrossRef]
71. Li, H.T.; He, L.; Qiu, J.B. Effects of the Chinese herb component phellopterin on the increase in cytosolic free calcium in PC12 cells. *Drug Dev. Res.* **2007**, *68*, 79–83. [CrossRef]
72. Ryu, S.Y.; Kim, J.C.; Kim, Y.S.; Kim, H.T.; Kim, S.K.; Chi, G.J.; Kim, J.S.; Lee, S.W.; Heor, J.H.; Cho, K.Y. Antifungal activities of coumarins isolated from *Angelica gigas* and *Angelica dahurica* against plant pathogenic fungi. *Korean J. Pestic. Sci.* **2001**, *5*, 26–35.

73. Kontogiorgis, C.A.; Hadjipavlou-Litina, D.J. Synthesis and antiinflammatory activity of coumarin derivatives. *J. Med. Chem.* **2005**, *48*, 6400–6408. [CrossRef]
74. Stein, G.S.; Lian, J.B.; Owen, T.A. Relationship of cell-growth to the regulation of tissue-specific gene-expression during osteoblast differentiation. *FASEB J.* **1990**, *4*, 3111–3123. [CrossRef] [PubMed]
75. Evans, D.B.; Bunning, R.A.D.; Russell, R.G.G. The effects of recombinant human interleukin-1 on cellular proliferation and the production of prostaglandin E2, plasminogen activator, osteocalcin and alkaline phosphatase by osteoblast-like cells derived from human bone. *Biochem. Biophys. Res. Commun.* **1990**, *166*, 208–216. [CrossRef]
76. Nash, L.A.; Ward, W.E. Comparison of black, green and rooibos tea on osteoblast activity. *Food Funct.* **2016**, *7*, 1166–1175. [CrossRef]
77. Ekeuku, S.O.; Pang, K.L.; Chin, K.Y. Effects of Caffeic Acid and Its Derivatives on Bone: A Systematic Review. *Drug Des. Dev. Ther.* **2021**, *15*, 259–275. [CrossRef] [PubMed]
78. Zhou, R.P.; Lin, S.J.; Wan, W.B.; Zuo, H.L.; Yao, F.F.; Ruan, H.B.; Xu, J.; Song, W.; Zhou, Y.C.; Wen, S.Y.; et al. Chlorogenic Acid Prevents Osteoporosis by Shp2/PI3K/Akt Pathway in Ovariectomized Rats. *PLoS ONE* **2016**, *11*, e0166751. [CrossRef] [PubMed]
79. Wong, S.K.; Chin, K.Y.; Ima-Nirwana, S. Quercetin as an Agent for Protecting the Bone: A Review of the Current Evidence. *Int. J. Mol. Sci.* **2020**, *21*, 6448. [CrossRef]
80. Jang, S.A.; Hwang, Y.H.; Kim, T.; Yang, H.; Lee, J.; Seo, Y.H.; Park, J.I.; Ha, H. Water Extract of *Agastache rugosa* Prevents Ovariectomy-Induced Bone Loss by Inhibiting Osteoclastogenesis. *Foods* **2020**, *9*, 1181. [CrossRef]
81. Goto, T.; Hagiwara, K.; Shirai, N.; Yoshida, K.; Hagiwara, H. Apigenin inhibits osteoblastogenesis and osteoclastogenesis and prevents bone loss in ovariectomized mice. *Cytotechnology* **2015**, *67*, 357–365. [CrossRef]
82. Kim, T.H.; Jung, J.W.; Ha, B.G.; Hong, J.M.; Park, E.K.; Kim, H.J.; Kim, S.Y. The effects of luteolin on osteoclast differentiation, function *in vitro* and ovariectomy induced bone loss. *J. Nutr. Biochem.* **2011**, *22*, 8–15. [CrossRef]
83. Nash, L.A.; Sullivan, P.J.; Peters, S.J.; Ward, W.E. Rooibos flavonoids, orientin and luteolin, stimulate mineralization in human osteoblasts through the Wnt pathway. *Mol. Nutr. Food Res.* **2015**, *59*, 443–453. [CrossRef]
84. Wattel, A.; Kamel, S.; Mentaverri, R.; Lorget, F.; Prouillet, C.; Petit, J.P.; Brazier, M. Potent inhibitory effect of naturally occurring flavonoids quercetin and kaempferol on in vitro osteoclastic bone resorption. *Biochem. Pharmacol.* **2003**, *65*, 35–42. [CrossRef]
85. Melguizo-Rodríguez, L.; Manzano-Moreno, F.J.; De Luna-Bertos, E.; Rivas, A.; Ramos-Torrecillas, J.; Ruiz, C.; García-Martínez, O. Effect of olive oil phenolic compounds on osteoblast differentiation. *Eur. J. Clin. Investig.* **2018**, *48*, e12904. [CrossRef]
86. Xu, B.; Wang, X.; Wu, C.; Zhu, L.; Chen, O.; Wang, X. Flavonoid compound icariin enhances BMP-2 induced differentiation and signalling by targeting to connective tissue growth factor (CTGF) in SAMP6 osteoblasts. *PLoS ONE* **2018**, *13*, e0200367. [CrossRef] [PubMed]
87. Liang, W.; Lin, M.; Li, X.; Li, C.; Gao, B.; Gan, H.; Yang, Z.; Lin, X.; Liao, L.; Yang, M. Icariin promotes bone formation via the BMP-2/Smad4 signal transduction pathway in the hFOB 1.19 human osteoblastic cell line. *Int. J. Mol. Med.* **2012**, *30*, 889–895. [CrossRef]
88. Wu, J.-B.; Fong, Y.-C.; Tsai, H.-Y.; Chen, Y.-F.; Tsuzuki, M.; Tang, C.-H. Naringin-induced bone morphogenetic protein-2 expression via PI3K, Akt, c-Fos/c-Jun and AP-1 pathway in osteoblasts. *Eur. J. Pharmacol.* **2008**, *588*, 333–341. [CrossRef] [PubMed]
89. Gaoli, X.; Yi, L.; Lili, W.; Qiutao, S.; Guang, H.; Zhiyuan, G. Effect of naringin combined with bone morphogenetic protein-2 on the proliferation and differentiation of MC3T3-E1 cells. Hua Xi Kou Qiang Yi Xue Za Zhi Huaxi Kouqiang Yixue Zazhi West China. *J. Stomatol.* **2017**, *35*, 275–280.
90. Hofbauer, L.C.; Kuhne, C.A.; Viereck, V. The OPG/RANKL/RANK system in metabolic bone diseases. *J. Musculoskelet. Neuronal Interact.* **2004**, *4*, 268–275.
91. Sugimoto, E.; Yamaguchi, M. Stimulatory effect of Daidzein in osteoblastic MC3T3-E1 cells. *Biochem. Pharmacol.* **2000**, *59*, 471–475. [CrossRef] [PubMed]
92. Eriksen, E.F.; Colvard, D.S.; Berg, N.J.; Graham, M.L.; Mann, K.G.; Spelsberg, T.C.; Riggs, B.L. Evidence of estrogen receptors in normal human osteoblast-like cells. *Science* **1998**, *241*, 84–86. [CrossRef]
93. Shin, C.S.; Cho, H.Y. Bone remodeling and mineralization. *J. Korean Soc. Endocrinol.* **2005**, *20*, 543–555. [CrossRef]
94. Bonjour, J.P. Calcium and phosphate: A duet of ions playing for bone health. *J. Am Coll. Nutr.* **2011**, *30*, 438S–448S. [CrossRef]
95. Mo, X.M.; Zeng, Y.; Hong, J. Biochemical characteristics of an ovariectomized female rat model of osteoporosis. *J. Tradit. Complement Med.* **1999**, 526–528.
96. Yun, J.H.; Hwang, E.S.; Kim, G.H. Effects of *Chrysanthemum indicum* L. extract on the growth and differentiation of osteoblastic MC3T3-E1 cells. *J. Korean Soc. Food Sci. Nutr.* **2011**, *40*, 1384–1390. [CrossRef]
97. Park, J.H.; Lee, J.W.; Kim, H.J.; Lee, I.S. Effects of *Solidago virga-aurea* var. *gigantea* Miq. root extracts on the activity and differentiation of MC3T3- E1 osteoblastic cell. *J. Korean Soc. Food Sci. Nutr.* **2005**, *34*, 929–936.
98. Akiko, M.; Tomoyo, K.Y.; Masato, H. Osteocalcin and its endocrine functions. *Biochem. Pharmacol.* **2017**, *132*, 1–8.
99. Manolagas, S.C. Osteocalcin promotes bone mineralization but is not a hormone. *PLoS Genet.* **2020**, *16*, e1008714. [CrossRef] [PubMed]
100. Komori, T. Signaling networks in RUNX2-dependent bone development. *J. Cell. Biochem.* **2011**, *112*, 750–755. [CrossRef]
101. Bronckers, A.L.; Sasaguri, K.; Engelse, M.A. Transcription and immunolocalization of Runx2/Cbfa1/Pebp2alphaA in developing rodent and human craniofacial tissues: Further evidence suggesting osteoclasts phagocytose osteocytes. *Microsc. Res. Tech.* **2003**, *61*, 540–548. [CrossRef]

102. Lorenzo, J.A.; Teitelbaum, S.; Faccio, R.; Takayanagi, H.; Choi, Y.; Horowitz, M.; Takayanagi, H. (Eds.) *Chapter 6: The Signaling Pathways Regulating Osteoclast Differentiation*; Academic Press: London, UK, 2011.
103. Bruderer, M.; Richards, R.G.; Alini, M.; Stoddart, M.J. Role and regulation of RUNX2 in osteogenesis. *Eur. Cell Mater.* **2014**, *28*, 269–286. [CrossRef]
104. Franceschi, R.T.; Xiao, G.; Zh Jiang, D.; Gopalakrishnan, R.; Yang Sh, Y.; Reith, E. Multiple signaling pathways converge on the Cbfa1/Runx2 transcription factor to regulate osteoblast differentiation. *Connect. Tissue Res.* **2003**, *44*, 109–116. [CrossRef] [PubMed]
105. Kim, J.E. Function of runx2 and osterix in osteogenesis and teeth. *J. Korean Assoc. Oral Maxillofac. Surg.* **2007**, *33*, 381–385.
106. Wu, C.F.; Lin, Y.S.; Lee, S.C.; Chen, C.Y.; Wu, M.C.; Lin, J.S. Effects of *Davallia formosana* Hayata Water and Alcohol Extracts on Osteoblastic MC3T3-E1 Cells. *Phytother. Res.* **2017**, *31*, 1349–1356. [CrossRef]
107. Hagiwara, K.; Goto, T.; Araki, M.; Miyazaki, H.; Hagiwara, H. Olive polyphenol hydroxytyrosol prevents bone loss. *Eur. J. Pharmacol.* **2011**, *662*, 78–84. [CrossRef] [PubMed]
108. Dai, Z.; Li, Y.; Quarles, L.D.; Song, T.; Pan, W.; Zhou, H.; Xiao, Z. Resveratrol enhances proliferation and osteoblastic differentiation in human mesenchymal stem cells via ER-dependent ERK1/2 activation. *Phytomedicine* **2007**, *14*, 806–814. [CrossRef] [PubMed]
109. Satué, M.; del Mar Arriero, M.; Monjo, M.; Ramis, J.M. Quercitrin and taxifolin stimulate osteoblast differentiation in MC3T3-E1 cells and inhibit osteoclastogenesis in RAW 264.7 cells. *Biochem. Pharmacol.* **2013**, *86*, 1476–1486. [CrossRef]
110. Srivastava, S.; Bankar, R.; Roy, P. Assessment of the role of flavonoids for inducing osteoblast differentiation in isolated mouse bone marrow derived mesenchymal stem cells. *Phytomedicine* **2013**, *20*, 683–690. [CrossRef] [PubMed]
111. Xiao, H.H.; Gao, Q.G.; Zhang, Y.; Wong, K.C.; Dai, Y.; Yao, X.S.; Wong, M.S. Vanillic acid exerts oestrogen-like activities in osteoblast-like UMR 106 cells through MAP kinase (MEK/ERK)-mediated ER signaling pathway. *J. Steroid Biochem. Mol. Biol.* **2014**, *144 Pt B*, 382–391. [CrossRef]
112. Kim, M.B.; Song, Y.; Hwang, J.K. Kirenol stimulates osteoblast differentiation through activation of the BMP and Wnt/β-catenin signaling pathways in MC3T3-E1 cells. *Fitoterapia* **2014**, *98*, 59–65. [CrossRef] [PubMed]
113. Xiao, H.H.; Fung, C.Y.; Mok, S.K.; Wong, K.C.; Ho, M.X.; Wang, X.L.; Yao, X.S.; Wong, M.S. Flavonoids from *Herba epimedii* selectively activate estrogen receptor alpha (ERα) and stimulate ER-dependent osteoblastic functions in UMR-106 cells. *J. Steroid Biochem. Mol. Biol.* **2014**, *143*, 141–151. [CrossRef]
114. Hu, B.; Yu, B.; Tang, D.; Li, S.; Wu, Y. Daidzein promotes osteoblast proliferation and differentiation in OCT1 cells through stimulating the activation of BMP-2/Smads pathway. *Genet. Mol. Res.* **2016**, *15*, 1–10. [CrossRef]
115. Isaac, J.; Erthal, J.; Gordon, J.; Gordon, J.; Duverger, O.; Sun, H.-W.; Lichtler, A.C.; Stein, G.S.; Lian, J.B.; Morasso, M.I. DLX3 regulates bone mass by targeting genes supporting osteoblast differentiation and mineral homeostasis in vivo. *Cell Death Differ.* **2014**, *21*, 1365–1376. [CrossRef] [PubMed]
116. Khrimian, L.; Obri, A.; Karsenty, G. Modulation of cognition and anxiety-like behavior by bone remodeling. *Mol. Metab.* **2017**, *6*, 1610–1615. [CrossRef] [PubMed]
117. Yang, Y.; Bai, Y.; He, Y.; Zhao, Y.; Chen, J.; Ma, L.; Pan, Y.; Hinten, M.; Zhang, J.; Karnes, R.J.; et al. PTEN Loss promotes intratumoral androgen synthesis and tumor microenvironment remodeling via aberrant activation of RUNX2 in castration-resistant prostate cancer. *Clin Cancer Res.* **2018**, *24*, 834–846. [CrossRef] [PubMed]

Disclaimer/Publisher's Note: The statements, opinions and data contained in all publications are solely those of the individual author(s) and contributor(s) and not of MDPI and/or the editor(s). MDPI and/or the editor(s) disclaim responsibility for any injury to people or property resulting from any ideas, methods, instructions or products referred to in the content.

Article

Novel Non-Toxic Highly Antibacterial Chitosan/Fe(III)-Based Nanoparticles That Contain a Deferoxamine—Trojan Horse Ligands: Combined Synthetic and Biological Studies

Omar M. Khubiev [1], Victoria E. Esakova [1], Anton R. Egorov [1], Artsiom E. Bely [1], Roman A. Golubev [1,2], Maxim V. Tachaev [1], Anatoly A. Kirichuk [1], Nikolai N. Lobanov [1], Alexander G. Tskhovrebov [1] and Andreii S. Kritchenkov [1,2,*]

[1] Faculty of Science, Peoples' Friendship University of Russia (RUDN University), 117198 Moscow, Russia; ihubievomar1@gmail.com (O.M.K.); 1032192994@rudn.ru (V.E.E.); sab.icex@mail.ru (A.R.E.); borutobirama@gmail.com (A.E.B.); asdfdss.asdasf@yandex.ru (R.A.G.); tatchaev@mail.ru (M.V.T.); kirichuk-aa@rudn.ru (A.A.K.); nnlobanov@mail.ru (N.N.L.); alexander.tskhovrebov@gmail.com (A.G.T.)
[2] Metal Physics Laboratory, Institute of Technical Acoustics NAS of Belarus, 210009 Vitebsk, Belarus
* Correspondence: kritchenkov-as@rudn.ru

Citation: Khubiev, O.M.; Esakova, V.E.; Egorov, A.R.; Bely, A.E.; Golubev, R.A.; Tachaev, M.V.; Kirichuk, A.A.; Lobanov, N.N.; Tskhovrebov, A.G.; Kritchenkov, A.S. Novel Non-Toxic Highly Antibacterial Chitosan/Fe(III)-Based Nanoparticles That Contain a Deferoxamine—Trojan Horse Ligands: Combined Synthetic and Biological Studies. *Processes* **2023**, *11*, 870. https://doi.org/10.3390/pr11030870

Academic Editors: Iliyan Ivanov and Stanimir Manolov

Received: 20 February 2023
Revised: 10 March 2023
Accepted: 13 March 2023
Published: 14 March 2023

Copyright: © 2023 by the authors. Licensee MDPI, Basel, Switzerland. This article is an open access article distributed under the terms and conditions of the Creative Commons Attribution (CC BY) license (https://creativecommons.org/licenses/by/4.0/).

Abstract: In this study, we prepared chitosan/Fe(III)/deferoxamine nanoparticles with unimodal size distribution (hydrodynamic diameter ca. 250 nm, zeta potential ca. 32 mV). The elaborated nanoparticles are characterized by outstanding in vitro and in vivo antibacterial activity, which exceeds even that of commercial antibiotics ampicillin and gentamicin. Moreover, the nanoparticles are non-toxic. We found that the introduction of iron ions into the chitosan matrix increases the ability of the resulting nanoparticles to disrupt the integrity of the membranes of microorganisms in comparison with pure chitosan. The introduction of deferoxamine into the obtained nanoparticles sharply expands their effect of destruction the bacterial membrane. The obtained antibacterial nanoparticles are promising for further preclinical studies.

Keywords: chitosan; iron(III); deferoxamine; nanoparticles; antibacterial activity; toxicity

1. Introduction

Infectious diseases of bacterial etiology are the cause of a large number of deaths and disability among the population [1–3]. In addition, infectious diseases affect not only people, but plants and animals, thereby causing significant damage to agriculture [4–6]. Therefore, infectious diseases represent both a medical and economic problem. Antibiotics have been used to treat bacterial infections for decades [7]. Undoubtedly, the introduction of antibiotics into clinical practice was one of the most important milestones in the history of medicine [8,9].

However, the use of antibiotics is associated with three major problems. The first problem is associated with the general systemic toxicity of antibiotics, which causes side effects during treatment and requires significant restrictions, especially in the cases of elderly patients, pregnant women, and children [10–12]. The second problem is related environmental issues. Antibiotics are often detected in wastewater, food, and even food packaging. Therefore, they harm the environment [13–16]. The third problem is related to the emergence of antibiotic resistance in bacteria, and this requires the prescription of large doses of antibiotics and/or the simultaneous use of several antibiotics [17–19]. All these problems have stimulated an intensive search for alternatives to traditional antibiotics, and such studies are among the most important tasks of medicinal chemistry and pharmacology.

Chitosan is a non-toxic, biocompatible and biodegradable polymer [20], which belongs to the most important eco-friendly macromolecular compounds. In addition, chitosan itself exhibits moderate antibacterial activity. The antibacterial effect of some of chitosan derivatives is comparable with that of common commercial antibiotics ampicillin and gentamicin, and its transfection activity is similar to that of commercially available vector lipofectamine. Meanwhile, the toxicity of these chitosan derivatives is much lower than that of the reference antibiotics or lipofectamine [21].

Iron(III) is one of the most non-toxic 3D transition metals in the periodic table [22,23]. Moreover, several non-toxic iron-based compounds with promising antibacterial activity (including in vivo) are described in the literature [24–27].

The non-toxic natural compound deferoxamine is part of a group of so-called siderophores [28,29]. These are small, high-affinity compounds that chelate iron. Siderophores are secreted by bacteria and help them to store iron. Antibacterial compounds conjugated to siderophores are actively taken up by bacteria and this is used in medicinal chemistry. Therefore, siderophores are often called the Trojan horse ligand [30,31].

In this study, we hypothesized that the non-covalent fusion of chitosan, iron(III) and deferoxamine (Trojan horse ligand) would lead to the formation of a novel antibacterial system. Such a system consists of natural non-toxic compounds and therefore should be a non-toxic and promising alternative to antibiotics. The results of the synthesis, characterization and investigation of the biological properties of this system are discussed in detail in the sections that follow below.

2. Materials and Methods

2.1. Materials

In this study, we used chitosan with a viscosity average molecular weight of 2.7 kDa and degree of acetylation of 10%, abbreviately named CH (Bioprogress, Losino-Petrovsky, Russia); iron(III) chloride hexahydrate and deferoxamine are abbreviately named DESF (Aldrich, St. Louis, MI, USA). All other chemicals and solvents were obtained from commercial sources and were used as received, without further purification.

2.2. Synthesis of Chitosan-Fe^{3+} Complex (POX-1)

A chitosan sample (1.0 g) was dispersed in 50 mL of a 1% solution of acetic acid and was stirred (400 rpm) for 24 h at room temperature. Then, 10 mL of the resultant solution was diluted to 200 mL with distilled water (solution A). Next, 0.1 g of ferric chloride(III) hexahydrate was dissolved in 100 mL of water (solution B). Solutions A and B were mixed and stirred for 1 h. The resultant solution was freeze-dried to give POX-1 as a soft orange cotton-like material.

2.3. Synthesis of Chitosan-Fe^{3+} Complex (POX-2 and POX-3)

First, 100 mg of POX-1 was dissolved in 100 mL of distilled water (solution C) and 100 mg of deferoxamine was dissolved in 100 mL of distilled water (solution D). Solutions C and D were mixed (solution E); one part of solution E was stirred for 2 h and freeze-dried to give POX-2, and another was stirred for 24 h and freeze-dried to give POX-3.

2.4. General Methods

The apparent hydrodynamic diameter and ζ-potential of nanoparticles in water were estimated at room temperature (approximately 25 °C) using a Photocor Compact-Z instrument (Russia) at λ = 659 nm and θ = 90°(10 scans, each one for 15 s).

IR spectroscopy was recorded on a Shimadzu IRSpirit at 4700 to 350 cm^{-1} (10 mg of sample without any specified sample preparation).

UV spectra were recorded using a Mettler UV5 spectrophotometer.

Differential thermal analysis (DTA) and thermogravimetric analysis (TGA) were performed on the SDT Q600 using a heating rate of 5 °C/min in the temperature range from 40 °C to 600 °C.

X-ray diffraction analysis was carried out on a Dron-7 X-ray diffractometer, using a 2θ angle interval from 7° to 40° with scanning step Δ2θ = 0.02° and exposure of 7 s per point.

Cu K$_\alpha$ radiation (Ni filter) was used, which was subsequently decomposed into K$_{\alpha 1}$ and K$_{\alpha 2}$ components during the processing of the spectra.

Loading efficiency (LE) was calculated using the following equation:

$$LE = [(m(\text{deferoxamine total}) - m(\text{c deferoxamine in supernatant}))/m(\text{deferoxamine total})] \times 100$$

The mass of deferoxamine in the supernatant was determined by UV spectroscopy at a wavelength of 252 nm (calibration curve method).

X-ray fluorescence analysis of the samples was performed on a Clever C-31 X-ray fluorescence spectrometer. The relative measurement error was ±7%. A rhodium tube with a voltage of 50 kV and a current of 100 μA acted as a generator of γ-rays. The samples were taken without filters for 2000 s.

High-resolution electrospray ionization mass spectrometry (positive ion mode) was carried out on a APEX-Qe ESI FT-ICR instrument (Bruker, Billerica, MA, USA) with CH$_3$CN as a solvent.

Antibacterial activity (in vitro and in vivo) and toxicity were evaluated completely as previously described by some of our group [32–35].

3. Results and Discussion

3.1. Preparation of Nanoparticles POX-1, POX-2 and POX-3

Treatment of the chitosan solution with iron(III) chloride immediately results in the generation of yellow-colored nanoparticles POX-1 of unimodal size distribution with hydrodynamic diameter ca. 285 nm and a high positive zeta potential (ca. 32 mV, see Table 1). The formed nanoparticles do not change their characteristic size and zeta potential values in a water nanosuspension for at least 10 days. Remarkably, when immersed in water after lyophilization, POX-1 is almost instantly redispersed, with the complete restoration of the starting values of the hydrodynamic diameter and zeta potential.

Table 1. Characteristics of the obtained nanoparticles.

Sample	D, nm	ζ, mV	Polydispersity Index
POX-1	285 ± 2	+31.8 ± 0.1	0.11 ± 0.02
POX-2	254 ± 1	+32.0 ± 0.3	0.10 ± 0.02
POX-3	260 ± 1	+32.5 ± 0.2	0.11 ± 0.03

The addition of deferoxamine to the POX-1 nanosuspension leads to the rapid formation of novel POX-2 nanoparticles of smaller size (hydrodynamic diameter ca. 254 nm) and the same zeta potential value as for POX-1 (ca. +32 mV, see Table 1). In water, POX-2, after 24 h, converts into POX-3 with hydrodynamic diameter ca. 260 nm, while the zeta potential value remains unchanged. It is likely that POX-2 and POX-3 are the same system, but this will be discussed in the following sections. It should be noted that both POX-2 and POX-3 are completely redispersible after lyophilization.

To confirm that deferoxamine is part of POX-2 and POX-3 and is not present in the solution in free form, we separated the resulting POX-2 and POX-3 from the supernatant after mixing the POX-1 and deferoxamine solutions. High-resolution mass spectrometry with electrospray ionization of the supernatant did not reveal any deferoxamine signals. Therefore, deferoxamine is included in the POX-2 and POX-3 nanoparticles. The loading efficiency (LE) of deferoxamine was ca. 100%.

X-ray fluorescence analysis of POX-1, POX-2 and POX-3 confirmed the presence of iron in the samples.

POX-2 also was characterized by scanning electron microscopy. The SEM image of POX-2 is presented in Figure 1.

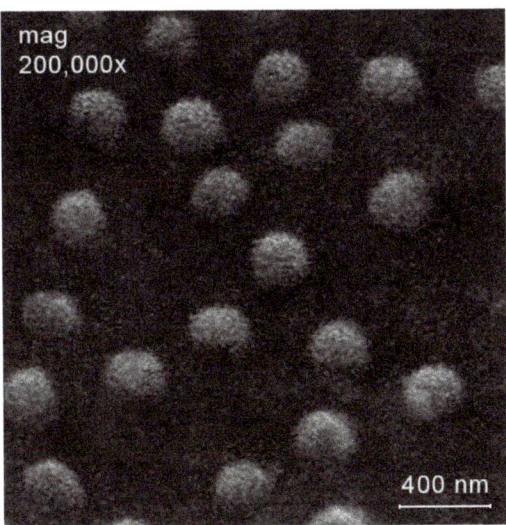

Figure 1. SEM image of POX-2.

3.2. Infrared Spectroscopy

The spectrum of $FeCl_3 \times 6H_2O$ (Figure 2A) displays small peaks of deformation vibrations of crystallized water at 1600 cm^{-1} and stretching vibrations of free water at 3530 cm^{-1}. A wide double band of stretching vibrations of crystallized water exhibits two pronounced maxima at 3000 and 3220 cm^{-1}, while coordinated water vibration bands arise at 840, 680, 600, 540 cm^{-1}. The spectrum of $FeCl_3 \times 6H_2O$ also displays stretching vibration bands attributed to Fe–Cl (360 cm^{-1}) and Fe–O (420 cm^{-1}) vibrations [36].

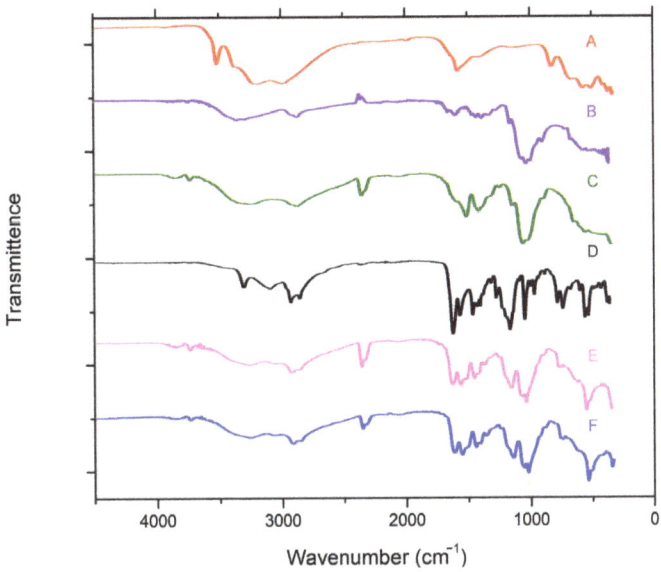

Figure 2. IR spectrum of $FeCl_3 \times 6H_2O$ (**A**); IR spectrum of chitosan (**B**); IR spectrum of POX-1 (**C**); IR spectrum of deferoxamine (**D**); IR spectrum of POX-3 (**E**); IR spectrum of POX-2 (**F**).

The spectrum of chitosan (Figure 2B) shows a wide band of O–H and N–H stretching (3440–3100 cm^{-1}), C–H stretching (2870 cm^{-1}) and bending (1460, 1420, 1380 cm^{-1}) vibrations, and N–H deformation vibrations (1590-1650 cm^{-1}). Absorption bands in the range of 900–1200 cm^{-1} are due to C–O–C, C–C and N–H deformation vibrations [37].

The spectrum of deferoxamine (Figure 2D) exhibits absorption bands of C=O, =O-H, -CONHR, -CH$_2$ stretching vibrations (2840–3240 cm^{-1}), peaks at 1620 cm^{-1} and 1560 cm^{-1} of stretching =N-CO and deformation NH vibration bands. Peaks of stretching vibrations -CH$_2$-CO are located in the interval 1400–1440 cm^{-1}, while deformation aliphatic amine vibration peaks are found at 1030–1220 cm^{-1}. In the range of 1040–500 cm^{-1}, we observe deformation vibrations -(CH$_2$)$_x$-, -NH$_2$.

The spectrum of POX-1 (Figure 2C) displays the C–H stretching vibration band at cm^{-1}, which is shifted to 2890 cm^{-1} as compared to the starting chitosan. The spectrum shows a wide band of O–H and N–H stretching (3440–3100 cm^{-1}), C–H stretching (2890–2870 cm^{-1}) and bending (1460, 1420, 1380 cm^{-1}) vibrations and characteristic bands due to C–O–C and C–C deformation vibrations (900–1200 cm^{-1}). The N–H deformation vibration band is shifted to 1560–1520 cm^{-1} in comparison with the chitosan.

The spectrum of POX-2 (Figure 2E) exhibits stretching vibration bands characteristic of POX-1 (Figure 2C), i.e., O–H and N–H peaks (3440–3100 cm^{-1}) and a characteristic band at 2360 cm^{-1}. C–H stretching vibration bands are located at 2920–2840 cm^{-1} and slightly shifted in comparison to POX-1. The spectrum of POX-2 exhibits stretching vibration bands characteristic of deferoxamine (Figure 2D), i.e., peaks of stretching =N–CO (1620 cm^{-1}) and deformation N–H (1560 cm^{-1}) vibration bands, stretching vibrations CH$_2$–CO (1400–1440 cm^{-1}) and a characteristic peak at 520–580 cm^{-1}.

The spectrum of POX-2 (Figure 2E) is identical to the spectrum of POX-3 (Figure 2F). This confirms our assumptions about their identical structures.

3.3. X-Ray Diffraction

The spectra of chitosan and POX-1 are very similar. Both diffractograms show similar stretched peaks at 10–30° 2θ, which are attributed to chitosan (Figure 3A) [38].

The maxima of the spectra of both chitosan and deferoxamine are in the same region (15–27° 2θ). The spectrum of chitosan displays a double wide peak with the left-shouldered maximum at 20° 2θ, while the diffractogram of deferoxamine at the same region exhibits four peaks with the main one at ca. 21.2° 2θ. Thus, the maximum and the shoulders of the deferoxamine-conditioned signals are right-shifted as compared to those of chitosan (Figure 3B).

The maxima of the spectra of both POX-2 and deferoxamine are located at ca. 21° 2θ. The spectrum of the POX-1 sample does not display any significant peak at this region. This fact indicates the presence of deferoxamine in POX-2 and its absence in POX-1 (Figure 3C).

The X-ray diffraction spectra of POX-2 and POX-3 samples are very similar. However, the diffractogram of POX-3 displays a broader maximum peak at 18–25°. 2θ. This indicates the presence of smaller, crystallographically active structural units in the main particles of the sample (for example, in the POX-3 nanoparticles, probably, there are smaller nanolevel particles) (Figure 3D). The described differences do not refute our assumption about the identity of the chemical structure of POX-2 and POX-3.

Generally, the X-ray diffraction study confirms the identical chemical structures of POX-2 and POX-3, and allows us to conclude that both POX-2 and POX-3 contain chitosan and deferoxamine.

3.4. Differential Thermal Analysis (DTA) and Thermogravimetric Analysis (TGA)

The pure chitosan TGA curve has two stages of thermal degradation (Figures 4 and 5). The first stage is associated with a small weight loss due to the evaporation of the adsorbed and weakly bound water (mass loss 7%, T_{max} = 121 °C). This stage is accompanied by a weak endothermic effect. The second stage is associated with the degradation of the polymer structure (mass loss 89%, T_{max} = 520 °C).

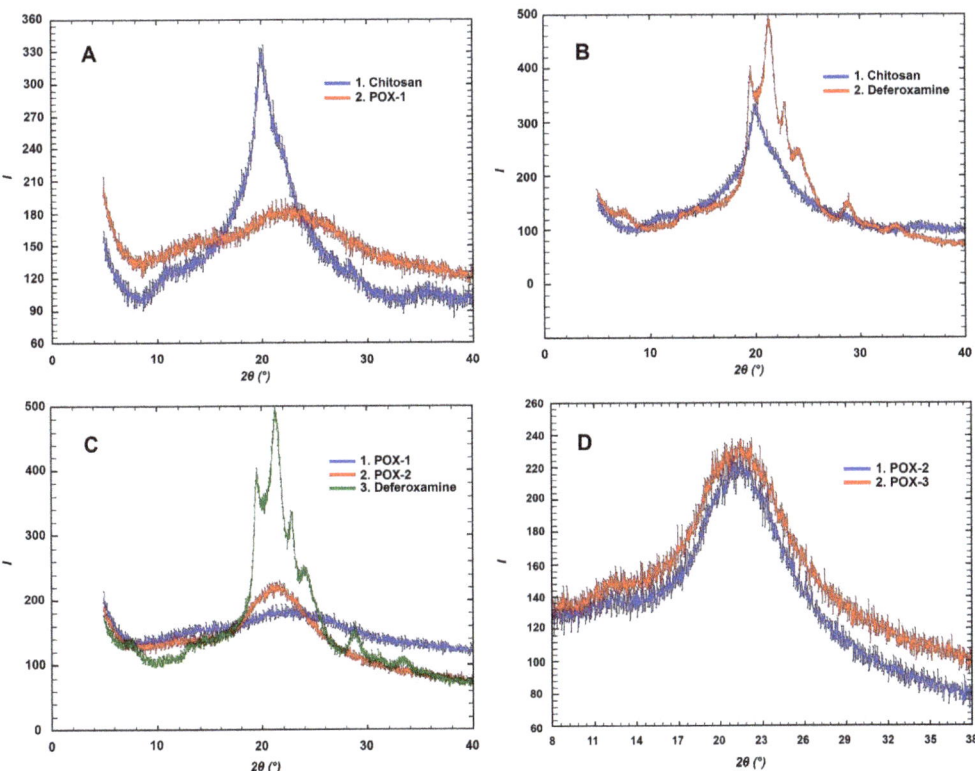

Figure 3. XRD patterns of chitosan and POX-1 (**A**); XRD patterns of chitosan and deferoxamine (**B**); XRD patterns of POX-1, POX-2 and deferoxamine (**C**); XRD patterns of POX-2 and POX-3 (**D**).

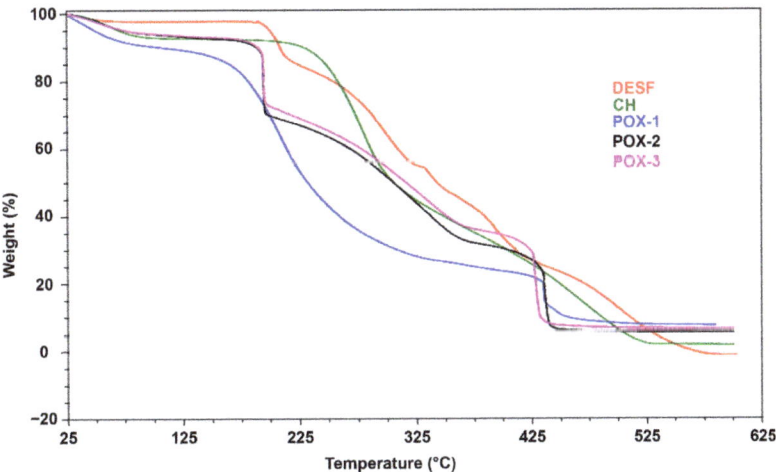

Figure 4. TGA curves of DESF, CH, POX-1, POX-2, POX-3.

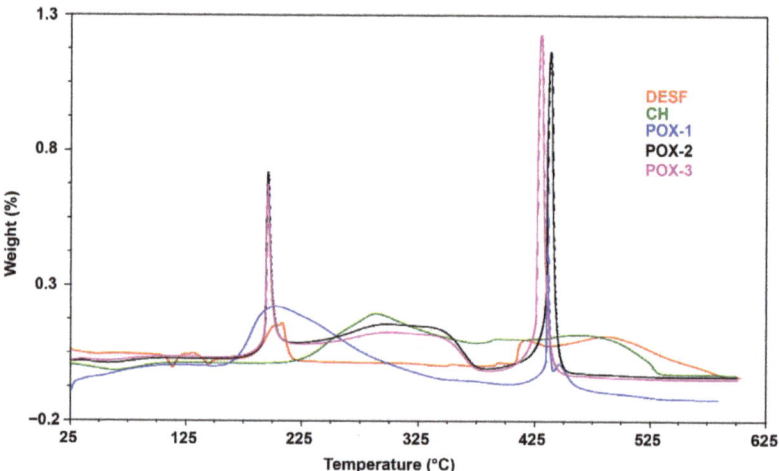

Figure 5. DTA curves of DESF, CH, POX-1, POX-2, POX-3.

POX-1 was obtained by the interaction of chitosan with Fe^{3+} ions. The thermal degradation curve of POX-1 is characterized by three stages of degradation. The first stage is associated with water loss, and it is accompanied by a slight endothermic effect (mass loss 10%, T_{max} = 122 °C). The second stage is associated with the degradation of the polymer structure, and it is accompanied by a pronounced, unsharp exothermic effect (mass loss 73%, T_{max} = 440 °C). The third stage of decomposition has a spasmodic, pronounced exothermic effect (mass loss 10%, T_{max} = 453 °C) (Figures 4 and 5).

POX-2 and POX-3 were prepared via the interaction of POX-1 with deferoxamine. Their thermal decomposition curves are almost identical, and the values of weight loss and maximum temperature differ by no more than 10% (Figures 4 and 5). The first stage of thermal degradation is associated with water loss (mass loss 7%, T_{max} = 146 °C). The second stage is accompanied by an acute exothermic effect (mass loss 25%, T_{max} = 200 °C) followed by a gradual loss of mass. The third stage is accompanied by a sharp, pronounced exothermic effect (T_{max} = 445 °C, weight loss 20%). The POX-1 and POX-2 curves are significantly different from those of the starting deferoxamine and POX-1, which indicates the formation of new systems. The systems include the characteristic features of the thermal decomposition of both starting deferoxamine and POX-1: (i) deferoxamine-like weight loss at 200 °C (for deferoxamine, at 202 °C); (ii) POX-1-like sharp weight loss at 445 °C (for POX-1, at 453 °C). These observations are in agreement with the results of the X-ray diffraction study.

3.5. Biological Studies

3.5.1. In Vitro Antibacterial Activity

The in vitro antibacterial activity of the prepared nanoparticles was compared with that of the starting chitosan, iron(III) chloride hexahydrate and deferoxamine. For the in vitro evaluation of the antibacterial effect, we used the conventional agar well diffusion method. This method allows one to directly estimate the diameter of the microbial colonies' growth inhibition zone. The compound that provokes the largest zone of inhibition of bacterial growth is considered the most active antibacterial agent. The results of the experiments are presented in Table 2.

Table 2. Antibacterial effects of the elaborated nanoparticles.

Sample	Inhibition Zone (mm) *	
	S. aureus	E. coli
Chitosan	13.1 ± 0.1	9.7 ± 0.3
Iron(III) chloride hexahydrate	16.0 ± 0.3	12.2 ± 0.2
Desferal	2.6 ± 0.1	1.3 ± 0.1
POX-1	22.4 ± 0.1	14.8 ± 0.2
POX-2	29.7 ± 0.2	22.5 ± 0.1
POX-3	29.9 ± 0.1	21.7 ± 0.3
Ampicillin	30.1 ± 0.3	
Gentamicin		22.1 ± 0.1

* Mean value ± SD, n = 3.

Both the starting chitosan and iron(III) hexahydrate are characterized by moderate antibacterial activity, while their composite POX-1 is ca. 1.5 times more effective toward the tested bacteria. This can be explained by the formation of highly positively charged nanoparticles of POX-1. Our previous studies showed that, in many instances, chitosan-based nanoparticles exhibit much stronger antibacterial effects than the starting chitosan in its native form of a molecular coil [39].

Deferoxamine practically does not have an antibacterial effect. In contrast, the blending of deferoxamine with POX-1 results in POX-2 with outstanding antibacterial activity, which is comparable to that of the reference antibiotics ampicillin and gentamicin. We speculate that the high antibacterial activity of POX-2 is associated with the symbatic action of POX-1/deferoxamine in their complex with POX-2, but a fuller understanding of their mechanism requires additional biological studies.

The antibacterial effect of POX-3 essentially does not differ from that of POX-2.

The most effective antibacterial nanoparticles proved to be POX-2 and POX-3. The minimum inhibitory concentration (MIC) values were 0.14 µg/mL (S. aureus) and 0.19 µg/mL (E. coli) (compare with MIC values of ampicillin 0.18 µg/mL (S. aureus), gentamicin 0.23 µg/mL (E. coli)). Thus, the most active elaborated antibacterial nanoparticles, POX-2 and POX-3, are not inferior in their in vitro effect to the conventional antibiotics ampicillin and gentamicin.

3.5.2. Effect of the Integrity of the Bacterial Membrane

The main recognized model of the mechanism of the antibacterial effect of chitosan is associated with its polycationic nature [40]. Due to the protonation of primary amino functionalities, the neutral chitosan (pK_a = 6.5) is converted into its polycation. The polycation interacts with the anionic moieties of the microbial cell, and this results in ionic pumps' dysfunction, osmotic imbalance, and overall membrane dysfunction, followed by cell membrane rupture and bacterial death [41].

To study the effects of the elaborated systems on the integrity of the microbial membrane, we used spectrophotometry of a suspension of bacterial cells in a 0.5% aqueous solution of sodium chloride in the UV region [42]. This approach is based on the fact that intracellular components are characterized by strong absorption at 260 nm [43]. In preliminary experiments, we found that both deferoxamine and iron(III) chloride did not damage the membranes of the tested bacteria. Thus, we compared the effect of starting chitosan and POX-1, POX-2 and POX-3 on the integrity of the cell membranes of S. aureus and E. coli. The results of the study are summarized in Figures 6 and 7.

In general, POX-1, POX-2 and POX-3 disrupt the integrity of the bacterial cell to a greater extent than the initial chitosan, and the effect of POX-2 and POX-3 is more pronounced than that of the starting chitosan. In addition, both in the case of Staphylococcus aureus and in the case of E. coli, POX-2 and POX-3 require less time to reach a plateau, i.e., their effect develops faster than that of chitosan of POX-1. Thus, the main mode of the antibacterial effect of POX-2 and POX-3 is the destruction of the integrity of the bacterial cell membrane.

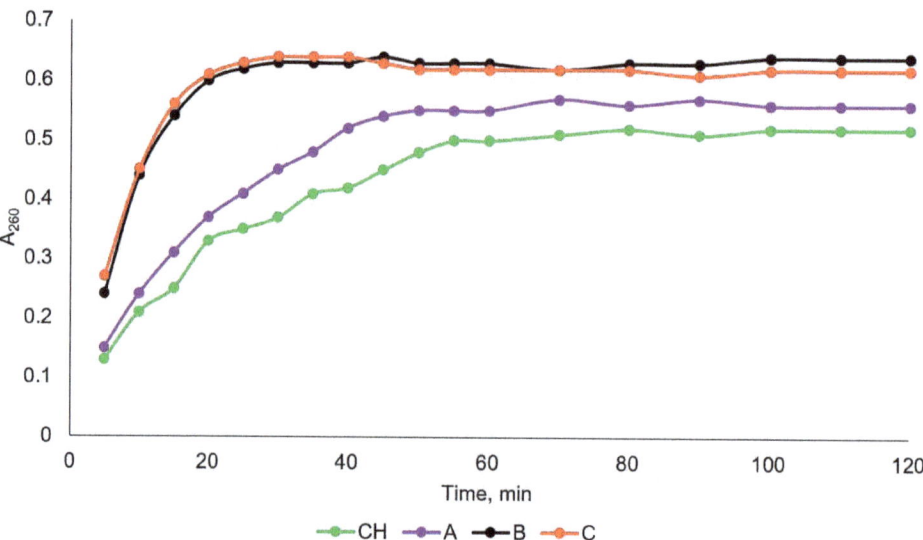

Figure 6. The effects of chitosan (CH), POX-1 (A), POX-2 (B), POX-3 (C) on the integrity of the cell membranes of *E. coli*.

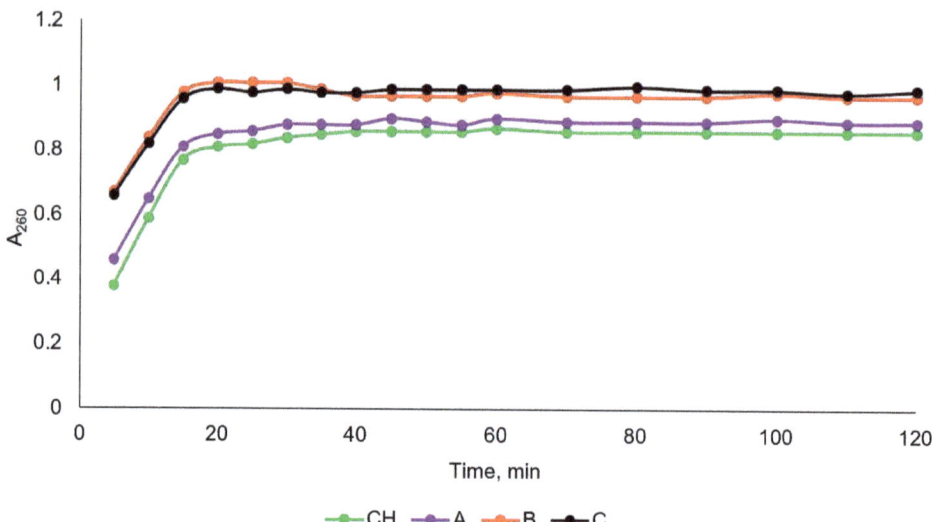

Figure 7. The effects of chitosan (CH), POX-1 (A), POX-2 (B), POX-3 (C) on the integrity of the cell membranes of *S. aureus*.

3.5.3. In Vitro Toxicity

The in vitro toxicity of the leading POX-2 and POX-3 was evaluated using the classic MTT test and compared with that of the starting chitosan, POX-1, deferoxamine and iron(III) chloride. As a quantitative measure of toxicity, we used the cell viability (CV, %) of the HEK-293 line after the cells were treated with a solution of the test substance (300 μg/mL).

The highest toxicity was found for iron(III) chloride (CV = 72%). The incorporation of iron(III) chloride into the chitosan polymeric matrix dramatically reduces its toxicity (CV = 94%). As a result, the formed nanoparticles POX-1 are characterized by the same toxicity as the starting chitosan (CV = 96%), which is considered a non-toxic polymer. The introduction of deferoxamine into POX-1 to give POX-2 or POX-3 did not lead to any noticeable changes in toxicity (for POX-1, CV = 93%; for POX-2, CV = 96%). However, it should be noted that the described nanoparticles are characterized by high positive values of the zeta potential; therefore, their intravenous administration into the general systemic circulation is undesirable, since it can cause the aggregation of negatively charged platelets. In this regard, in further experiments in vivo, we decided to use the intracavitary method of administration for these nanoparticles.

3.5.4. In Vivo Antibacterial Activity

At the next stage of the current work, we evaluated the in vivo antibacterial activity and toxicity of the leading systems, i.e., POX-2 and POX-3, in rats and compared their effects with those of antibiotics ampicillin and gentamicin.

The rats were subjected to the so-called model peritonitis. To imitate peritonitis, we infected the rats with a microbial mixture containing hospital strains of S. aureus and E. coli. Six hours after the introduction of microorganisms, all rats showed the conventional symptoms of peritonitis: lethargy, food refusal, rapid breathing and abdominal distention. In the control groups, exudate (200 µL) was collected with a sterile syringe 24 h after infection. A day later, all other rats were injected with a solution of the tested POX-2 or POX-3, or ampicillin or gentamicin. Then, 200 µL of exudate was taken after 7 h. To 200 µL of exudate, 1000 µL of 0.9% NaCl aqueous solution was added. Next, 100 µL of the resulting solution was evenly applied to a Petri dish with meat peptone agar. Colonies were counted 24 h after incubation at 37 °C. Subsequently, colony-forming units (CFU) were recalculated per 1 mL of exudate (Table 3).

Table 3. In vivo antibacterial effect.

Tested Sample	CFU per 1 mL of Exudate (7 h after Treatment or 31 h after Infection)
Control without treatment (24 h after infection)	2690
POX-2	0
POX-3	0
Ampicillin	540
Gentamicin	370

All tested samples demonstrated high in vivo antibacterial effects. However, POX-2 and POX-3 showed extremely high antimicrobial effects: no growth of colonies was found after collection of the exudate. The lower efficacy of the antibiotics can be explained by their rapid elimination after intracavitary administration. The elimination of polymer-based nanoparticles is much slower, and this leads to an increase in antibacterial activity compared low-molecular compounds, i.e., antibiotics, ampicillin and gentamicin. It should be noted that the results of this in vivo study in rats may not be directly applicable to human patients, and further studies, including clinical trials, are necessary to confirm the efficacy and safety of these nanoparticles in humans.

4. Conclusions

The results of this work can be considered from the following main perspectives.

Firstly, we elaborated the following types of chitosan-based nanoparticles: chitosan/Fe(III) (POX-1) and chitosan/Fe(III)/deferoxamine (POX-2 and POX-3, which differs from POX-2 in that it has been stirred in water for 24 h rather than 1 h). The resulting nanoparticles are characterized by close values of the hydrodynamic diameter and zeta potential. Characterization of the obtained POX-2 and POX-3 by physicochemical methods of analysis clearly demonstrated that POX-2 and POX-3 are practically identical nanoparticles. The equal biological effects of POX-2 and POX-3 support this conclusion. Therefore, POX-2 is stable in water for at least 24 h, retaining all the characteristic features of its structure and biological effects.

Secondly, we evaluated the in vitro and in vivo antibacterial activity of the prepared POX-1, POX-2 and POX-3. The most effective antibacterial species are POX-2 and POX-3 and their antimicrobial efficiency is equal, which is not surprising since POX-2 and POX-3 are the same system. POX-2 and POX-3 are characterized by outstanding in vivo antibacterial effects, and their activity exceeds even that of commercial antibiotics ampicillin and gentamicin. Moreover, POX-2 and POX-3 are non-toxic.

Thirdly, we concluded that the introduction of iron ions into the chitosan matrix increases the ability of the resulting nanoparticles to disrupt the integrity of the membranes of microorganisms in comparison with pure chitosan. The introduction of deferoxamine into the obtained nanoparticles sharply expands their effect of destruction of the bacterial membrane.

Finally, we provide a fast, simple and convenient route to obtain highly effective, non-toxic antibacterial systems. This route is based on non-covalent chemistry and does not require laborious and sophisticated organic synthesis methods. The obtained antibacterial nanoparticles are promising for further preclinical studies, and this project is underway in our group.

Author Contributions: Conceptualization, O.M.K. and A.S.K.; methodology, O.M.K. and A.R.E.; software, A.E.B.; validation, A.A.K., A.G.T. and A.S.K.; formal analysis, N.N.L.; investigation, O.M.K. and V.E.E.; resources, R.A.G.; data curation, V.E.E.; writing—original draft preparation, A.R.E. and O.M.K.; writing—review and editing, A.G.T.; visualization, M.V.T.; supervision, A.S.K.; project administration, A.S.K.; funding acquisition, A.R.E. All authors have read and agreed to the published version of the manuscript.

Funding: This study was supported by the Russian Science Foundation, grant № 23-23-00021.

Institutional Review Board Statement: The animal study protocol was approved by the Ethics Committee and followed the recommendations of European Directive 2010/63/EU of 22 September 2010.

Data Availability Statement: Not applicable.

Conflicts of Interest: The authors declare no conflict of interest.

References

1. Hyman, P.; Abedon, S.T. Chapter 7—Bacteriophage Host Range and Bacterial Resistance. In *Advances in Applied Microbiology*; Academic Press: Cambridge, MA, USA, 2010; Volume 70, pp. 217–248.
2. Międzybrodzki, R.; Borysowski, J.; Weber-Dąbrowska, B.; Fortuna, W.; Letkiewicz, S.; Szufnarowski, K.; Pawełczyk, Z.; Rogóż, P.; Kłak, M.; Wojtasik, E.; et al. Chapter 3—Clinical Aspects of Phage Therapy. In *Advances In Virus Research*; Łobocka, M., Szybalski, W., Eds.; Academic Press: Cambridge, MA, USA, 2012; Volume 83, pp. 73–121.
3. Mandell, L.A.; Wunderink, R.G.; Anzueto, A.; Bartlett, J.G.; Campbell, G.D.; Dean, N.C.; Dowell, S.F.; File, T.M., Jr.; Musher, D.M.; Niederman, M.S.; et al. Infectious Diseases Society of America/American Thoracic Society consensus guidelines on the management of community-acquired pneumonia in adults. *Clin. Infect. Dis.* **2007**, *44* (Suppl. S2), S27–S72. [CrossRef] [PubMed]
4. Jones, J.D.G.; Dangl, J.L. The plant immune system. *Nature* **2006**, *444*, 323–329. [CrossRef] [PubMed]
5. Nazarov, P.A.; Baleev, D.N.; Ivanova, M.I.; Sokolova, L.M.; Karakozova, M.V. Infectious Plant Diseases: Etiology, Current Status, Problems and Prospects in Plant Protection. *Acta Nat.* **2020**, *12*, 46–59. [CrossRef]
6. Prescott, J.F.; MacInnes, J.I.; Van Immerseel, F.; Boyce, J.D.; Rycroft, A.N.; Vázquez-Boland, J.A. *Pathogenesis of Bacterial Infections in Animals*; Wiley: Hoboken, NJ, USA, 2022.
7. Hutchings, M.I.; Truman, A.W.; Wilkinson, B. Antibiotics: Past, present and future. *Curr. Opin. Microbiol.* **2019**, *51*, 72–80. [CrossRef]
8. Aminov, R.I. A brief history of the antibiotic era: Lessons learned and challenges for the future. *Front Microbiol.* **2010**, *1*, 134. [CrossRef]
9. Gualerzi, C.O.; Brandi, L.; Fabbretti, A.; Pon, C.L. *Antibiotics: Targets, Mechanisms and Resistance*; Wiley: Hoboken, NJ, USA, 2013.
10. Jourdan, A.; Sangha, B.; Kim, E.; Nawaz, S.; Malik, V.; Vij, R.; Sekhsaria, S. Antibiotic hypersensitivity and adverse reactions: Management and implications in clinical practice. *Allergy Asthma Clin. Immunol.* **2020**, *16*, 6. [CrossRef] [PubMed]
11. Cunha, B.A. Antibiotic side effects. *Med. Clin. N. Am.* **2001**, *85*, 149–185. [CrossRef] [PubMed]

12. Mohsen, S.; Dickinson, J.A.; Somayaji, R. Update on the adverse effects of antimicrobial therapies in community practice. *Can. Fam. Physician* **2020**, *66*, 651–659. [PubMed]
13. Bombaywala, S.; Mandpe, A.; Paliya, S.; Kumar, S. Antibiotic resistance in the environment: A critical insight on its occurrence, fate, and eco-toxicity. *Environ. Sci. Pollut. Res.* **2021**, *28*, 24889–24916. [CrossRef]
14. Roy, B.; Suresh, P.K. Toxic Effect of Antibiotics on Freshwater Algal Systems and the Mechanisms of Toxicity: A Review. *Nat. Environ. Pollut. Technol.* **2021**, *20*, 1611–1619. [CrossRef]
15. Yang, Q.; Gao, Y.; Ke, J.; Show, P.L.; Ge, Y.; Liu, Y.; Guo, R.; Chen, J. Antibiotics: An overview on the environmental occurrence, toxicity, degradation, and removal methods. *Bioengineered* **2021**, *12*, 7376–7416. [CrossRef] [PubMed]
16. Kovalakova, P.; Cizmas, L.; McDonald, T.J.; Marsalek, B.; Feng, M.; Sharma, V.K. Occurrence and toxicity of antibiotics in the aquatic environment: A review. *Chemosphere* **2020**, *251*, 126351. [CrossRef] [PubMed]
17. Darby, E.M.; Trampari, E.; Siasat, P.; Gaya, M.S.; Alav, I.; Webber, M.A.; Blair, J.M.A. Molecular mechanisms of antibiotic resistance revisited. *Nat. Rev. Microbiol.* **2022**, 1–16. [CrossRef]
18. Zaman, S.B.; Hussain, M.A.; Nye, R.; Mehta, V.; Mamun, K.T.; Hossain, N. A Review on Antibiotic Resistance: Alarm Bells are Ringing. *Cureus* **2017**, *9*, e1403. [CrossRef]
19. Urban-Chmiel, R.; Marek, A.; Stępień-Pyśniak, D.; Wieczorek, K.; Dec, M.; Nowaczek, A.; Osek, J. Antibiotic Resistance in Bacteria-A Review. *Antibiotics* **2022**, *11*, 1079. [CrossRef] [PubMed]
20. Varlamov, V.P.; Il'ina, A.V.; Shagdarova, B.T.; Lunkov, A.P.; Mysyakina, I.S. Chitin/Chitosan and Its Derivatives: Fundamental Problems and Practical Approaches. *Biochemistry* **2020**, *85*, 154–176. [CrossRef] [PubMed]
21. Guarnieri, A.; Triunfo, M.; Scieuzo, C.; Ianniciello, D.; Tafi, E.; Hahn, T.; Zibek, S.; Salvia, R.; De Bonis, A.; Falabella, P. Antimicrobial properties of chitosan from different developmental stages of the bioconverter insect Hermetia illucens. *Sci. Rep.* **2022**, *12*, 8084. [CrossRef]
22. Mitra, S.; Chakraborty, A.J.; Tareq, A.M.; Emran, T.B.; Nainu, F.; Khusro, A.; Idris, A.M.; Khandaker, M.U.; Osman, H.; Alhumaydhi, F.A.; et al. Impact of heavy metals on the environment and human health: Novel therapeutic insights to counter the toxicity. *J. King Saud Univ. Sci.* **2022**, *34*, 101865. [CrossRef]
23. Kritchenkov, A.S.; Bokach, N.A.; Starova, G.L.; Kukushkin, V.Y. A palladium(II) Center activates nitrile ligands toward 1,3-dipolar cycloaddition of nitrones substantially more than the corresponding platinum(II) center. *Inorg. Chem.* **2012**, *51*, 11971–11979. [CrossRef]
24. Guo, D.; Xia, Q.; Zeng, Q.; Wang, X.; Dong, H. Antibacterial Mechanisms of Reduced Iron-Containing Smectite–Illite Clay Minerals. *Environ. Sci. Technol.* **2021**, *55*, 15256–15265. [CrossRef]
25. Kritchenkov, A.S.; Luzyanin, K.V.; Bokach, N.A.; Kuznetsov, M.L.; Gurzhiy, V.V.; Kukushkin, V.Y. Selective Nucleophilic Oxygenation of Palladium-Bound Isocyanide Ligands: Route to Imine Complexes That Serve as Efficient Catalysts for Copper-/Phosphine-Free Sonogashira Reactions. *Organometallics* **2013**, *32*, 1979–1987. [CrossRef]
26. Abdulsada, F.M.; Hussein, N.N.; Sulaiman, G.M.; Al Ali, A.; Alhujaily, M. Evaluation of the Antibacterial Properties of Iron Oxide, Polyethylene Glycol, and Gentamicin Conjugated Nanoparticles against Some Multidrug-Resistant Bacteria. *J. Funct. Biomater.* **2022**, *13*, 138. [CrossRef]
27. Gudkov, S.V.; Burmistrov, D.E.; Serov, D.A.; Rebezov, M.B.; Semenova, A.A.; Lisitsyn, A.B. Do Iron Oxide Nanoparticles Have Significant Antibacterial Properties? *Antibiotics* **2021**, *10*, 884. [CrossRef] [PubMed]
28. Reedijk, J. *Reference Module in Chemistry, Molecular Sciences and Chemical Engineering*; Elsevier: Amsterdam, The Netherlands, 2013.
29. Codd, R. Siderophores and iron transport. In *Reference Module in Chemistry, Molecular Sciences and Chemical Engineering*; Elsevier: Amsterdam, The Netherlands, 2021.
30. De Carvalho, C.C.C.R.; Fernandes, P. Siderophores as "Trojan Horses": Tackling multidrug resistance? *Front. Microbiol.* **2014**, *5*, 290. [CrossRef] [PubMed]
31. Saha, M.; Sarkar, S.; Sarkar, B.; Sharma, B.K.; Bhattacharjee, S.; Tribedi, P. Microbial siderophores and their potential applications: A review. *Environ. Sci. Pollut. Res.* **2016**, *23*, 3984–3999. [CrossRef] [PubMed]
32. Kritchenkov, A.S.; Egorov, A.R.; Dysin, A.P.; Volkova, O.V.; Zabodalova, L.A.; Suchkova, E.P.; Kurliuk, A.V.; Shakola, T.V. Ultrasound-assisted Cu(I)-catalyzed azide-alkyne click cycloaddition as polymer-analogous transformation in chitosan chemistry. High antibacterial and transfection activity of novel triazol betaine chitosan derivatives and their nanoparticles. *Int. J. Biol. Macromol.* **2019**, *137*, 592–603. [CrossRef]
33. Kritchenkov, A.S.; Egorov, A.R.; Volkova, O.V.; Kritchenkov, I.S.; Kurliuk, A.V.; Shakola, T.V.; Khrustalev, V.N. Ultrasound-assisted catalyst-free phenol-yne reaction for the synthesis of new water-soluble chitosan derivatives and their nanoparticles with enhanced antibacterial properties. *Int. J. Biol. Macromol.* **2019**, *139*, 103–113. [CrossRef]
34. Kritchenkov, A.S.; Zhaliazniak, N.V.; Egorov, A.R.; Lobanov, N.N.; Volkova, O.V.; Zabodalova, L.A.; Suchkova, E.P.; Kurliuk, A.V.; Shakola, T.V.; Rubanik, V.V.; et al. Chitosan derivatives and their based nanoparticles: Ultrasonic approach to the synthesis, antimicrobial and transfection properties. *Carbohydr. Polym.* **2020**, *242*, 116478. [CrossRef]
35. Kritchenkov, A.S.; Kletskov, A.V.; Egorov, A.R.; Tskhovrebov, A.G.; Kurliuk, A.V.; Zhaliazniak, N.V.; Shakola, T.V.; Khrustalev, V.N. New water-soluble chitin derivative with high antibacterial properties for potential application in active food coatings. *Food Chem.* **2020**, *343*, 128696. [CrossRef]
36. Inam, M.A.; Khan, R.; Park, D.R.; Lee, Y.-W.; Yeom, I.T. Removal of Sb(III) and Sb(V) by Ferric Chloride Coagulation: Implications of Fe Solubility. *Water* **2018**, *10*, 418. [CrossRef]

37. Fernandes, L.; Resende, C.; Tavares, D.; Soares, G.; de Oliveira Castro, L.; Granjeiro, J. Cytocompatibility of Chitosan and Collagen-Chitosan Scaffolds for Tissue Engineering. *Polímeros* **2010**, *21*, 1–6. [CrossRef]
38. Kumar, S.; Koh, J. Physiochemical, Optical and Biological Activity of Chitosan-Chromone Derivative for Biomedical Applications. *Int. J. Mol. Sci.* **2012**, *13*, 6102–6116. [CrossRef] [PubMed]
39. Ravi Kumar, M.N.V. A review of chitin and chitosan applications. *React. Funct. Polym.* **2000**, *46*, 1–27. [CrossRef]
40. Li, Q.; Mahendra, S.; Lyon, D.Y.; Brunet, L.; Liga, M.V.; Li, D.; Alvarez, P.J.J. Antimicrobial nanomaterials for water disinfection and microbial control: Potential applications and implications. *Water Res.* **2008**, *42*, 4591–4602. [CrossRef] [PubMed]
41. Dash, M.; Chiellini, F.; Ottenbrite, R.M.; Chiellini, E. Chitosan—A versatile semi-synthetic polymer in biomedical applications. *Prog. Polym. Sci.* **2011**, *36*, 981–1014. [CrossRef]
42. Kong, M.; Chen, X.G.; Xing, K.; Park, H.J. Antimicrobial properties of chitosan and mode of action: A state of the art review. *Int. J. Food Microbiol.* **2010**, *144*, 51–63. [CrossRef]
43. Okamoto, T.; Okabe, S. Ultraviolet absorbance at 260 and 280 nm in RNA measurement is dependent on measurement solution. *Int. J. Mol. Med.* **2000**, *5*, 657–659. [CrossRef]

Disclaimer/Publisher's Note: The statements, opinions and data contained in all publications are solely those of the individual author(s) and contributor(s) and not of MDPI and/or the editor(s). MDPI and/or the editor(s) disclaim responsibility for any injury to people or property resulting from any ideas, methods, instructions or products referred to in the content.

Article

Procainamide Charge Transfer Complexes with Chloranilic Acid and 2,3-Dichloro-5,6-dicyano-1,4-benzoquinone: Experimental and Theoretical Study

A. F. M. Motiur Rahman [1,*], Ahmed H. Bakheit [1], Shofiur Rahman [2], Gamal A. E. Mostafa [1,*] and Haitham Alrabiah [1]

1. Department of Pharmaceutical Chemistry, College of Pharmacy, King Saud University, Riyadh 11451, Saudi Arabia
2. Biological and Environmental Sensing Research Unit, King Abdullah Institute of Nanotechnology, King Saud University, Riyadh 11451, Saudi Arabia
* Correspondence: afmrahman@ksu.edu.sa (A.F.M.M.R.); gmostafa@ksu.edu.sa (G.A.E.M.)

Abstract: The formation of charge transfer (CT) complexes between bioactive molecules and/or organic molecules is an important aspect in order to understand 'molecule-receptor' interactions. Here, we have synthesized two new CT complexes, procainamide-chloranilic acid (PA-ChA) and procainamide-2,3-dichloro-5,6-dicyano-1,4-benzoquinone (PA-DDQ), from electron donor procainamide (PA), electron acceptor chloranilic acid (ChA), and 2,3-dichloro-5,6-dicyano-1,4-benzoquinone (DDQ). The structures of these two CT complexes were elucidated/characterized using FTIR, NMR, and many other spectroscopic methods. A stability study of each complex was conducted for the first time using various spectroscopic parameters (e.g., formation constant, molar extinction coefficient, ionization potential oscillator strength, dipole moment, and standard free energy). The formation of CT complexes in solution was confirmed by spectrophotometric determination. The molecular composition of each complex was determined using the spectrophotometric titration method and gave a 1:1 (donor:acceptor) ratio. In addition, the formation constant was determined using the Benesi–Hildebrand equation. To understand the noncovalent interactions of the complexes, density functional theory (DFT) calculations were performed using the ωB97XD/6-311++G(2d,p) level of theory. The DFT-computed interaction energies (ΔIEs) and the Gibbs free energies (ΔGs) were in the same order as observed experimentally. The DFT-calculated results strongly support our experimental results.

Keywords: procainamide; chloranilic acid; DDQ; charge transfer complex; DFT

Citation: Rahman, A.F.M.M.; Bakheit, A.H.; Rahman, S.; Mostafa, G.A.E.; Alrabiah, H. Procainamide Charge Transfer Complexes with Chloranilic Acid and 2,3-Dichloro-5,6-dicyano-1,4-benzoquinone: Experimental and Theoretical Study. *Processes* **2023**, *11*, 711. https://doi.org/10.3390/pr11030711

Academic Editors: Iliyan Ivanov, Stanimir Manolov and Donatella Aiello

Received: 19 January 2023
Revised: 8 February 2023
Accepted: 26 February 2023
Published: 27 February 2023

Copyright: © 2023 by the authors. Licensee MDPI, Basel, Switzerland. This article is an open access article distributed under the terms and conditions of the Creative Commons Attribution (CC BY) license (https://creativecommons.org/licenses/by/4.0/).

1. Introduction

Procainamide, 4-amino-N-(2-(diethylamino)ethyl)benzamide (PA, Figure 1), is used to treat abnormal heart rhythms, namely Wolff–Parkinson–White syndrome (WPWS)-associated arrhythmias [1,2]. According to the Vaughan Williams classification, it is classified as a class IA agent with a sodium-channel-blocking effect [1,2].

Recently, extensive care has been given to the development of the formation of charge transfer (CT) complexes derived from an electron donor and an acceptor molecule. Due to the intriguing structures; physical and chemical properties; applications in different fields, especially optical materials [2,3]; drug–receptor interactions [4]; solar energy and surface chemistry [5]; field-effect transistors; light-emitting devices, lasers, and sensors and stimuli-responsive behavior [6]; organic semiconductor properties [7,8]; and various biological applications [9,10], the synthesis of CT complexes is of interest. CT complexes are formed with unique types of interactions, which are incorporated in the formation of π-π stacking of the aromatic complexes, hydrogen bonds, and/or electron transfer from a donor to an acceptor [5,11–13]. The reaction mechanism of electron transfer from the donor to the acceptor of the CT complexes was reported by Mullikan [14,15]. It should be noted that molecular acceptors can be used for the determination of drugs in dosage forms [16]. Nevertheless, the formation of the CT complexes is echo-friendly, inexpensive, simple, and easy compared to other techniques [17]. On the other hand, CT complexes are used to remove and utilize discarded drugs from the environment [18]. Here, we report the synthesis and characterization of two new CT complexes 'procainamide-chloranilic acid' (PA-ChA) and 'procainamide-2,3-dichloro-5,6-dicyano-1,4-benzoquinone' (PA-DDQ). The structures of the CT complexes were elucidated using UV, fluorescence, FTIR, and NMR spectrometry. Electronic properties and conductometry were evaluated. A spectroscopic study was carried out to determine the formation constant and stoichiometry of the CT complexes. In addition, different physicochemical properties of the CT complexes were assessed to determine their stability. Furthermore, electronic structures were examined by DFT calculations to determine the frontier molecular orbitals and the relocation of the electron density.

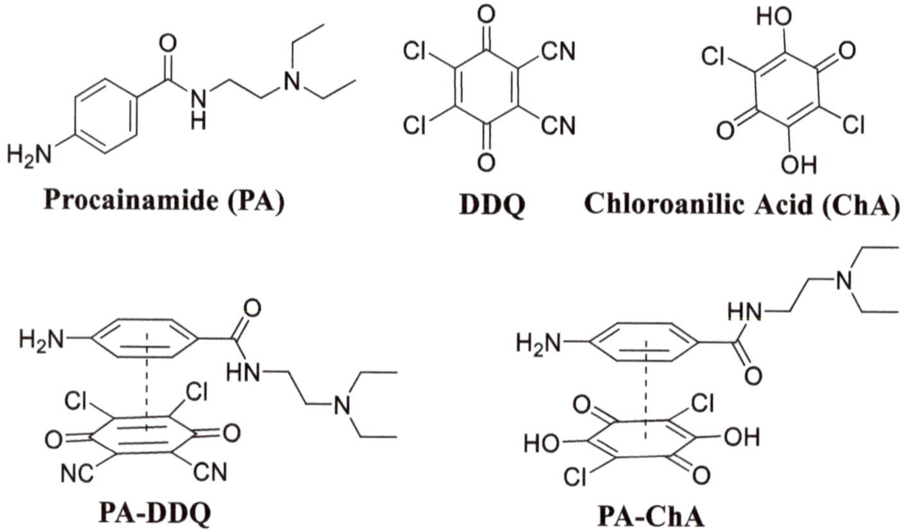

Figure 1. Chemical structures of procainamide and its complexes.

2. Experimental Section

2.1. General

All the chemicals were reagent grade. Procainamide (purity > 99.5) was purchased from Sigma-Aldrich (Darmstadt, Germany), and chloranilic acid and 2,3-dichloro-5,6-dicyano-1,4-benzoquinone were purchased from Merck (Darmstadt, Germany). Melting points were determined on a Barnstead electrothermal digital melting point apparatus, model IA9100, BIBBY scientific limited Stone (Staffordshire, UK). IR spectra were recorded on a Jasco FT/IR-6600 spectrometer (Tokyo, Japan) in KBr. NMR spectra were taken from 400 MHz premium shielded NMR spectroscopy in deuterated dimethyl sulfoxide (DMSO-d_6) (Agilent Technologies, Santa Clara, CA, USA). UV-vis spectra were measured using a Genesis G10S UV-Vis spectrophotometer (Thermo Fisher Scientific, Pleasanton, CA, USA), using acetonitrile as the solvent. All the measurements were performed using various solvents and quartz cells with a path length of 1 cm. The UV-Vis spectra in the solution were measured over the range of 200–700 nm. The conductivities of procainamide and its CT complexes (PA-ChA and PA-DDQ) were measured on an Orion conductometer (Beverly, MA, USA).

2.2. Synthesis of CT Complexes PA-ChA and PA-DDQ

2.2.1. Preparation of PA-ChA Complex

A solution of PA (235 mg, 1 mmol) in 20 mL of methanol was added to a solution of ChA (209 mg, 1 mmol) in 20 mL of methanol. The reaction mixture was stirred for 3 h at 25 °C, filtered, washed with ice-cold acetonitrile (20 mL), and dried over anhydrous $CaCl_2$. Purple solid (yield > 95%). Melting point, 193 °C. IR: ν (cm^{-1}): 3523, 3440, 3270, 3151, 3002, 2715, 2670, 2600, 2529, 1637, 1576, 1530, 1500, 1436, 1381, 1344, 1288, 1173, 1124. 1030, 982, 875, 839, 777, 753, 653, and 571 cm^{-1}. ^{1}H-NMR (DMSO-d_6, 400 MHz) δ: 9.10 (br, -NH), 8.32 (t, -NH), 7.59 (d, J = 8.8 Hz, 2H), 6.55 (d, J = 8.4 Hz, 2H), 3.54–3.50 (q, -N-CH$_2$, 2H), 3.21–3.15 (m, 3 × -CH$_2$, 6H), and 1.18 (t, J = 7.2 Hz, 2 × -CH$_3$, 6H) ppm.

2.2.2. Preparation of PA-DDQ Complex

A solution of PA (235 mg, 1 mmol) in 20 mL of methanol was added to a solution of DDQ (227 mg, 1 mmol) in 20 mL of methanol. The reaction mixture was stirred for 3 h at 25 °C. The solid formation was filtered, washed with ice-cold acetonitrile (20 mL), and dried over anhydrous $CaCl_2$. Brown solid (yield > 95%). Melting point, 287 °C. IR: ν (cm^{-1}): 3430, 3248, 2987, 2944, 2581, 2471, 2217, 1653, 1610, 1562, 1507, 1482, 1407, 1320, 1246, 1178, 1145, 1018, 961, 892, 867, 757, 680, and 625 cm^{-1}. ^{1}H-NMR (DMSO-d_6, 400 MHz) δ: 11.00 (s, -OH), 10.05 (brs,-NH), 9.56 (s, -OH), 8.90 (t,-NH), 8.88 (t,-NH), 7.98 (d, J = 8.8 Hz, 1H), 7.88 (d, J = 8.4 Hz, 1H), 7.44 (d, J = 8.8 Hz, 1H), 7.20 (d, J = 8.4 Hz, 1H), 3.64 (quint, -N-CH$_2$, 2H), 3.25–3.16 (m, 3 × -CH$_2$, 6H), and 1.22 (t, J = 6.8 Hz, 2 × -CH$_3$, 6H) ppm.

2.3. Stoichiometry

Spectrophotometric titration measurements were carried out for the reactions of PA with both ChA and DDQ, where acetonitrile was used as a blank solvent at wavelengths of 515 and 490 nm, respectively. A 0.25, 0.50, 0.75, 1.00, 1.50, 2.0, 2.50, 3.00, 3.50, and 4.00 mL aliquot of a standard solution (1 × 10^{-3} M) of the appropriate acceptor in acetonitrile was added to 1.00 mL of 1 × 10^{-3} M PA (in acetonitrile), and the total volume of the mixture was 5 mL. The concentration of PA (CD) in the reaction mixture was maintained at 1 × 10^{-3} M, whereas the concentration of the acceptors (CA) changed over a wide range of concentrations (from 0.25 × 10^{-3} M to 4 × 10^{-3} M) to produce solutions with an acceptor molar ratio that varied from 1:4 to 4:1. The stoichiometry of the CT complexes was obtained from the determination of the conventional spectrophotometric titration method.

2.4. Formation Constant

The formation constant of the investigated CT complexes was determined using the modified Benesi–Hildebrand method [19] using the following Equation (1):

$$[A_o]/A = 1/\varepsilon^{AD} + 1/K_c \cdot \varepsilon \times 1/D_o \qquad (1)$$

where '[A$_o$]' is the molar concentration of the acceptor ChA or DDQ, 'D$_o$' is the molar concentration of the donor (PA); 'A' is the absorbance of the formed CT complexes at λ max; 'ε' is the molar absorptivity of the complexes; and 'K$_c$' is the association constant of the complexes (L mol^{-1}).

2.5. Spectroscopic Physical Parameters

2.5.1. Oscillator Strength (f) Transition Dipole Moment (−μe)

Oscillator strength and transition dipole moment were measured using Equations (2) and (3), respectively.

$$f = 4.32 \times 10^{-9} \left[\varepsilon_{max} \Delta v_{\frac{1}{2}} \right] \qquad (2)$$

$$\mu = 0.0958 \left[\frac{\varepsilon_{max} \Delta v_{\frac{1}{2}}}{\overline{v}_{max}} \right]^{1/2} \qquad (3)$$

2.5.2. Ionization Potential (I_P)

Ionization potential (I_P) values were recorded using the following equation [20].

$$I_P \text{ (eV)} = 5.76 + 1.53 \times 10^{-4} \, v_{CT} \quad (4)$$

2.5.3. Energy of the CT Complexes (E_{CT})

The energies of the CT complexes (E_{CT}) (eV) were calculated using Equation (5).

$$E_{CT} = hv_{CT} = 1243.667/\lambda_{CT} \quad (5)$$

2.5.4. Resonance Energy (R_N)

The resonance energy (R_N) of the CT complexes was estimated according to Equation (6) presented by Briegleb and Czekalla [21].

$$\varepsilon_{max} = 7.7 \times 10^{-4}/[hv_{CT}/R_N] - 3.5] \quad (6)$$

2.5.5. Dissociation Energy (W) (eV)

Further evidence of the nature of CT interactions of the synthesized CT complexes was calculated according to Equation (7).

$$W = I_P - E^A - E_{CT} \quad (7)$$

2.5.6. Gibbs Free Energy Change ($\Delta G°$)

Finally, the nature of the interaction of CT complexes was examined using Gibbs free energy calculation as shown in $\Delta G°$ Equation (8).

$$\Delta G° = -RT \ln K_{CT} \quad (8)$$

2.6. DFT Calculations

Single-point density functional theory (DFT) calculations were performed using the long-range corrected hybrid functional ωB97XD in conjunction with the 6-311++G(2d,p) basis set. All the DFT calculations were performed using Gaussian 16, Revision C.01 [22] in the gas phase first, and then the optimized structures were further calculated in the acetonitrile solvent system using the polarized continuum model (PCM). A vibrational analysis was carried out for each optimized molecule to ensure that they were in a vibrational energy minimum and had no imaginary frequencies (Supplementary Materials).

3. Results and Discussion

3.1. Synthesis of CT Complexes PA-ChA and PA-DDQ

The interaction of PA (donor) with ChA/DDQ (as π acceptors) in methanol produced colored CT complexes with high molar absorptivity. The synthesis of CT complexes was straightforward as explained in Scheme 1. Electron donor PA in methanol was simply added to the acceptor molecules (ChA/DDQ) to form colored CT complexes (Scheme 1).

3.2. Electronic Absorption Spectra

CT complexes were characterized using the spectrophotometric method. As shown in Figure 2, the absorption intensity of the CT complexes was related to the formation of the radical anion. PA showed peaks at 206 (n→σ* for C-N), 264 (n→π* for C=O), and 299 (n→π* for C=O) nm at the concentration of 8×10^{-5} M; ChA at 230 (n→π* for C=O), 282 (n→π* for C=O), and 431 (n→π* for C=O) nm; and DDQ at 249 (n→π* for C=O) and 380 (n→π* for C≡N) nm at the concentration of 1×10^{-4} M. Absorption maxima for PA-ChA were found at 246 (n→π* for C=O), 316 (n→π* for C=O), and 525 (n→π* for the visible region) nm and those for PA-DDQ were found at 246 (n→π* for C=O), 306 (n→π* for C=O), 458 (n→π* for the visible region), 543 (n→π* for the visible region), and 590 (n→π* for the visible region) nm at the concentration of 4×10^{-4} M (Figure 2). Acetonitrile was used as solvent blank for all measurements. The resulting color was stable for more than five hours, indicating the high stability of these complexes. The band gaps of the formed complex were calculated from the formula Eg (eV) = hυ = hc/λae (nm) (2) where "h" is Planck's constant, c is the velocity of light, and λ is the wavelength. The energy gaps were 2.41 and 2.53 eV for PA-ChA and PA-DDQ complexes, respectively.

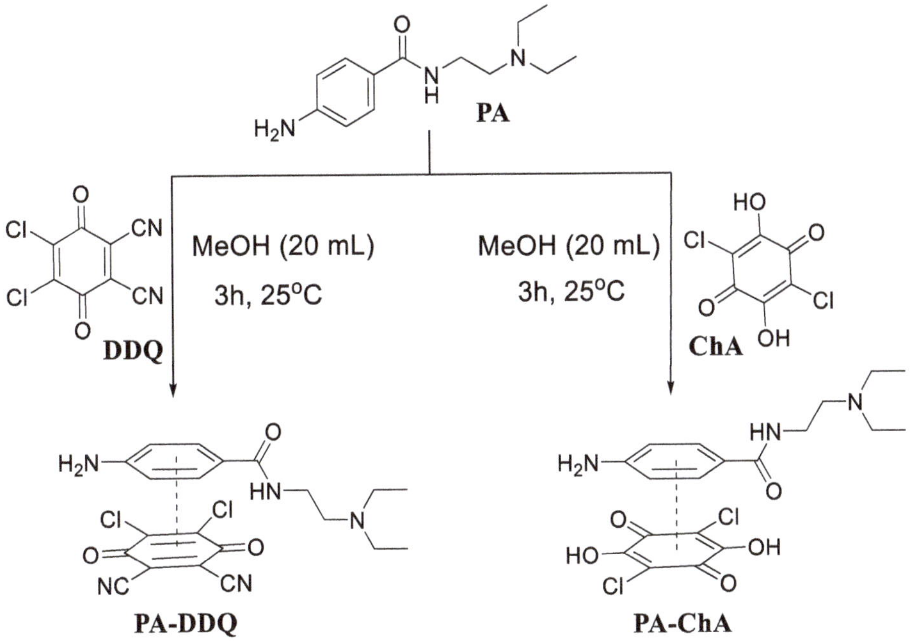

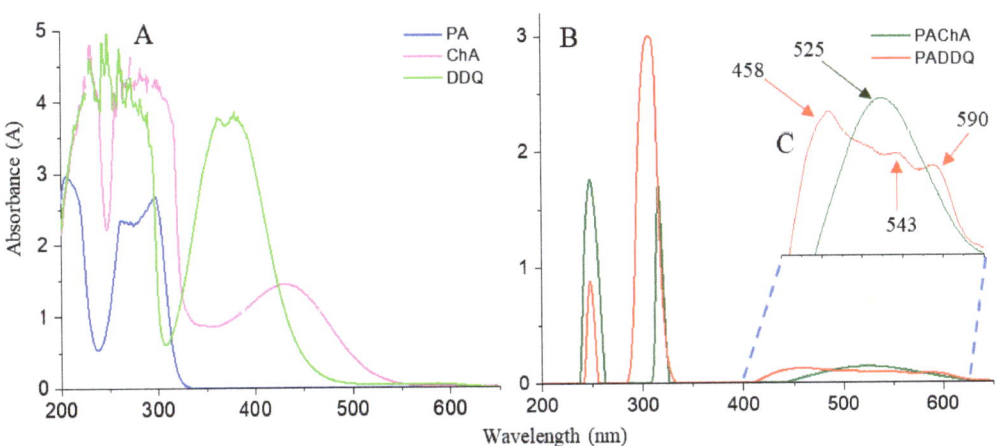

Scheme 1. Synthesis of CT complexes PA-ChA and PA-DDQ.

Figure 2. Absorption curve of CT complex: (A) absorption maximum of PA, ChA, and DDQ; (B) absorption maximum of PA-ChA and PA-DDQ complexes; and (C) expanded spectra shown at 400–650 nm for PA-ChA and PA-DDQ complexes (all spectra were taken in acetonitrile).

3.3. Molecular Composition of the PA Complexes

The stoichiometry of the investigated complexes was monitored and determined spectrophotometrically using spectrophotometric titration (Figure 3). As shown in Figure 4, PA reacts with CHA/DDQ in a 1:1 ratio.

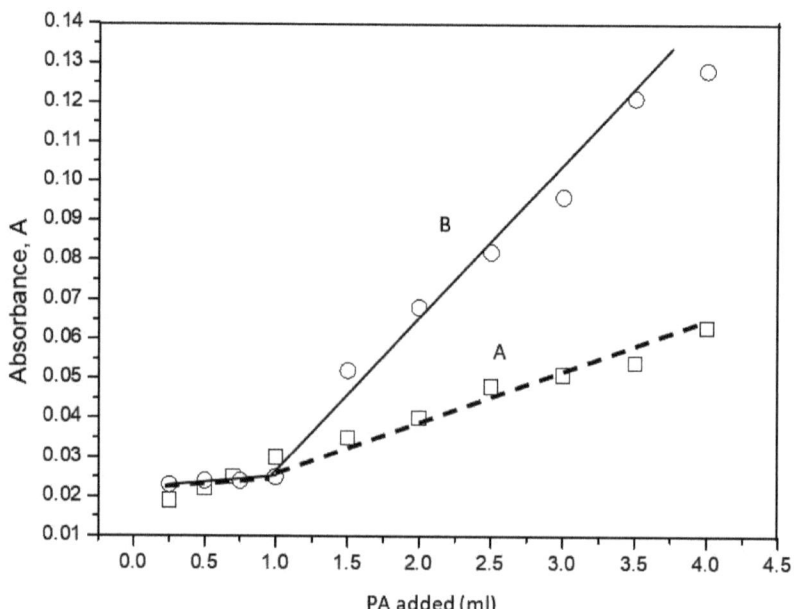

Figure 3. Spectrophotometric titration curves of A PA-ChA and B PA-DDQ CT complexes in acetonitrile at the detectable peaks (donor and acceptor, 1×10^{-3} M).

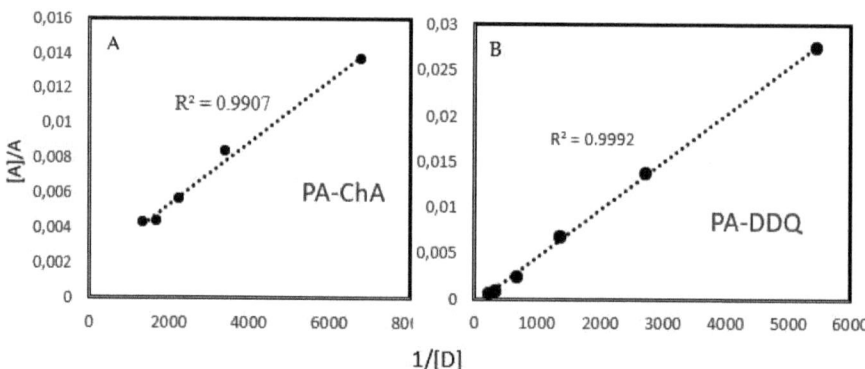

Figure 4. The 1:1 Benesi–Hildebrand plot of procainamide CT complexes at detectable peaks: (**A**) for PA-ChA and (**B**) PA-DDQ.

3.4. Formation Constant

Upon plotting the values [A]/A against $1/[D_o]$, a straight line was obtained (Figure 4). The molar absorptivity and formation constant in acetonitrile were found to be 0.6×10^3 L mol^{-1} cm^{-1} and 1×10^3 L mol^{-1}cm^{-1} for the PA-CHA complex and 14×10^3 and 0.1×10^3 L mol^{-1} for the DDQ complex, respectively. Those values indicate the stability of the CT complexes. The formation constant values show that the stability of the PA-ChA complex is higher than that of the PA-DDQ complex.

3.5. Spectroscopic Physical Parameters

3.5.1. Oscillator Strength (*f*) Transition Dipole Moment (Debye)

As shown in Table 1, relatively high '*f*' values of 0.212 and 0.31 indicate the strong interactions between PA and ChA/DDQ, and therefore, the CT complexes PA-ChA and PA-DDQ have high stability. In addition, the transition dipole moment was recorded as very high (4.82 and 5.62 Debye for PA-ChA and PA-DDQ, respectively), supporting the formation of strong bonded CT complexes between the donor (PA) and acceptors (ChA/DDQ).

Table 1. Spectroscopic characteristics of procainamide CT complexes.

Parameters	PA-ChA	PA-DDQ
Wavelength, nm	515	490
Molar absorptivity (ε), L mol^{-1} cm^{-1}	0.6×10^3	1×10^3
Formation constant K = L mol^{-1}	1.4×10^3	0.1×10^3
Oscillator strength (*f*)	0.212	0.31
Transition dipole moment (Debye)	4.82	5.65
Ionization potential: I_P (eV)	18.2	16.61
Energy: hv (eV)	2.41	2.53
Resonance energy: R_N (eV)	1.2	1.11
Dissociation energy: W(eV)	14.69	12.18
Gibbs free energy: ΔG (kJ mol^{-1})	-18 kJ mol^{-1}	-11 kJ mol^{-1}

3.5.2. Ionization Potential (I_P)

Ionization potential (I_P) values were recorded at 18.2 eV and 16.6 eV for PA-ChA and PA-DDQ complexes, respectively (Equation (4)). A low Ip value of PA (8.7 eV) [23] compared to the CT complexes indicated its better stability [20]. Therefore, the formation of CT complexes is preferable while PA is reacting with ChA and DDQ.

3.5.3. Energy of the CT Complexes (E_{CT})

As presented in Table 1, the values of the energy (E_{CT}) of the CT complexes were found to be 2.41 and 2.53 eV for PA-ChA and PA-DDQ, respectively (Equation (5)). These very low E_{CT} values of the CT complexes indicate that the PA-ChA and PA-DDQ complexes' stability is very high.

3.5.4. Resonance Energy (R_N)

The resonance energies (R_N) of the CT complexes were found to be 1.2 and 1.11 eV for PA-ChA and PA-DDQ, respectively (Table 1). The lower resonance energies of the CT complexes indicates their higher stability.

3.5.5. Dissociation Energy (W)

Further evidence of the nature of CT interactions of the synthesized CT complexes was found through calculations using Equation (7). The dissociation energy (W) values were found at 14.69 and 12.18 eV for PA-ChA and PA-DDQ, respectively (Table 1). These values indicated that the synthesized complexes have strong CT interactions and, therefore, high stability.

3.5.6. Gibbs Free Energy Change ($\Delta G°$)

As listed in Table 1, the higher negative values suggest that the CT complexes formed between PA and the acceptors are exergonic. Generally, the values of $\Delta G°$ become more negative as the value of K_{CT} increases, where the CT interaction between the donor and acceptors becomes strong. Thus, the complex composition is subject to a lower degree of freedom, and the values of ΔG° become higher negative values. The negative value of ΔG° pointed out that the interaction between the donor (PA) and the acceptors (ChA and DDQ) is spontaneous. Their values are −18 and −11 kJmol^{-1} for PA-ChA and PA-DDQ, respectively.

3.6. Spectroscopy

3.6.1. Infrared (IR) Spectra

IR Spectra of PA-ChA Complex

From the comparison of the FTIR spectra of PA, ChA, and the PA-ChA complex (Figure 5), a characteristic C-Cl band at 571 cm^{-1} for ChA and 572 cm^{-1} for PA-ChA was observed, which confirms the complex formation. Furthermore, a red shift of 1664 cm^{-1} for ChA was observed at 1530 cm^{-1} for PA-ChA [24]. Other important peaks are summarized in Table 2 [25,26]. It should be noted that the vibrational bands for O-H, C-H, aromatic C=O, and C-O for ChA to PA-ChA have been shifted from 3560 to 3523, 3235 to 3151, 1664 to 1637, and 1207 to 1173, respectively.

Table 2. IR spectral bands of PA, ChA, and their complex (PA-ChA).

PA	ChA	PA-ChA Complex	Possible Assignments
3402	3560	3523	ν(O-H)
		3440	ν(N-H)
3320		3270	ν(CONH)
3215	3235	3151	ν(C-H) (aromatic)
2938		3002	
2576		2715	ν($^+$N-H)
		2670	
		2600	
2465		2529	ν($^+$N-H)
1637	1664	1637	ν(C=O)
1599	1631	1576	
1542		1530	ν(C=C) (aromatic ring)
1512		1500	
1467		1436	ν(C-H) (alkanes)
1392	1369	1381	ν(C-C) (alkanes)
1323		1344	ν(C-C), ν(C-N) (alkanes)
1295	1264	1288	ν(C-N) (alkanes)
1185	1207	1173	ν(C-O)
1145		1124	ν(C-H) (bending)
1027		1030	ν(NH)
964	983	982	ν(C-H) (bending)
839	854	875	
806		839	
768	752	777	
702	690	753	ν(N-H)
652		653	ν(C-N-C)
	572	571	ν(C-Cl)

IR Spectra of PA-DDQ Complex

As shown in Figure 6, from the comparison of the IR spectra of PA, DDQ, and the PA-DDQ complex, a characteristic C≡N band at 2233 cm^{-1} for DDQ and 2216 cm^{-1} for the PA-DDQ complex was observed, which confirms the complex formation [25,26]. Similarly, a red shift of 1674 cm^{-1} for DDQ was observed at 1481 cm^{-1} for DDQ-ChA [24,27]. Other important peaks are summarized in Table 3.

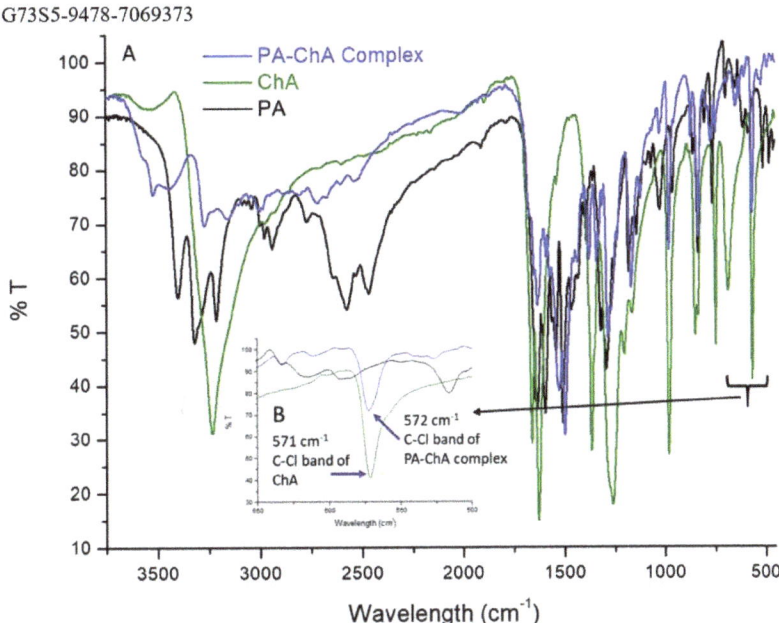

Figure 5. Comparison of IR spectra of procainamide (PA), chloranilic acid (ChA), and PA-ChA complex: (**A**) whole spectra; (**B**) showing the characteristic C-Cl band of ChA and PA-ChA complex.

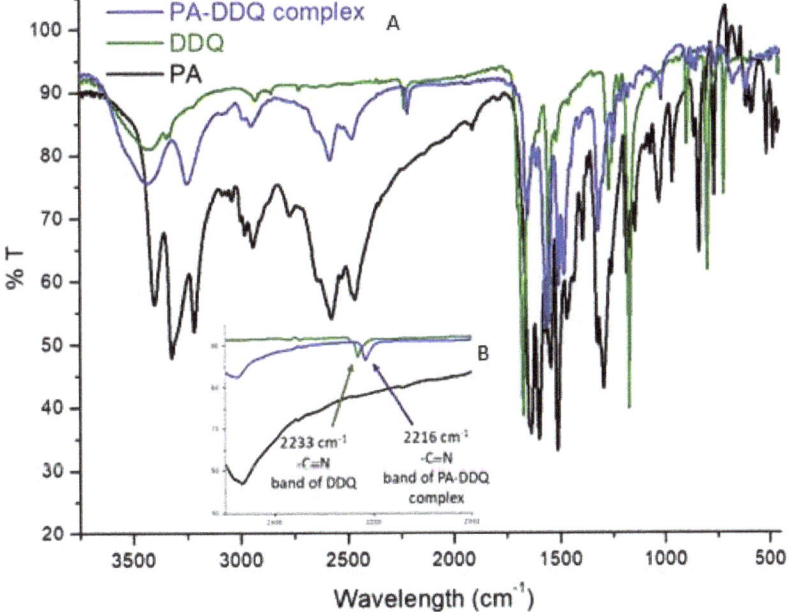

Figure 6. Comparison of IR spectra of PA, DDQ, and PA-DDQ complex: (**A**) whole spectra; (**B**) showing the characteristic -C≡N band of DDQ and PA-DDQ complex.

Table 3. IR spectral bands of PA, DDQ, and their complex (PA-DDQ).

Procainamide	DDQ	PA-DDQ Complex	Possible Assignments
3402	3430	3430	ν(N-H)
3320		3248	ν(O-H)
3215		2987	ν(C-H) (aromatic)
2938		2944	ν(O-H)
2576		2581	ν(⁺N-H)
2465		2471	ν(⁺N-H)
	2232	2217	ν(CN)
1637	1674	1653	ν(CO)
1599		1610	ν(N-H)
1542	1554	1562	ν(C=C) (aromatic ring)
1512		1507	
1467		1482	
1392		1407	
1323		1320	ν(C-C), ν(C-O) (alkanes)
1295	1269	1246	ν(C-N) (alkanes)
1185	1173	1178	ν(C-O) (alkanes)
1145		1145	ν(C-N) (alkanes)
1027		1018	ν(NH)
964		961	ν(C-H) (alkanes)
840	897	892	
806	801	867	
768		757	ν(N-H)
702	722	680	ν(C-N-C)
652		625	

3.6.2. NMR Spectra

NMR Spectra of PA-ChA Complex

From the comparison of the proton NMR (^{1}H-NMR) spectra of procainamide (PA) and the PA-Chloranilic acid (PA-ChA) complex (Figure 7), the formation of the PA-ChA complex is confirmed. Aromatic protons in positions 2 and 6 of the PA-ChA complex slightly (0.13 ppm) shifted downfield from 6.52 to 6.65 ppm, and another two aromatic protons in positions 3 and 5 slightly (0.01 ppm) shifted upfield from 7.59 to 7.60 ppm.

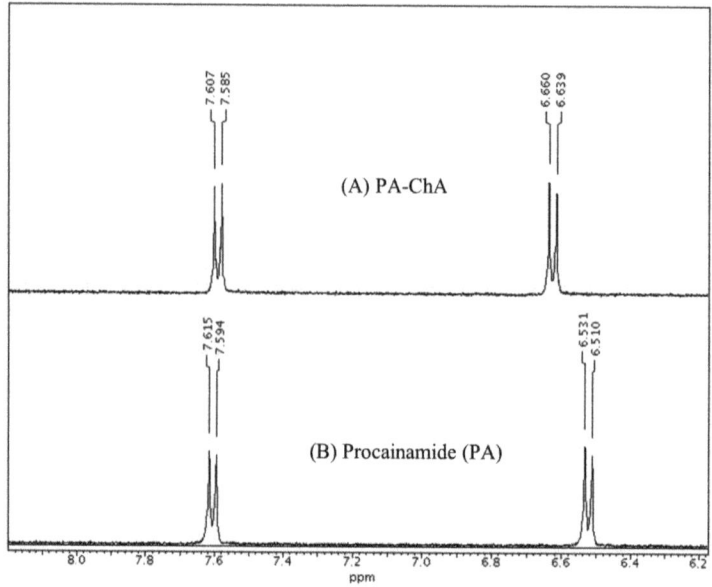

Figure 7. Comparison of proton-NMR (^{1}H-NMR) spectra of Procainamide and Procainamide-ChA complex: (**A**) aromatic region of Procainamide-ChA complex; (**B**) aromatic region of Procainamide (PA).

On the other hand, as shown in Figure 8, -CH$_2$ protons adjacent to -CONH in the aliphatic region were upfield-shifted (0.04 ppm) from 3.56 to 3.52 ppm, -CH$_2$ protons adjacent to tertiary amine were downfield-shifted (0.06 ppm), and methyl protons (-CH$_3$) were also upfield-shifted (0.02 ppm) (Figure 8). In addition to this, -NH$_2$ peaks were upfield-shifted from 10.29 to 9.10 ppm. These changes in chemical shifts might be due to changes in the structural configuration of the complex formation.

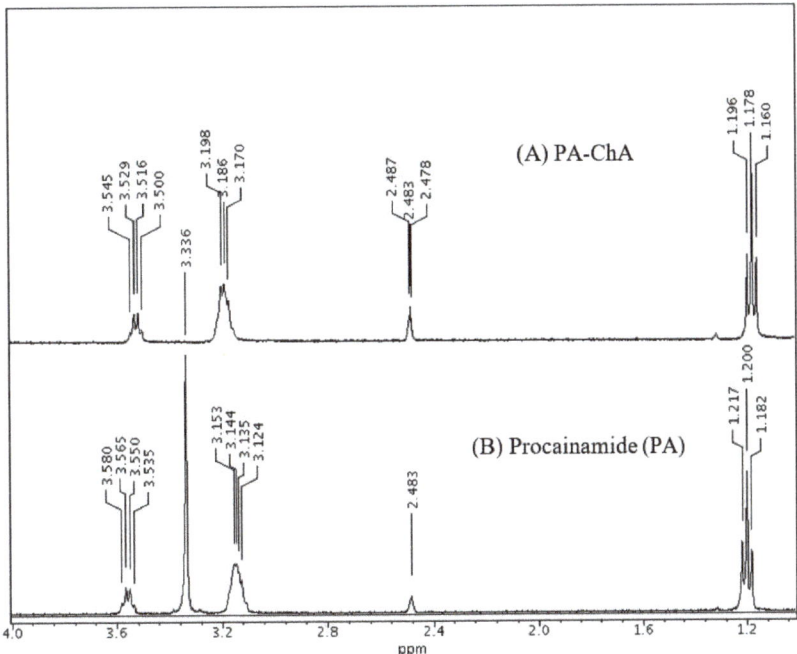

Figure 8. Comparison of proton-NMR (^{1}H-NMR) spectra of PA and PA-ChA complex: (**A**) aliphatic region of PA-ChA complex; (**B**) aliphatic region of PA.

NMR Spectra of PA-DDQ Complex

In the case of the PA-DDQ complex, the chemical shifts of aromatic protons were dramatically downfield-shifted and split into four different chemical shifts. As shown in Figure 9, aromatic protons of PA were given the chemical shifts at 7.60 ppm as a doublet for the protons in positions 3 and 5 and 6.52 ppm as a doublet for the protons in positions 2 and 6; on the other hand, aromatic protons of the PA-DDQ complex were given the chemicals shifts at 7.98, 7.88, 7.45, and 7.20 ppm for the protons in positions 3, 5, 2, and 6, respectively. It should be noted that the protons in positions 3 and 5 were downfield-shifted to 0.38 and 0.28 ppm, and the protons in positions 2 and 6 were downfield-shifted to 0.92 and 0.88 ppm, respectively.

On the other hand, in the aliphatic region, there is little change in chemical shifts, similar to the PA-ChA complex. As shown in Figure 10, -CH$_2$ protons adjacent to -CONH in the aliphatic region were downfield-shifted (0.08 ppm) from 3.56 to 3.64 ppm, -CH$_2$ protons adjacent to tertiary amine were downfield-shifted (0.09 ppm), and in the case of methyl protons (-CH$_3$), they were also downfield-shifted (0.02 ppm). Interestingly, -NH$_2$/-OH peaks were given in 11.01, 10.05, 9.56, 9.00, and 8.89 ppm. These changes in chemical shifts are obviously due to changes in the structural configuration of the complex formation.

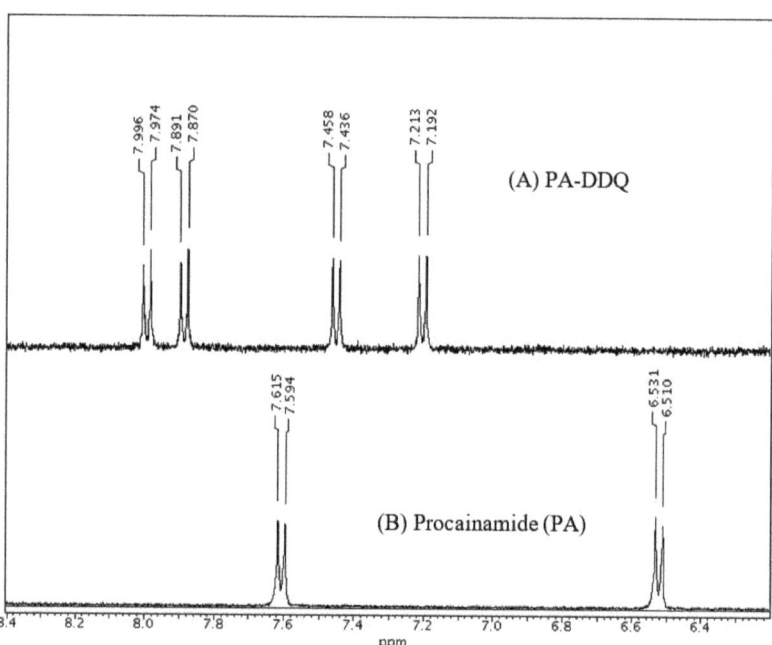

Figure 9. Comparison of proton-NMR (^{1}H-NMR) spectra of PA and PA-DDQ complex. (**A**) aromatic region of PA-DDQ complex; (**B**) aromatic region of PA.

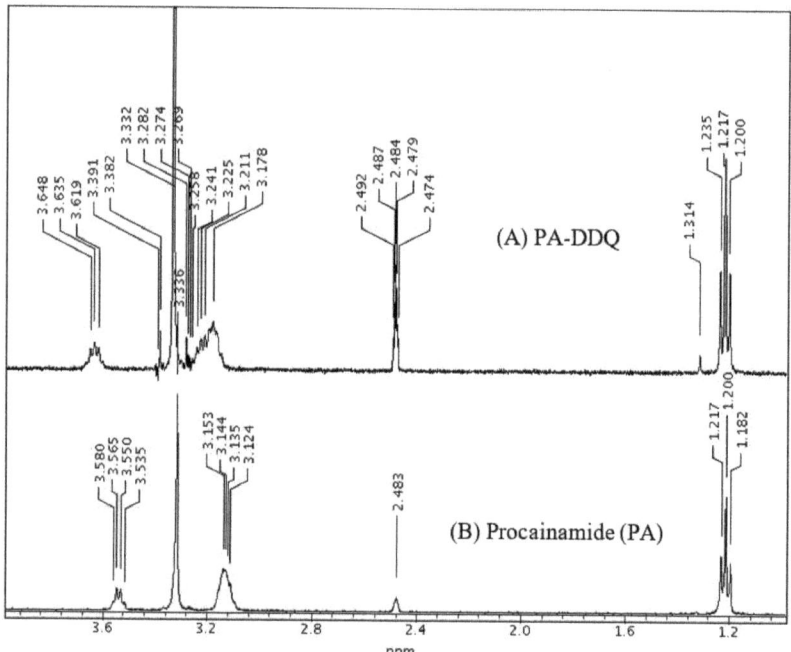

Figure 10. Comparison of proton-NMR (^{1}H-NMR) spectra of PA and PA-DDQ complex. (**A**) Aliphatic region of PA-DDQ complex; (**B**) aliphatic region of PA.

3.7. DFT/TD-DFT Calculations

3.7.1. Optimized Geometrical Structures

The DFT interaction energy (ΔIE) values were calculated by using Equations (9) and (10) in the acetonitrile solvent system for the hypothetical modeled complexes (Figure 11, and the calculated values are summarized in Table 4.

$$\Delta IE_{int} = E_{[PA] \supset \cap [DDQ]} - (E_{[PA]} + E_{[DDQ]}) \tag{9}$$

$$\Delta IE = E_{[PA] \supset \cap [ChA]} - (E_{[PA]} + E_{[ChA]}) \tag{10}$$

where $E_{[PA] \supset \cap [DDQ]}$ and $E_{[PA] \supset \cap [ChA]}$ represent the electronic energy of the optimized structures of PA-DDQ and PA-ChA complexes, respectively, and $E_{[PA]}$, $E_{[ChA]}$, and $E_{[ChA]}$ represent the optimized energy of free PA, ChA, and DDQ, respectively.

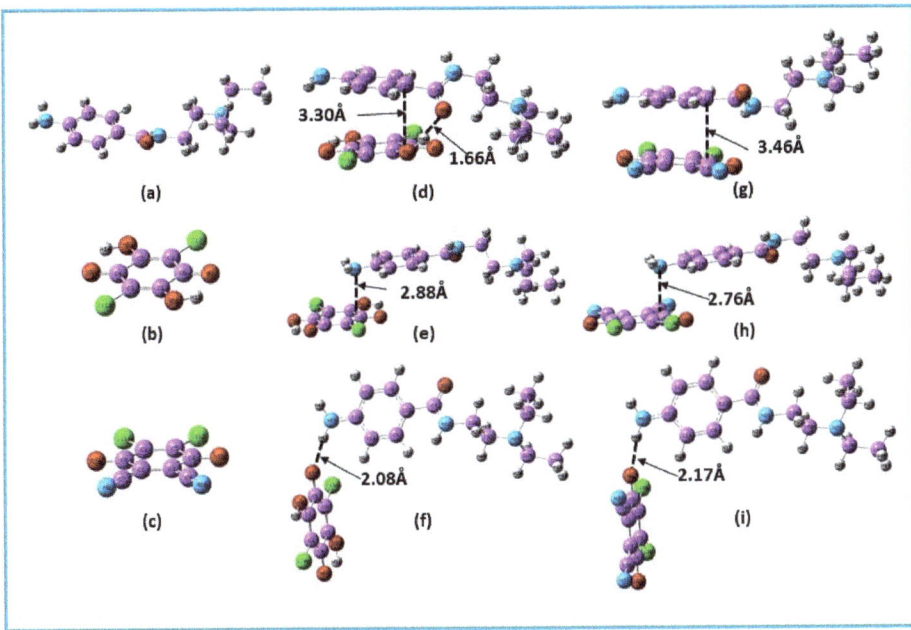

Figure 11. Optimized structures of PA, ChA, DDQ, and their complexes: (**a**) PA; (**b**) ChA; (**c**) DDQ; (**d**) face-to-face I fashion complex of PA with ChA; (**e**) face-to-face II fashion complex of PA complex with ChA; (**f**) edge-to-edge fashion complex of PA with ChA; (**g**) face-to-face I fashion complex of PA with DDQ; (**h**) face-to-face II fashion complex of PA complex with DDQ; and (**i**) edge-to-edge fashion complex of PA with DDQ at the ωB97XD/6-311++G(2d,p) level of theory in acetonitrile solvent system. Close nonbonded contact distances are highlighted in Å.

On the other hand, the Gibbs interaction energy (ΔGint) values were obtained by using Equations (11) and (12) in the acetonitrile solvent system.

$$\Delta G_{int} = E_{[PA] \supset \cap [DDQ]} - (G_{[PA]} + G_{[DDQ]}) \tag{11}$$

$$\Delta G_{int} = G_{[PA] \supset \cap [ChA]} - (G_{[PA]} + G_{[ChA]}) \tag{12}$$

where $G_{[PA] \supset \cap [DDQ]}$ and $G_{[PA] \supset \cap [ChA]}$ represent the Gibbs free energy of the optimized molecular complex of PA with DDQ and ChA, respectively, and $G_{[PA]}$ and $G_{[ChA]}$ represent the Gibbs free energy of the optimized free PA and ChA, respectively.

The calculated ΔIE and ΔG values are represented in the following order: face-to-face I fashion > face-to-face II fashion > edge-to-edge fashion. Negative ΔIE and ΔG values correlate with favorable interactions. These results strongly suggest that the π–π* interactions between the two aromatic rings and the hydrogen bond play an important role in the interaction of the PA with ChA and DDQ, in agreement with the findings of others and also strongly supporting the results of our experimental free energies changes (ΔG) ([PA]⊃ ∩[ChA] = -4.3×10^3 and [PA]⊃ ∩[DDQ] = -4.08×10^3 kJ mol^{-1}). The face-to-face I fashion of the PA-ChA and PA-DDQ complexes are thus considered to be stable structures based on the experimental and theoretical findings.

Table 4. DFT-calculated electronic binding interaction energies (ΔIE kJ/mole) and ΔG (kJ/mole) for the PA complex with ChA and DDQ compounds at the ωB97XD/6-311++G(d,2p) level of theory in acetonitrile solvent.

Complexation Mode	Interaction Energy ΔIE (kJ mol^{-1})	Gibbs Interaction Energy ΔG (kJ mol^{-1})
Face-to-face I fashion for PA⊃ ∩ChA	−64.21	−7.70
Face-to-face II fashion for PA⊃ ∩ChA	−56.90	18.05
Edge-to-edge fashion for PA⊃ ∩ChA	−30.08	39.89
Face-to-face I fashion for PA⊃ ∩DDQ	−61.40	−3.03
Face-to-face II fashion for PA⊃ ∩DDQ	−55.90	14.74
Edge-to-edge fashion for PA⊃ ∩DDQ	−30.55	32.36

3.7.2. HOMO–LUMO Analysis

The structures were drawn in *GaussView 6.0.16* program. The highest occupied molecular orbitals (HOMOs) in which electrons are located and the lowest unoccupied molecular orbitals (LUMOs) were calculated based on the most stable geometry of the complexes. The HOMO of a chemical species is therefore nucleophilic or electron-donating, and the LUMO is electrophilic or electron-accepting. According to Koopmans' theorem [28], the energy of the HOMO (E_{HOMO}), which is indicative of nucleophilic components, is correlated with the ionization potential's negative value (IP = $-E_{HOMO}$).

The energy of the LUMO (E_{LUMO}) is related to the electron affinity's negative value (EA = $-E_{LUMO}$) and is a measure of the susceptibility of the molecule or species toward the reaction with nucleophiles. A large HOMO-LUMO gap signifies that the chemical species is extremely stable and has low reactivity. The HOMO-LUMO energy values shown in Table 5 can be used to calculate a number of other significant and valuable quantum chemical properties. These include global hardness (η), global softness (S), electrophilicity index (ω), electronegativity (χ), and chemical potential (μ), all of which give a measure of chemical reactivity. The hardness value (η) is a qualitative indication of its low polarizability and can be computed using Equation (13):

$$\eta = \left[\frac{E_{LUMO} - E_{HOMO}}{2}\right] = \left[\frac{IP - EA}{2}\right] \quad (13)$$

On the other hand, "soft" molecules are highly polarizable, have modest HOMO-LUMO energy gaps, and can be calculated by using Equation (14):

$$S = \left[\frac{2}{E_{LUMO} - E_{HOMO}}\right] = \left[\frac{2}{IP - EA}\right] = \frac{1}{\eta} \quad (14)$$

Electronegativity

The ability to attract electrons is a characteristic of a chemical's electronegativity (χ) and determines how chemically reactive it is, which can be computed using Equation (15):

$$\chi = \left[\frac{E_{LUMO} - E_{HOMO}}{2}\right] = \left[\frac{IP + EA}{2}\right] \quad (15)$$

Chemical Potential

The chemical potential (μ) is the ability for an electron to be taken out of a molecule, and it can be determined using Equation (16):

$$\mu = \left[\frac{E_{HOMO} - E_{LUMO}}{2}\right] = -\left[\frac{IP + EA}{2}\right] \quad (16)$$

Electrophilicity Index

The electrophilicity index (ω) measures the strength of the electron flow between a donor and an acceptor in a substance's electron acceptors. The mathematical expression for ω is as follows in Equation (17).

$$\omega = \frac{\chi^2}{2\eta} = \left[\frac{\left(\frac{E_{HOMO} - E_{LUMO}}{2}\right)^2}{(E_{LUMO} E_{HOMO})}\right] = \left[\frac{\left(\frac{IP + EA}{2}\right)^2}{IP - EA}\right] \quad (17)$$

Table 5. HOMO-LUMO gap (ΔE_{gap}), ionization potential (IP), electron affinity (EA), electronegativity (χ), chemical potential (μ), hardness (η), softness (S), electrophilicity index (ω), dipole moments (dm), and polarizability (α) of the PA, ChA, DDQ, and their complexes at the ωB97XD/6−311++G(d,2p) level of theory in acetonitrile solvent.

	E_{HOMO} (eV)	E_{LUMO} (eV)	ΔE_{gap} (eV)	IP (eV)	EA (eV)	χ (eV)	μ (eV)	η (eV)	S (eV)	ω (eV)	dm (Debye)	Polarizability (α)
PA	−7.97	−0.81	7.16	7.97	0.81	4.39	−4.39	3.58	0.28	2.70	5.58	248.81
ChA	−10.18	−3.18	7.00	10.18	3.18	6.68	−6.68	3.50	0.29	6.37	0.01	151.11
DDQ	−8.02	−3.03	4.99	8.02	3.03	5.52	−5.52	2.50	0.40	6.11	10.27	463.23
Face-to-face I fashion for PA⊃∩ChA complex	−8.04	−1.95	6.09	8.04	1.95	4.99	−4.99	3.04	0.33	4.10	6.58	408.18
Face-to-face II fashion for PA⊃∩ChA complex	−8.05	−2.01	6.04	8.05	2.01	5.03	−5.03	3.02	0.33	4.18	5.75	418.10
Edge-to-edge fashion for PA⊃∩ChA complex	−7.86	−2.21	5.65	7.86	2.21	5.03	−5.03	2.83	0.35	4.48	10.13	403.93
Face-to-face I fashion for PA⊃∩DDQ complex	−8.08	−3.06	5.02	8.08	3.06	5.57	−5.57	2.51	0.40	6.18	10.56	467.96
Face-to-face II fashion for PA⊃∩DDQ complex	−8.08	−2.96	5.12	8.08	2.96	5.52	−5.52	2.56	0.39	5.94	10.35	465.30
Edge-to-edge fashion for PA⊃∩DDQ complex	−7.91	−3.26	4.66	7.91	3.26	5.59	−5.59	2.33	0.43	6.70	6.05	431.41

The molecular electrostatic surface potentials [29] of PA, ChA, DDQ, and PA complexes with ChA and DDQ are shown in Figures 12 and 13. The relative polarities and reactive sites of the species-negative ESP are shown in red, and the order of increasing electrostatic potential (i.e., highest negative value) is red > orange > yellow > green > blue. The carbonyl oxygen (-C=O) atom of PA, which is illustrated in red in Figures 12 and 13, has a high electron density and is the preferred site for electrophilic attack and interaction with the nucleophilic partly positive charged hydrogen atoms (blue color). The yellow color indicates the slightly rich electron regions, and the green reflects more neutral zones. The HOMO and LUMO properties and the quantum chemical properties of PA, ChA, DDQ, and PA complexes with ChA and DDQ are summarized in Table 5.

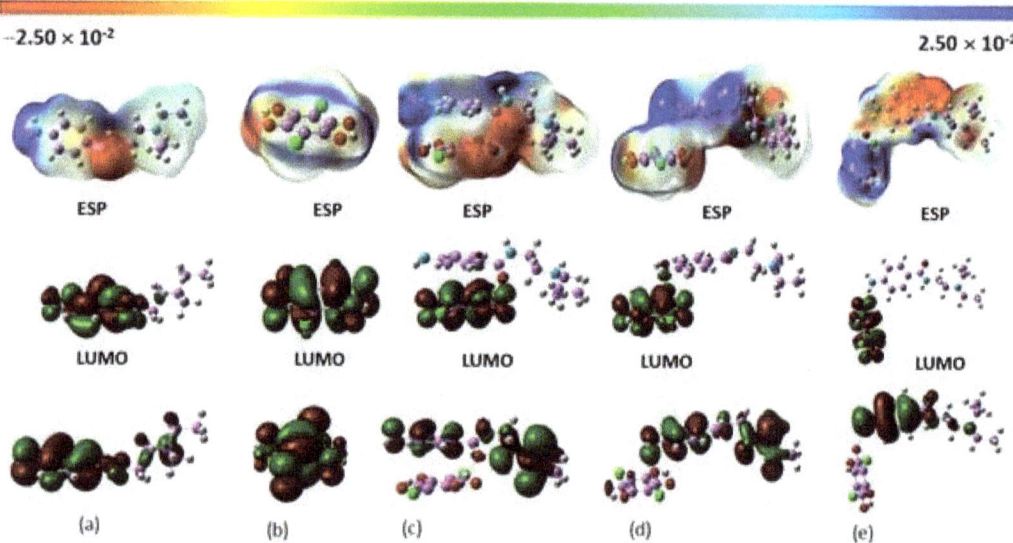

Figure 12. Molecular electrostatic potential (MEP) maps of HOMO and LUMO structures of PA, ChA, and their 1:1 complex: (**a**) PA; (**b**) ChA; (**c**) face-to-face I fashion complex of PA with ChA; (**d**) face-to-face II fashion complex of PA complex with ChA; and (**e**) edge-to-edge fashion complex of PA with ChA at the ωB97XD/6-311++G(2d,p) level of theory in acetonitrile solvent system.

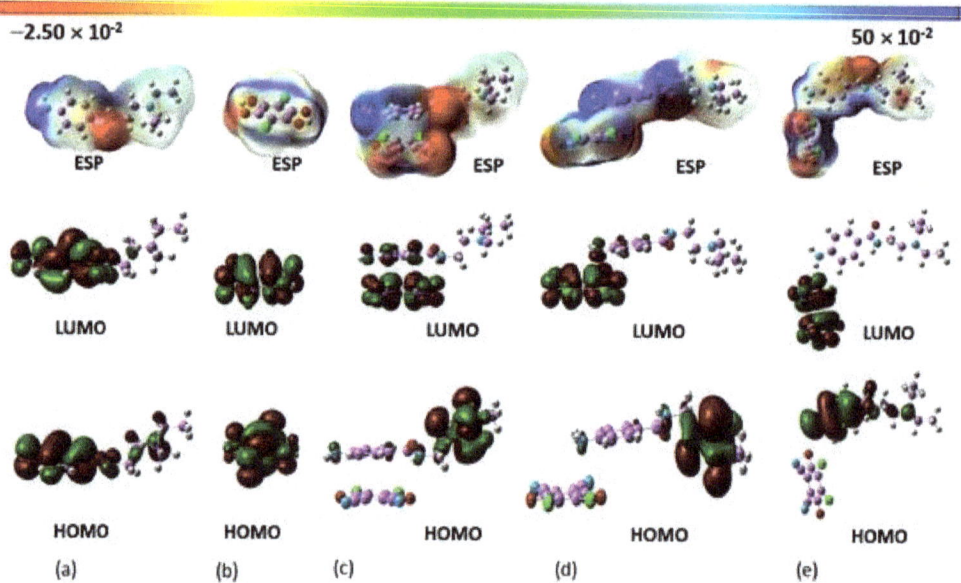

Figure 13. Molecular electrostatic potential (MEP) maps of HOMO and LUMO structures of PA, DDQ, and their 1:1 complex: (**a**) PA; (**b**) DDQ; (**c**) face-to-face I fashion complex of PA with DDQ; (**d**) face-to-face II fashion complex of PA complex with DDQ; and (**e**) edge-to-edge fashion complex of PA with DDQ at the ωB97XD/6-311++G(2d,p) level of theory in acetonitrile solvent system.

As shown in Table 5, the electronegativity (χ) for ChA is 6.68 eV and for DDQ is 5.52 eV, which indicates that ChA and DDQ have the ability to form CT complexes with PA. In addition, the electrophilicity index (ω) of ChA is 6.37 eV and for DDQ is 6.11 eV, which also suggests the formation of CT complexes with PA. The stability between the PA-ChA and PA-DDQ complexes was measured using their HOMO-LUMO gaps (ΔE_{gap}), and the face-to-face I fashion for PA⊃∩ChA complex was found to be 6.09 eV, while the Face-to-face II fashion for PA⊃∩DDQ complex was found to be 5.12 eV.

3.7.3. Theoretical Electronic Absorption Spectra

The predicted electronic spectra of the resultant complexes of PA with ChA and DDQ after being calculated using the first six single-point calculations in the acetonitrile solvent system at the TD-DFT/wB97XD/6-31+G(d,2p) basis set level of theory are shown in Figure 14. The spectra were plotted by applying a Gaussian broadening of 0.333 eV half-width at half height. The theoretical electronic absorption spectra of the donor (PA) and acceptors (ChA/DDQ) as well as synthesized CT complexes (PA-ChA/PA-DDQ) are shown in Figure 14. A strong absorption band at 250 nm was observed for PA, while weak absorption maxima for ChA or DDQ were observed at 291 and 276 nm, respectively. On the other hand, two broad bands at 426 nm (excitation energy of 2.91 eV and oscillator strength of 0.041) and 622 nm (excitation energy of 1.99 eV and oscillator strength of 0.142) were observed for the two new CT complexes PA-ChA and PA-DDQ, respectively, in the acetonitrile solvent system at 298 K.

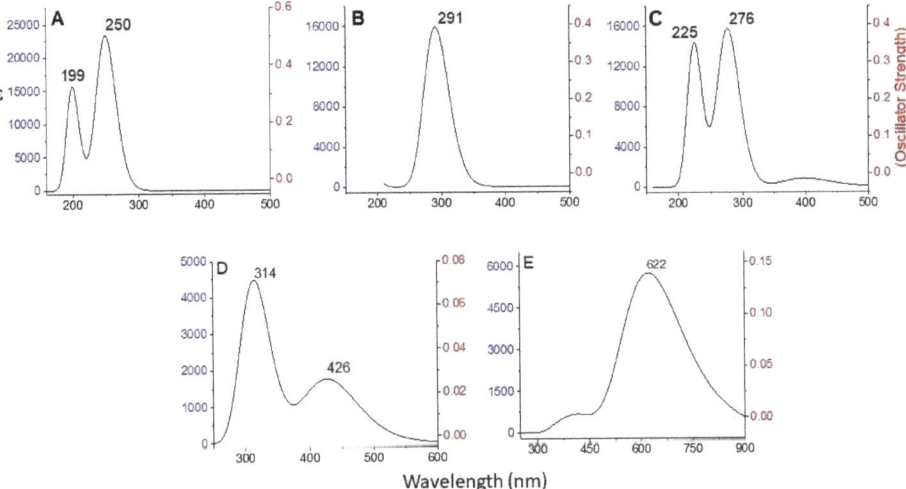

Figure 14. DFT calculated UV-Vis spectra of PA, ChA, DDQ, and their complexes: (**A**) PA; (**B**) ChA; (**C**) DDQ; (**D**) face-to-face I fashion complex of PA with ChA; and (**E**) face-to-face I fashion complex of PA complex with DDQ at the ωB97XD/6-311++G(2d,p) level of theory in acetonitrile solvent system.

4. Conclusions

Two new CT complexes (PA-ChA and PA-DDQ) have been synthesized from an electron donor PA and an electron acceptor ChA/DDQ using a simple, easy, and economically inexpensive synthetic method. The formation of the complexes was confirmed by various spectroscopic analysis techniques. The DFT-computed calculation strongly supports our experimental results.

Supplementary Materials: The following data are available online at https://www.mdpi.com/article/10.3390/pr11030711/s1, DFT/TD-DFT calculation files.

Author Contributions: Conceptualization, A.F.M.M.R. and G.A.E.M.; methodology, A.F.M.M.R. and G.A.E.M.; validation, A.F.M.M.R. and G.A.E.M.; formal analysis, A.F.M.M.R., S.R., A.H.B. and G.A.E.M.; investigation, A.F.M.M.R. and G.A.E.M.; resources; A.F.M.M.R., H.A. and G.A.E.M.; Software, S.R. and A.H.B.; data curation, A.F.M.M.R., S.R., A.H.B. and G.A.E.M.; writing—original draft preparation, A.F.M.M.R., S.R., A.H.B. and G.A.E.M.; writing—review and editing, A.F.M.M.R., S.R., A.H.B., H.A. and G.A.E.M.; supervision, A.F.M.M.R. and G.A.E.M.; project administration G.A.E.M.; funding acquisition, H.A. and G.A.E.M. All authors have read and agreed to the published version of the manuscript.

Funding: This research was funded by the Researchers Supporting Project, King Saud University, through grant No. RSP-2023R501.

Institutional Review Board Statement: Not applicable.

Informed Consent Statement: Not applicable.

Data Availability Statement: All the Supporting Data for DFT/TD-DFT calculation can be obtained upon request.

Acknowledgments: The authors extend their appreciation to the Researchers Supporting Project, King Saud University.

Conflicts of Interest: The authors clarify that this study has no competing interest.

References

1. Mian, M.S.; El-Obeid, H.A.; Al-Badr, A.A. Procainamide Hydrochloride. In *Analytical Profiles of Drug Substances and Excipients*; Brittain, H.G., Ed.; Academic Press: Cambridge, MA, USA, 2001; Volume 28, pp. 251–332.
2. Zamponi, G.W.; Sui, X.; Codding, P.W.; French, R.J. Dual actions of procainamide on batrachotoxin-activated sodium channels: Open channel block and prevention of inactivation. *Biophys. J.* **1993**, *65*, 2324–2334. [CrossRef]
3. Shen, D.; Chen, W.-C.; Lo, M.-F.; Lee, C.-S. Charge-transfer complexes and their applications in optoelectronic devices. *Mater. Today Energy* **2021**, *20*, 100644. [CrossRef]
4. Adam, A.M.A. Application of charge-transfer complexation for evaluation of the drug-receptor mechanism of interaction: Spectroscopic and structure morphological properties of procaine and pilocarpine complexes with chloranilic acid acceptor. *Russ. J. Gen. Chem.* **2014**, *84*, 1225–1236. [CrossRef]
5. Adam, A.M.A. Nano-structured complexes of reserpine and quinidine drugs with chloranilic acid based on intermolecular H-bond: Spectral and surface morphology studies. *Spectrochim. Acta A Mol. Biomol. Spectrosc.* **2014**, *127*, 107–114. [CrossRef]
6. Wang, W.; Luo, L.; Sheng, P.; Zhang, J.; Zhang, Q. Multifunctional Features of Organic Charge-Transfer Complexes: Advances and Perspectives. *Chem. A Eur. J.* **2021**, *27*, 464–490. [CrossRef]
7. Zhu, L.; Yi, Y.; Fonari, A.; Corbin, N.S.; Coropceanu, V.; Brédas, J.-L. Electronic Properties of Mixed-Stack Organic Charge-Transfer Crystals. *J. Phys. Chem. C* **2014**, *118*, 14150–14156. [CrossRef]
8. McGuire, M.A. Crystal and Magnetic Structures in Layered, Transition Metal Dihalides and Trihalides. *Crystals* **2017**, *7*, 121. [CrossRef]
9. Khan, I.M.; Islam, M.; Shakya, S.; Alam, K.; Alam, N.; Shahid, M. Synthesis, characterization, antimicrobial and DNA binding properties of an organic charge transfer complex obtained from pyrazole and chloranilic acid. *Bioorganic. Chem.* **2020**, *99*, 103779. [CrossRef]
10. Karmakar, A.; Bandyopadhyay, P.; Mandal, N. Synthesis, spectroscopic, theoretical and antimicrobial studies on molecular charge-transfer complex of 4-(2-thiazolylazo)resorcinol (TAR) with 3, 5-dinitrosalicylic acid, picric acid, and chloranilic acid. *J. Mol. Liq.* **2019**, *299*, 112217. [CrossRef]
11. Singh, N.; Khan, I.M.; Ahmad, A.; Javed, S. Synthesis, crystallographic and spectrophotometric studies of charge transfer complex formed between 2,2′-bipyridine and 3,5-dinitrosalicylic acid. *J. Mol. Liq.* **2014**, *191*, 142–150. [CrossRef]
12. Refat, M.S.; Saad, H.A.; Adam, A.M. Spectral, thermal and kinetic studies of charge-transfer complexes formed between the highly effective antibiotic drug metronidazole and two types of acceptors: σ- and π-acceptors. *Spectrochim. Acta A Mol. Biomol. Spectrosc* **2015**, *141*, 202–210. [CrossRef]
13. Nampally, V.; Palnati, M.K.; Baindla, N.; Varukolu, M.; Gangadhari, S.; Tigulla, P. Charge Transfer Complex between O-Phenylenediamine and 2, 3-Dichloro-5, 6-Dicyano-1, 4-Benzoquinone: Synthesis, Spectrophotometric, Characterization, Computational Analysis, and its Biological Applications. *ACS Omega* **2022**, *7*, 16689–16704. [CrossRef] [PubMed]
14. Mulliken, R.S. Structures of Complexes Formed by Halogen Molecules with Aromatic and with Oxygenated Solvents1. *J. Am. Chem. Soc.* **1950**, *72*, 600–608. [CrossRef]
15. Mulliken, R.S. Molecular Compounds and their Spectra. III. The Interaction of Electron Donors and Acceptors. *J. Phys. Chem.* **1952**, *56*, 801–822. [CrossRef]

16. Alanazi, A.; Abounassif, M.; Alrabiah, H.; Mostafa, G. Development of two charge transfer complex spectrophotometric methods for determination of tofisopam in tablet dosage form. *Trop. J. Pharm. Res.* **2016**, *15*, 995. [CrossRef]
17. Refat, M.S.; El-Korashy, S.A.; El-Deen, I.M.; El-Sayed, S.M. Experimental and spectroscopic studies of charge transfer reaction between sulfasalazine antibiotic drug with different types of acceptors. *Drug Test Anal.* **2011**, *3*, 116–131. [CrossRef]
18. Adam, A.M.A.; Refat, M.S.; Hegab, M.S.; Saad, H.A. Spectrophotometric and thermodynamic studies on the 1:1 charge transfer interaction of several clinically important drugs with tetracyanoethylene in solution-state: Part one. *J. Mol. Liq.* **2016**, *224*, 311–321. [CrossRef]
19. Khan, I.M.; Shakya, S.; Islam, M.; Khan, S.; Najnin, H. Synthesis and spectrophotometric studies of CT complex between 1,2-dimethylimidazole and picric acid in different polar solvents: Exploring antimicrobial activities and molecular (DNA) docking. *Phys. Chem. Liq.* **2021**, *59*, 753–769. [CrossRef]
20. Aloisi, G.G.; Pignataro, S. Molecular complexes of substituted thiophens with σ and π acceptors. Charge transfer spectra and ionization potentials of the donors. *J. Chem. Soc. Faraday Trans. 1 Phys. Chem. Condens. Phases* **1973**, *69*, 534–539. [CrossRef]
21. Briegleb, G.; Czekalla, J. Die Intensität von Elektronenüberführungsbanden in Elektronen-Donator-Akzeptor-Komplexen1. *Z. Für Phys. Chem.* **1960**, *24*, 37–54. [CrossRef]
22. Frisch, M.J.; Trucks, G.W.; Schlegel, H.B.; Scuseria, G.E.; Robb, M.A.; Cheeseman, J.R.; Scalmani, G.; Barone, V.; Mennucci, B.; Petersson, G.A.; et al. *Gaussian 16, Revision C.01*; Gaussian, Inc.: Wallingford, CT, USA, 2019.
23. Siraki, A.G.; Deterding, L.J.; Bonini, M.G.; Jiang, J.; Ehrenshaft, M.; Tomer, K.B.; Mason, R.P. Procainamide, but not N-Acetylprocainamide, Induces Protein Free Radical Formation on Myeloperoxidase: A Potential Mechanism of Agranulocytosis. *Chem. Res. Toxicol.* **2008**, *21*, 1143–1153. [CrossRef] [PubMed]
24. Ranzieri, P.; Masino, M.; Girlando, A. Charge-Sensitive Vibrations in p-Chloranil: The Strange Case of the CC Antisymmetric Stretching. *J. Phys. Chem. B* **2007**, *111*, 12844–12848. [CrossRef] [PubMed]
25. IR Spectroscopy Tutorial: Alkyl Halides. Available online: https://orgchemboulder.com/Spectroscopy/irtutor/alkhalidesir.shtml (accessed on 17 October 2022).
26. IR Spectrum Table & Chart. 2022. Available online: https://www.sigmaaldrich.com/SA/en/technical-documents/technical-article/analytical-chemistry/photometry-and-reflectometry/ir-spectrum-table (accessed on 17 October 2022).
27. Fukui, K.-i.; Yonezawa, T.; Shingu, H. A Molecular Orbital Theory of Reactivity in Aromatic Hydrocarbons. *J. Chem. Phys.* **1952**, *20*, 722–725. [CrossRef]
28. Luo, J.; Xue, Z.Q.; Liu, W.M.; Wu, J.L.; Yang, Z.Q. Koopmans' Theorem for Large Molecular Systems within Density Functional Theory. *J. Phys. Chem. A* **2006**, *110*, 12005–12009. [CrossRef] [PubMed]
29. Murray, J.S.; Politzer, P. The electrostatic potential: An overview. *WIREs Comput. Mol. Sci.* **2011**, *1*, 153–163. [CrossRef]

Disclaimer/Publisher's Note: The statements, opinions and data contained in all publications are solely those of the individual author(s) and contributor(s) and not of MDPI and/or the editor(s). MDPI and/or the editor(s) disclaim responsibility for any injury to people or property resulting from any ideas, methods, instructions or products referred to in the content.

Article

Evaluation of Antiaging Effect of Sheep Placenta Extract Using SAMP8 Mice

Ming-Yu Chou [1], Chi-Pei Ou Yang [2], Wen-Ching Li [3], Yao-Ming Yang [4], Yu-Ju Huang [2], Ming-Fu Wang [3,*] and Wan-Teng Lin [5,*]

1. International Aging Industry Research and Development Center (AIC), Providence University, Taichung 43301, Taiwan
2. Genius Bull International Ltd., No. 220, Sec. 2, Taiwan Blvd., West Dist., Taichung 403, Taiwan
3. Department of Food and Nutrition, Providence University, Taichung 43301, Taiwan
4. Dong Wu Zhu Mu Qin Qi Yue Yi Biological and Technology Co., Ltd., Industry Area, Wu Li Ya Si Tai Town, Xiligol League, Inner Mongolia, China
5. Department of Hospitality Management, College of Agriculture, Tunghai University, Taichung 407224, Taiwan
* Correspondence: mfwang@pu.edu.tw (M.-F.W.); 040770@thu.edu.tw (W.-T.L.); Tel.: +886-4-2359-0121 (ext. 37709) (W.-T.L.)

Citation: Chou, M.-Y.; Yang, C.-P.O.; Li, W.-C.; Yang, Y.-M.; Huang, Y.-J.; Wang, M.-F.; Lin, W.-T. Evaluation of Antiaging Effect of Sheep Placenta Extract Using SAMP8 Mice. *Processes* **2022**, *10*, 2242. https://doi.org/10.3390/pr10112242

Academic Editors: Hah Young Yoo and Tao Sun

Received: 18 August 2022
Accepted: 25 October 2022
Published: 1 November 2022

Publisher's Note: MDPI stays neutral with regard to jurisdictional claims in published maps and institutional affiliations.

Copyright: © 2022 by the authors. Licensee MDPI, Basel, Switzerland. This article is an open access article distributed under the terms and conditions of the Creative Commons Attribution (CC BY) license (https://creativecommons.org/licenses/by/4.0/).

Abstract: Widely used in traditional medicine, sheep placenta extract (SPE) is known for its physiological effects such as wound healing, antioxidant, and anti-inflammatory properties. However, the effect of SPE on antiaging is still unclear. In this study, we investigated the effect of SPE on aging through the senescence-accelerated mouse prone 8 (SAMP8) strain. We designed an experiment using both male and female mice randomly divided into 4 groups (n = 10) as follows: Group A—control group; Group B—low-dose SPE (61.5 mg/kg BW/day); Group C—medium-dose SPE (123 mg/kg BW/day); and Group D—high-dose SPE (184.5 mg/kg BW/day). As a result of measuring the aging index parameters such as skin glossiness, spine lordosis, and kyphosis, it was found that the treatment of SPE lowered the aging index. In addition, we found that biochemical parameters such as lactic acid, glucose, ketone bodies, free fatty acids, tumor necrosis factor-alpha (TNF-α), and interleukin 6 (IL-6) were not changed in the experimental group treated with SPE for 13 weeks. Finally, we found that lipid peroxidation (LPO) was decreased, while the activities of catalase and superoxide dismutase (SOD) were significantly increased in the brain tissues of SPE-treated male and female mice. Supplementation of SPE lowered the oxidative stress caused by the aging process in mice without toxicity and decreased the aging index, suggesting the value of SPE as an effective antiaging treatment.

Keywords: antiaging; sheep placenta; antioxidant; SAMP8; histology

1. Introduction

The placenta joins the fetus to the mother and functions as a means of transport to transfer nutrients such as glucose, amino acids, fatty acids, vitamins, and other minerals essential for fetus growth and development [1]. The nutrients supplied by the placenta may be retained after the birth of the fetus [2,3]. Collagen, elastin, laminin, vitamins, trace minerals, nucleic acids, amino acids, peptides, cytokines, and growth factors have all been identified in human placental extracts [4–7].

With its high nutritional content and biologically active components, the placenta has historically been valued as a traditional folk medicine in China and other areas of the world. Placentophagy has a long history and has been practiced in many regions of the world. However, few scientific studies of the medicinal value of placental extract have been conducted [8]. Dried human placenta, also known as "Zi He Che", was used in sixteenth-century China as a cure for impotence and for infertility, liver, and kidney issues, according to Li Shizen's "Compendium of Materia Medica" [9]. It was also used to boost energy and stamina. The Araucanian Indians of Argentina use dried umbilical cord ground into powder to heal sick children. The Kol tribe in Central India consumes the placenta to enhance reproductive function [10].

In mainland China, since the Tang dynasty, practitioners of traditional Chinese medicine (TCM) have valued the placenta as a transitory organ that sustains the growth and development of a fetus in utero [11]. TCM practitioners have used the placenta to reduce mental tension and anxiety, enrich the blood, tonify Qi, increase essences, reduce fatigue and spasm, and detoxify the body [12]. In recent times, increased interest in the pharmacological effects and therapeutic value of the human placenta has led to more scientific studies, as well as increased usage in clinical settings [13].

Placenta treatment is now used for a variety of conditions, including adrenocortical hypofunction, lumbago, compromised immunity, infertility, depression, lack of lactation, and hair loss, as well as for antiaging purposes [9]. Within conventional medicine, practitioners are increasingly acknowledging the effectiveness of human placenta treatment, although restricted availability has prevented its widespread use to date. The benefits of treatment with sheep placenta (SP) extract include blood and skin nourishment, sedation effects, and increased longevity [14]. SP extract is one of the best medications for boosting essential vitality. Practitioners favor the use of sheep placenta because it shares the same nutritional composition and pharmacological properties as the human placenta [15]. However, fatigue in mice was also relieved by preparations made from goat placenta [16]. SP extract (SPE) is rich in tiny molecular peptides, nucleic acids, enzymes, amino acids, growth factors, collagen, and other active ingredients that may boost immunity, and reduce the effects of hypoxia [17].

The senescence-accelerated mouse (SAM) is a naturally occurring experimental mouse breed that exhibits early aging characteristics such as senile amyloidosis, senile osteoporosis, cataracts, a suppressed immune system, and deficiencies in learning and memory [18,19]. Breeders have developed a number of senescence-prone (SAMP) and senescence-resistant (SAMR) lines. Due to the early onset of brain shrinkage, which is accompanied by learning and/or memory impairment and affective disturbances, SAMP8 is a commonly utilized strain, particularly in dementia- and aging-related studies [20]. The longevity of SAMP8 is roughly half that of SAMR1 [21].

Integrative biological medicine is important for managing and restoring health and fitness. Placental extracts that still retain active components may promote rejuvenation, revitalization, and the restoration of youth and vitality by delaying aging processes. The placental extract is a powerful therapeutic agent with effective regeneration abilities that counter the aging process. In the present study, we evaluated the antiaging effect of SPE in SAMP8 mice.

2. Materials and Methods

2.1. Animals

We kept animals in transparent plastic cages of 30(W) × 20(D) × 10(H) cm^3 in a dust-free automatic control room. We maintained room temperature at 22 ± 2 °C, with a relative humidity of 65 ± 5%. We controlled the light cycle using an automatic timer, with daily light and dark periods from 07:00 to 19:00 and 19:00 to 07:00 h, respectively. We provided animals with food and water ad libitum.

2.2. Sheep Placenta Extract Preparation and Dosage

For our experimental samples, we used dried sheep placenta powder provided by Zhenyuebo International Co., Ltd., produced from the placentas of Ujimqin sheep from Inner Mongolia, manufactured by East Ujimqin Banner Yueyi Biotechnology Co., Ltd., Inner Mongolia, China.

We determined dose conversion using the experimental evaluation method of the Ministry of Health and Welfare. First, we determined the recommended daily human intake per kilogram of body weight. We then expanded the recommended human dose by a factor of 12.3 to obtain the dose for mice in the medium-dose group. We then multiplied this dose by a factor of 0.5 for the low-dose group, and by 1.5 for the high-dose group.

Low-dose group (0.5 times)

For this group, we first calculated the recommended daily intake of human adults as 600 mg/60 kg BW/day $\times$ 1/2 = 5 mg/kg BW/day. We then calculated daily intake for mice based on the proportion of their body weight, as follows: 5 mg/kg BW/day $\times$ 12.3 = 61.5 mg/kg BW/day.

Medium-dose group (1 time)

For this group, we first calculated the recommended daily intake of human adults as 600 mg/60 kg BW/day $\times$ 1 = 10 mg/kg BW/day. We then calculated daily intake for mice based on the proportion of their body weight, as follows: 10 mg/kg BW/day $\times$ 12.3 = 123 mg/kg BW/day.

High-dose group (1.5 times)

For this group, we first calculated the recommended daily intake of human adults as 600 mg/60 kg BW/day $\times$ 1.5 = 15 mg/kg BW/day. We then calculated daily intake for mice based on the proportion of their body weight, as follows: 15 mg/kg BW/day $\times$ 12.3 = 184.5 mg/kg BW/day.

2.3. Experimental Animal Grouping

We used 3-month-old male and female SAMP8 mice as our experimental animals, and we randomly divided them into one control group and three experimental groups (low, medium, and high doses) with 10 mice in each group (n = 10), and 80 mice in total. For both male and female animals, Group A was the control group, Group B was the low-dose SPE group (61.5 mg/kg BW/day), Group C was the medium-dose SPE group (123 mg/kg BW/day), and Group D was the high-dose SPE group (184.5 mg/kg BW/day). We obtained approval for our animal experiment procedures from the Institutional Animal Care and Use Committee (IACUC) of Providence University (Approval No: 20201218 A008).

At the end of the treatment, we euthanized all mice using 95% CO_2. We then collected blood and separated serum using centrifugation. This was then stored at -80 °C. We collected and weighed organs such as the liver, lung, heart, kidney, adipose tissue, and skeletal muscles, and stored them at -80 °C. We stored small portions of vital organ tissue sections in formalin for histological analysis. We used brain tissue to estimate LPO, catalase activities, and SOD activities.

2.4. Aging Index

We included behavioral aspects in the evaluation items of the aging index. For 30 s, we observed the reactivity of mice (reactivity) and also their escape response (passivity) following pinching of the skin on the nape of the neck. We also noted appearance aspects such as the glossiness of the skin (glossiness) and its roughness (coarseness), as well as any loss of hair (hair loss). We also sought to identify any skin ulceration (ulcer). With regard to animal eyes (eyes), we looked for mucositis around the eyes or edema of the eyelid (periophthalmic lesion). We also inspected animals using touch to assess any changes in spine lordokyphosis (lordokyphosis). We assessed each evaluation item by means of 5 grades, namely, 0, 1, 2, 3, and 4. The higher the score, the more serious the aging phenomenon observed.

2.5. Serum Biochemical Parameters Analysis

After 13 weeks of SPE treatment, we anesthetized all the mice and collected their blood. We then separated serum to analyze various biochemical parameters using an autoanalyzer (Hitachi 7060, Hitachi, Tokyo, Japan). These biochemical parameters were as follows: glucose, total protein, albumin, triglyceride, total cholesterol, HDL, LDL, AST, ALT, BUN, creatinine, sodium, potassium, uric acid, creatine kinase, ketones, free fatty acids, IL-6, and TNF-α.

2.6. Histological Analysis

We collected the liver, heart, lung, kidney, and brain from all the treatment groups. We carried out hematoxylin and eosin (H&E) staining on the collected organs, following previously reported methods [22].

2.7. Statistical Analysis

We analyzed the data obtained using the SPSS statistical software package. We subjected the experimental data to a one-way analysis of variance (one-way ANOVA) to identify differences between multiple groups. To compare the differences between groups, we used Duncan's multiple-range test. A significant difference was indicated when $p < 0.05$.

3. Results

3.1. Effects of SPE on Food Intake, Water Intake, Body Weight, and Organ Weight of SAMP8 Mice

Tables 1 and 2 show the effects of feeding 3-month-old SAMP8 mice with SPE in terms of changes in body weight, food intake, and water intake for male and female animals, respectively. For groups of both sexes, we found no significant changes in body weight, food intake, and water intake after the administration of SPE. Tables 3 and 4 show weight changes in the brain, heart, liver, spleen, lung, and kidney of male and female mice, respectively. We found no significant changes in organ weights in the animal groups that received SPE treatment.

Table 1. Effects of SPE treatment on body weight, food intake, and water intake in male SAMP8 mice.

Group	Body Weight (g)			Food Intake (g/Day)	Water Intake (mL/Day)
	Initial	Final	Gain		
A	29.81 ± 0.58	31.49 ± 2.02	1.69 ± 0.43	4.61 ± 0.08	6.54 ± 0.09
B	30.24 ± 0.67	32.01 ± 0.65	1.70 ± 0.56	4.73 ± 0.10	6.93 ± 0.14
C	29.52 ± 0.67	32.76 ± 0.63	2.80 ± 0.52	4.74 ± 0.08	6.65 ± 0.14
D	29.68 ± 0.85	30.76 ± 0.80	1.08 ± 0.31	4.87 ± 0.03	6.95 ± 0.14

Group A—control group, Group B—low-dose SPE (61.5 mg/kg BW/day), Group C—medium-dose SPE (123 mg/kg BW/day), and Group D—high-dose SPE (184.5 mg/kg BW/day). Data are expressed as the mean ± SEM and analyzed by one-way ANOVA.

Table 2. Effects of SPE treatment on body weight, food intake, and water intake in female SAMP8 mice.

Group	Body Weight (g)			Food Intake (g/Day)	Water Intake (mL/Day)
	Initial	Final	Gain		
A	25.58 ± 0.53	26.73 ± 0.54	1.16 ± 0.24	4.28 ± 0.09	5.01 ± 0.10
B	25.85 ± 0.57	27.39 ± 0.43	1.59 ± 0.62	4.08 ± 0.09	5.00 ± 0.06
C	25.80 ± 0.55	27.17 ± 0.56	1.23 ± 0.27	4.16 ± 0.08	5.04 ± 0.04
D	25.55 ± 0.42	27.36 ± 0.50	1.82 ± 0.49	4.34 ± 0.09	5.18 ± 0.06

Group A—control group, Group B—low dose SPE (61.5 mg/kg BW/day), Group C—medium dose SPE (123 mg/kg BW/day), and Group D—high dose SPE (184.5 mg/kg BW/day). Data are expressed as the mean ± SEM and analyzed by one-way ANOVA.

Table 3. Effects of SPE treatment on organ weights of male SAMP8 mice.

Group	Relative Organ Weights (g/100 g Body Weight)					
	Brain	Heart	Liver	Spleen	Lung	Kidney
A	1.41 ± 0.03	0.63 ± 0.02	4.46 ± 0.14	0.29 ± 0.03	0.67 ± 0.02	1.60 ± 0.04
B	1.38 ± 0.04	0.61 ± 0.02	4.58 ± 0.07	0.33 ± 0.04	0.66 ± 0.01	1.62 ± 0.04
C	1.30 ± 0.04	0.58 ± 0.02	4.40 ± 0.14	0.27 ± 0.02	0.66 ± 0.07	1.67 ± 0.11
D	1.42 ± 0.03	0.63 ± 0.01	4.81 ± 0.10	0.32 ± 0.02	0.72 ± 0.04	1.75 ± 0.06

Group A—control group, Group B—low-dose SPE (61.5 mg/kg BW/day), Group C—medium-dose SPE (123 mg/kg BW/day), and Group D—high-dose SPE (184.5 mg/kg BW/day). Data are expressed as the mean ± SEM and analyzed by one-way ANOVA.

Table 4. Effects of SPE treatment on organ weights of male SAMP8 mice.

Group	Relative Organ Weights (g/100 g Body Weight)					
	Brain	Heart	Liver	Spleen	Lung	Kidney
A	1.56 ± 0.06	0.65 ± 0.01	4.98 ± 0.11	0.42 ± 0.02	0.82 ± 0.02	1.42 ± 0.06
B	1.58 ± 0.07	0.60 ± 0.02	4.66 ± 0.14	0.42 ± 0.03	0.79 ± 0.04	1.40 ± 0.03
C	1.55 ± 0.07	0.62 ± 0.02	4.65 ± 0.15	0.37 ± 0.01	0.77 ± 0.02	1.42 ± 0.02
D	1.61 ± 0.08	0.59 ± 0.02	4.57 ± 0.13	0.40 ± 0.02	0.86 ± 0.05	1.38 ± 0.04

Group A—control group, Group B—low-dose SPE (61.5 mg/kg BW/day), Group C—medium-dose SPE (123 mg/kg BW/day), and Group D—high-dose SPE (184.5 mg/kg BW/day). Data are expressed as the mean ± SEM and analyzed by one-way ANOVA.

3.2. Effect of SPE on Aging Index of Senescence-Accelerated Mice

Tables 5 and 6 show the effects of SPE administration on aging index assessment scores after 10 weeks for male and female mice, respectively. The skin glossiness, spinal curvature, and total aging index assessment scores were significantly lower for mice in the treatment groups ($p < 0.05$) compared with animals in the control group. SPE treatment produced the same effects in both male and female mice.

Table 5. Effects of SPE administration on aging index assessment scores of male SAMP8 mice.

Group	A	B	C	D
Behavior				
Reactivity	0.10 ± 0.10	0.00 ± 0.00	0.11 ± 0.11	0.00 ± 0.00
Passivity	0.40 ± 0.16	0.11 ± 0.11	0.22 ± 0.15	0.00 ± 0.00
Skin				
Glossiness	1.60 ± 0.16 [a]	1.00 ± 0.00 [b]	0.78 ± 0.15 [b]	0.67 ± 0.17 [b]
Coarseness	1.30 ± 0.15	1.67 ± 0.17	1.44 ± 0.18	1.44 ± 0.18
Hair loss	1.50 ± 0.22	1.44 ± 0.18	1.22 ± 0.15	1.44 ± 0.18
Ulcer	0.00 ± 0.00	0.00 ± 0.00	0.00 ± 0.00	0.00 ± 0.00
Eyes				
Periophthalmic lesion	0.20 ± 0.13	0.11 ± 0.11	0.00 ± 0.00	0.11 ± 0.11
Spine				
Lordokyphosis	1.60 ± 0.16 [a]	1.00 ± 0.00 [b]	1.00 ± 0.00 [b]	1.00 ± 0.00 [b]
Total	6.70 ± 0.26 [a]	5.33 ± 0.24 [b]	4.78 ± 0.32 [b]	4.67 ± 0.17 [b]

Group A—control group, Group B—low-dose SPE (61.5 mg/kg BW/day), Group C—medium-dose SPE (123 mg/kg BW/day), and Group D—high-dose SPE (184.5 mg/kg BW/day). Data are expressed as the mean ± SEM and analyzed by one-way ANOVA. Groups with different letters indicate significant differences among each group ($p < 0.05$).

Table 6. Effects of SPE administration on aging index assessment scores of female SAMP8 mice.

Group	A	B	C	D
Behavior				
Reactivity	0.78 ± 0.22	0.78 ± 0.22	0.56 ± 0.18	0.67 ± 0.17
Passivity	1.11 ± 0.11	0.78 ± 0.15	0.89 ± 0.11	0.56 ± 0.24
Skin				
Glossiness	0.89 ± 0.11	0.33 ± 0.17	0.56 ± 0.18	0.67 ± 0.17
Coarseness	0.78 ± 0.15	0.56 ± 0.18	0.89 ± 0.11	0.67 ± 0.24
Hair loss	0.78 ± 0.22	0.67 ± 0.24	0.67 ± 0.17	0.67 ± 0.17
Ulcer	0.44 ± 0.18	0.56 ± 0.18	0.11 ± 0.11	0.22 ± 0.15
Eyes				
Periophthalmic lesion	0.78 ± 0.22	0.78 ± 0.15	0.44 ± 0.18	0.67 ± 0.24
Spine				
Lordokyphosis	1.00 ± 0.00 [a]	0.78 ± 0.15 [ab]	0.56 ± 0.18 [b]	0.44 ± 0.18 [b]
Total	6.56 ± 0.18 [a]	5.22 ± 0.32 [b]	4.67 ± 0.17 [b]	4.56 ± 0.18 [b]

Group A—control group, Group B—low-dose SPE (61.5 mg/kg BW/day), Group C—medium-dose SPE (123 mg/kg BW/day), and Group D—high-dose SPE (184.5 mg/kg BW/day). Data are expressed as the mean ± SEM and analyzed by one-way ANOVA. Groups with different letters indicate significant differences among each group ($p < 0.05$).

3.3. Effects of SPE on Serum Biochemical Parameters

Tables 7 and 8 show the effects of SPE on the serum biochemical parameters glucose, total protein, albumin, triglyceride, total cholesterol, HDL, LDL, AST, ALT, BUN, creatinine, sodium, potassium, uric acid, creatine kinase, ketones, free fatty acids, IL-6, and TNF-α in male and female SAMP8 mice, respectively. For mice of both sexes, we found no significant changes between the treatment groups after SPE administration for 13 weeks.

Table 7. Effects of SPE on serum biochemical parameters of male SAP8 mice.

Group	A	B	C	D
Glucose (mg/dL)	112.02 ± 0.99	109.56 ± 1.22	114.21 ± 0.64	109.86 ± 1.95
Total protein (g/dL)	5.29 ± 0.05	5.40 ± 0.06	5.38 ± 0.07	5.36 ± 0.03
Albumin (g/dL)	2.95 ± 0.11	3.12 ± 0.08	3.06 ± 0.06	2.88 ± 0.08
Triglyceride (mg/dL)	108.65 ± 0.56	103.41 ± 0.54	104.43 ± 1.95	106.41 ± 0.11
Total cholesterol (mg/dL)	116.48 ± 1.81	114.91 ± 2.64	118.00 ± 2.99	115.49 ± 1.83
HDL (mg/dL)	53.57 ± 1.65	54.22 ± 1.72	56.35 ± 1.53	56.21 ± 1.03
LDL (mg/dL)	7.20 ± 0.14	7.06 ± 0.13	7.24 ± 0.09	7.29 ± 0.15
AST (U/L)	89.14 ± 0.75	87.06 ± 0.86	88.87 ± 0.42	88.09 ± 0.38
ALT (U/L)	59.51 ± 0.91	61.05 ± 0.87	59.49 ± 0.64	59.57 ± 0.42
BUN (mg/dL)	25.91 ± 1.22	26.81 ± 0.39	26.99 ± 0.46	27.72 ± 0.70
Creatinine (mg/dL)	0.30 ± 0.02	0.32 ± 0.01	0.29 ± 0.02	0.33 ± 0.02
Sodium (mg/dL)	6.46 ± 0.10	6.77 ± 0.11	6.52 ± 0.13	6.68 ± 0.08
Potassium (mg/dL)	7.71 ± 0.13	7.94 ± 0.12	7.86 ± 0.10	7.75 ± 0.15
Uric acid (mg/dL)	4.41 ± 0.05	4.38 ± 0.08	4.54 ± 0.07	4.57 ± 0.08
Creatine kinase (U/L)	260.52 ± 2.26	259.76 ± 3.23	264.30 ± 1.75	262.96 ± 1.88
Ketones (mmol/L)	0.66 ± 0.04	0.65 ± 0.02	0.68 ± 0.04	0.62 ± 0.07
Free fatty acids (mmol/L)	0.72 ± 0.08	0.79 ± 0.02	0.76 ± 0.01	0.73 ± 0.04
IL-6 (pg/mL)	<1.5	<1.5	<1.5	<1.5
TNF-α (pg/mL)	<0.106	<0.106	<0.106	<0.106

Group A—control group, Group B—low-dose SPE (61.5 mg/kg BW/day), Group C—medium-dose SPE (123 mg/kg BW/day), and Group D—high-dose SPE (184.5 mg/kg BW/day). Data are expressed as the mean ± SEM and analyzed by one-way ANOVA.

Table 8. Effects of SPE on serum biochemical parameters of female SAP8 mice.

Group	A	B	C	D
Glucose (mg/dL)	116.12 ± 1.43	115.66 ± 1.15	113.40 ± 1.84	114.51 ± 2.01
Total Protein (g/dL)	5.36 ± 0.02	5.46 ± 0.11	5.35 ± 0.04	5.37 ± 0.02
Albumin (g/dL)	3.02 ± 0.04	2.93 ± 0.07	3.09 ± 0.11	2.90 ± 0.08
Triglyceride (mg/dL)	101.27 ± 1.66	103.22 ± 0.68	103.18 ± 1.00	104.65 ± 0.78
Total cholesterol (mg/dL)	117.99 ± 2.10	120.29 ± 2.20	116.88 ± 1.96	120.43 ± 2.39
HDL (mg/dL)	55.14 ± 2.03	54.16 ± 1.84	56.99 ± 1.13	57.03 ± 1.70
LDL (mg/dL)	7.36 ± 0.11	7.28 ± 0.12	7.22 ± 0.16	7.42 ± 0.09
AST (U/L)	88.94 ± 0.51	88.33 ± 0.33	87.36 ± 0.55	88.09 ± 0.63
ALT (U/L)	59.33 ± 0.97	61.31 ± 1.18	60.41 ± 0.50	58.87 ± 1.38
BUN (mg/dL)	27.98 ± 0.46	28.18 ± 0.47	25.62 ± 1.38	28.07 ± 0.73
Creatinine (mg/dL)	0.31 ± 0.01	0.28 ± 0.02	0.30 ± 0.01	0.33 ± 0.03
Sodium (mg/dL)	6.59 ± 0.21	6.64 ± 0.18	6.67 ± 0.11	6.75 ± 0.19
Potassium (mg/dL)	7.89 ± 0.10	7.83 ± 0.07	7.75 ± 0.09	7.79 ± 0.11
Uric acid (mg/dl)	4.64 ± 0.04	4.56 ± 0.07	4.73 ± 0.10	4.68 ± 0.02
Creatine kinase (U/L)	259.32 ± 1.10	262.86 ± 0.71	261.66 ± 0.63	260.96 ± 1.10
Ketone (mmol/L)	0.72 ± 0.07	0.69 ± 0.05	0.67 ± 0.04	0.70 ± 0.06
Free fatty acids (mmol/L)	0.73 ± 0.03	0.70 ± 0.06	0.76 ± 0.03	0.74 ± 0.02
IL-6 (pg/mL)	<1.5	<1.5	<1.5	<1.5
TNF-α (pg/mL)	<0.106	<0.106	<0.106	<0.106

Group A—control group, Group B—low-dose SPE (61.5 mg/kg BW/day), Group C—medium-dose SPE (123 mg/kg BW/day), and Group D—high-dose SPE (184.5 mg/kg BW/day). Data are expressed as the mean ± SEM and analyzed by one-way ANOVA.

3.4. Effects of SPE on SOD Activities, Catalase Activities, and Lipid Peroxidation in Aging Mice

We quantified lipid peroxidation, SOD activities, and catalase activities in the brain tissues of male and female SAMP8 mice, as shown in Figure 1. For aging mice of both sexes, SPE treatment significantly reduced TBARS content ($p < 0.05$). At the same time, the activities of cellular antioxidant enzymes SOD and catalase significantly ($p < 0.05$) increased after SPE treatment in both male and female aging mice.

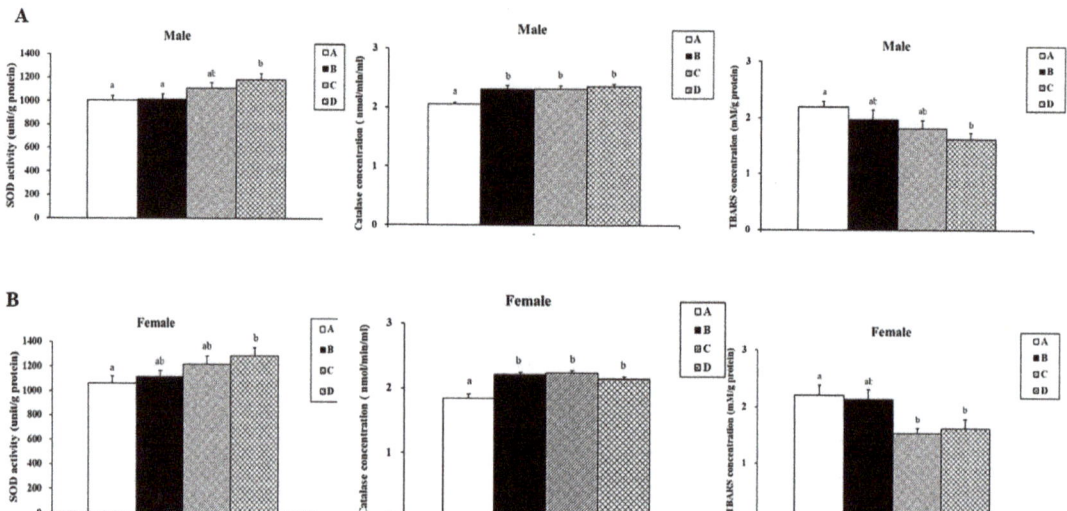

Figure 1. Effects of SPE treatment on SOD activities, catalase activities, and LPO in brain tissues of SAMP8 mice. (**A**) Male SAMP8 mice. (**B**) Female SAMP8 mice. Data are expressed as the mean ± SEM and analyzed by one-way ANOVA. Groups with different letters indicate significant differences among each group ($p < 0.05$).

3.5. Effects of SPE on Histology of Vital Organs in Aging Mice

Figures 2 and 3 show the effects of SPE treatment on changes in the histology of vital organs such as the brain, heart, kidney, liver, and lung in male and female mice, respectively. We found that a higher dose of SPE (184.5 mg/kg BW) administration to aging mice of either sex did not result in any significant structural changes in these vital organs. All organs displayed normal morphological architecture in the SPE-treated groups.

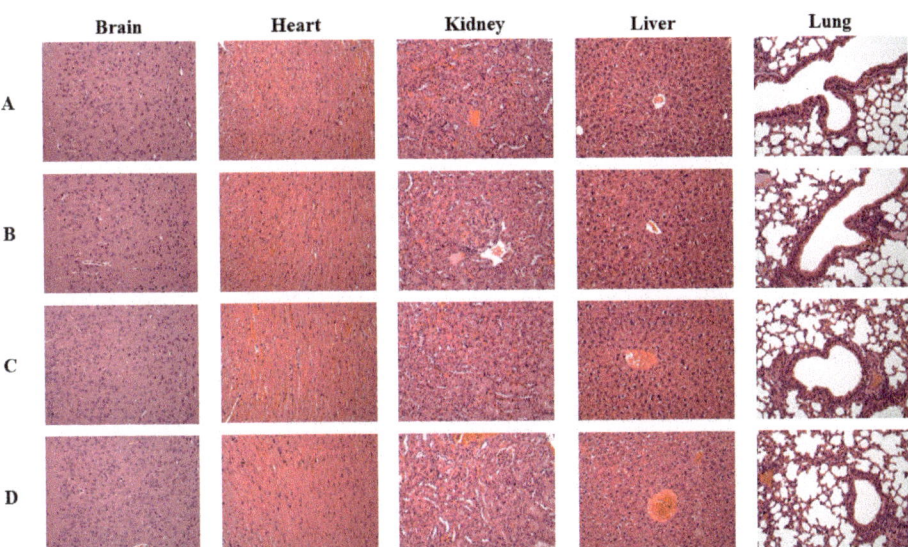

Figure 2. Effects of SPE treatment on histology of the brain, heart, liver, kidney, and lung of male SAMP8 mice (100×). Group (**A**)—control group, Group (**B**)—low-dose SPE (61.5 mg/kg BW/day), Group (**C**)—medium-dose SPE (123 mg/kg BW/day), and Group (**D**)—high-dose SPE (184.5 mg/kg BW/day).

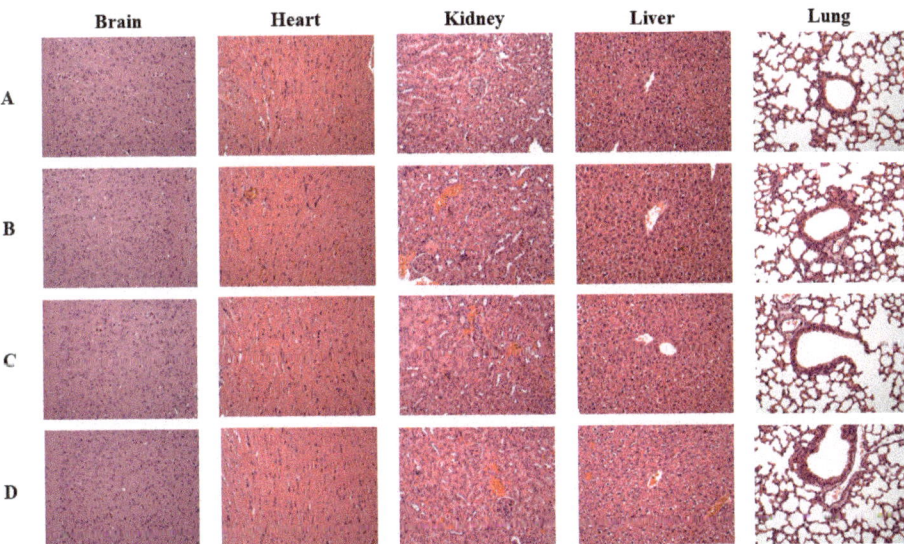

Figure 3. Effects of SPE treatment on histology of the brain, heart, liver, kidney, and lung of female SAMP8 mice (100×). Group (**A**)—control group, Group (**B**)—low-dose SPE (61.5 mg/kg BW/day), Group (**C**)—medium-dose SPE (123 mg/kg BW/day), and Group (**D**)—high-dose SPE (184.5 mg/kg BW/day).

4. Discussion

Practitioners of traditional and folk medicine have used the placenta to treat a wide range of ailments. Along with its medicinal uses, it has also been used as an anesthetic in various clinical settings, including internal medicine, general surgery, ENT, ophthalmology, orthopedic surgery, plastic surgery, dermatology, obstetrics, and gynecology [9]. The main amino acids in SPE are glutamic acid, aspartic acid, lysine, and leucine. Eight essential amino acids constitute 31.20 g/100 g of SPE or 39.48 percent of total amino acids. Additionally, several cytokines and growth factors present in placenta extract contribute to their pharmacological activities [23].

Reactive oxygen species (ROS), which cause oxidative damage and lipid peroxidative injury, are produced throughout the aging process [24]. Free radicals and their metabolic byproducts must be removed by enzymes of antioxidant systems. The key enzymes in antioxidant systems are SOD and catalase [25]. Superoxide anion free radicals and hydrogen ions are dismutated into H_2O_2 and O_2 by the antioxidant enzyme SOD, which is crucial for cleaning up free radicals [26]. Catalase eliminates oxidative risks by accelerating the synthesis of H_2O and O_2 from H_2O_2. One of the byproducts of lipid peroxidation is TBARS. During the aging process, TBARS levels increase [27]. In this study, we used these three defining indices to assess the antioxidant mechanism of SPE. We found that in SAMP8 mice, SOD and catalase activities decreased and the level of TBARS increased; however, SPE treatment for 13 weeks increased SOD and catalase antioxidant enzyme activities and lowered TABRAS levels in brain tissue. Antiaging processes may include regulating the activity of these antioxidant enzymes. There is strong evidence to support the original free radical hypothesis of aging, which states that free radical damage caused by numerous endogenous ROS is related to aging but is most likely the cause of it. Uracil, tyrosine, phenylalanine, and tryptophan are the main antioxidant substances found in placenta extract. These elements are responsible for 59% of PE's anti-oxidative actions. Freeze-dried placenta powder delays D-galactose-induced aging in female KM mice by improving immunity and reducing LPO in the brain [28], which is in line with our present findings. Preparations from deer fetuses and placenta produce an antiaging effect by reducing monoamine oxidase activities in the brain and liver and lipofuscin activities in the heart and brain of aging rats [29].

High doses (1000 mg/kg) of swine placenta extract powder administered for 28 days do not produce any adverse effects in rats. Male and female rats treated with swine placenta powder showed no changes in hematological parameters and no changes to vital organs and tissues including the kidneys, heart, lung, liver, spleen, brain, adrenals, epididymides, prostate, seminal vesicles, thymus, testes, or ovaries. Uterus histology and organ weights were also unaffected [30]. In our study, we checked both male and female SAMP8 mice every day for symptoms of toxicity and death. In addition, we periodically checked the body weight, food and water consumption, and physical and visual health of the animals. We also carried out regular urinalysis and performed clinical biochemistry assessments of blood and plasma samples. We found that a higher dose of SPE (184.5 mg/kg BW) did not induce any changes in serum biochemical parameters or any histological changes in vital organs. This is in line with the previous findings and suggests that SPE is safe for human consumption.

Manufacturers employ a variety of techniques to produce placental extracts using raw materials from human, bovine, and porcine placentas. As a result, the components of placental extracts can be quite complex and may widely vary between products. Placental extracts include N-acetylneuraminic acid, glucosamine, omega-3 fatty acids, various fatty acids, and many other amino acids and nucleotides that have antioxidative or anti-inflammatory and wound-healing activities [31–33]. We postulated that most of these compounds/molecules would also be present in our sheep placenta extract. However, we were unable to validate the crucial elements and signaling pathways important for the prevention of aging. More research is needed to understand the molecular mechanism underpinning the antiaging activity of SPE, and this may be a challenging task.

As well as the aging index, serum biochemical parameters might be used as an indicator of aging. The liver function markers, ALT and AST, and the kidney function markers, creatinine and uric acid, are key indicators of tissue damage during the aging process [34].

However, in our study, these biochemical parameters did not significantly change after SPE administration. Inflammation plays a key role in the process of senescence [35]. Levels of cytokine IL-6 increase in elderly individuals, in comparison with young people, but levels of IL-1β and TNF-α do not increase [36]. In our study, we found no significant changes in levels of IL-6 or TNF-α.

In this study, we investigated the effect of different doses of SPE for 13 weeks on male and female SAMP8 in terms of the aging index, serum biochemical levels, and antioxidant status. We found that medium and high doses of sheep placenta reduced the appearance changes in aging, lowered levels of brain lipid peroxides, and increased SOD and catalase activities, thereby enhancing the antioxidant capacity of the body and helping to reduce the damage of oxidative stress, with the overall effect of delaying the aging process. SPE is rich in peptides and amino acids, whose strong antioxidant activities might contribute to antiaging effects. However, further characterization and molecular studies of SPE are needed to better understand its antiaging effects. In conclusion, SPE supplementation reduces the aging index and lowers the oxidative stress caused by the aging process in male and female mice, without toxicity, suggesting the value of SPE as an effective antiaging treatment.

Author Contributions: Conceptualization, W.-T.L.; methodology, C.-P.O.Y. and M.-Y.C.; validation, W.-C.L. and Y.-M.Y.; formal analysis, W.-C.L.; investigation, M.-Y.C. and Y.-J.H.; resources, M.-F.W.; data curation, M.-F.W.; writing—original draft preparation, M.-F.W. and W.-T.L.; writing—review and editing, M.-F.W. and W.-T.L.; supervision, W.-T.L.; project administration, M.-F.W.; funding acquisition, W.-T.L. All authors have read and agreed to the published version of the manuscript.

Funding: This research was funded by Providence University, fund number: 109: PU109-11150-A044, and Tunghai University Fund number: THU 111611.

Institutional Review Board Statement: All the animal treatment procedures were performed in accordance with the Guide for Care and Use of Laboratory Animals (National Institutes of Health Publication No. 85-23, raised 1996). Animal experiment procedures in the present study were approved by the Institutional Animal Care and Use Committee (IACUC) of Providence University (Approval No.: 20201218 A008).

Informed Consent Statement: Not applicable.

Data Availability Statement: Data are contained within the article.

Conflicts of Interest: The authors declare no conflict of interest.

References

1. Sun, C.; Groom, K.M.; Oyston, C.; Chamley, L.W.; Clark, A.R.; James, J.L. The placenta in fetal growth restriction: What is going wrong? *Placenta* **2020**, *96*, 10–18. [CrossRef] [PubMed]
2. Brett, K.E.; Ferraro, Z.M.; Yockell-Lelievre, J.; Gruslin, A.; Adamo, K.B. Maternal–fetal nutrient transport in pregnancy pathologies: The role of the placenta. *Int. J. Mol. Sci.* **2014**, *15*, 16153–16185. [CrossRef] [PubMed]
3. Rasool, A.; Alvarado-Flores, F.; O'Tierney-Ginn, P. Placental impact of dietary supplements: More than micronutrients. *Clin. Ther.* **2021**, *43*, 226–245. [CrossRef] [PubMed]
4. Ghoneum, M.; El-Gerbed, M.S. Human placental extract ameliorates methotrexate-induced hepatotoxicity in rats via regulating antioxidative and anti-inflammatory responses. *Cancer Chemother. Pharmacol.* **2021**, *88*, 961–971. [CrossRef] [PubMed]
5. Goswami, S.; Sarkar, R.; Saha, P.; Maity, A.; Sarkar, T.; Das, D.; Chakraborty, P.D.; Bandyopadhyay, S.; Ghosh, C.K.; Karmakar, S. Effect of human placental extract in the management of biofilm mediated drug resistance–A focus on wound management. *Microb. Pathog.* **2017**, *111*, 307–315. [CrossRef]
6. Kaneko, Y.; Sano, M.; Seno, K.; Oogaki, Y.; Takahashi, H.; Ohkuchi, A.; Yokozawa, M.; Yamauchi, K.; Iwata, H.; Kuwayama, T. Olive leaf extract (OleaVita) suppresses inflammatory cytokine production and NLRP3 inflammasomes in human placenta. *Nutrients* **2019**, *11*, 970. [CrossRef]
7. Mumtaz, S.M.; Goyal, R.K.; Ameen, A.; Alexandrovich, B.I.; Gupta, M. Animal Placental Therapy: An Emerging Tool for Health Care. *Curr. Tradit. Med.* **2022**, *8*, 20–30. [CrossRef]
8. Pogozhykh, O.; Prokopyuk, V.; Figueiredo, C.; Pogozhykh, D. Placenta and placental derivatives in regenerative therapies: Experimental studies, history, and prospects. *Stem Cells Int.* **2018**, *2018*, 4837930. [CrossRef]
9. Pan, S.Y.; Chan, M.K.; Wong, M.B.; Klokol, D.; Chernykh, V. Placental therapy: An insight to their biological and therapeutic properties. *Blood* **2017**, *4*, 12.

10. Ober, W.B. Notes on placentophagy. *Bull. N. Y. Acad. Med.* **1979**, *55*, 591.
11. Hayes, E.H. Consumption of the placenta in the postpartum period. *J. Obstet. Gynecol. Neonatal. Nurs.* **2016**, *45*, 78–89. [CrossRef]
12. Koike, K.; Yamamoto, Y.; Suzuki, N.; Yamazaki, R.; Yoshikawa, C.; Takano, F.; Takuma, K.; Sugiura, K.; Inoue, M. Efficacy of porcine placental extract on climacteric symptoms in peri-and postmenopausal women. *Climacteric* **2012**, *16*, 28–35. [CrossRef] [PubMed]
13. Silini, A.R.; Cargnoni, A.; Magatti, M.; Pianta, S.; Parolini, O. The long path of human placenta, and its derivatives, in regenerative medicine. *Front. Bioeng. Biotechnol.* **2015**, *3*, 162. [CrossRef] [PubMed]
14. Liu, W.; Hou, Y.; Cheng, Y. Polypeptide preparation of sheep placenta and its free radicals scavenging activity. *China Brew.* **2014**, *33*, 89–93.
15. Wu, Y. Study on the new method of extraction placental peptide and analysis of nutritive composition in sheep placenta. *Food Sci.* **2005**, *26*, 295–297.
16. Zhang, S.; Fang, F.; Gu, Y.; Zhang, Y.; Tao, Y.; Zhang, X. Effects of goat placental preparations on relieving fatigue and anti-oxidation in mice. *J. Tradit. Chin. Vet. Med.* **2008**, *6*, 8–10.
17. Hou, Y.; Zhou, J.; Liu, W.; Cheng, Y.; Wu, L.; Yang, G. Preparation and characterization of antioxidant peptides from fermented goat placenta. *Korean J. Food Sci. Anim. Resour.* **2014**, *34*, 769. [CrossRef] [PubMed]
18. Lai, P.-F.; Baskaran, R.; Kuo, C.-H.; Day, C.H.; Chen, R.-J.; Ho, T.-J.; Yeh, Y.-L.; Padma, V.V.; Lai, C.-H.; Huang, C.-Y. Bioactive dipeptide from potato protein hydrolysate combined with swimming exercise prevents high fat diet induced hepatocyte apoptosis by activating PI3K/Akt in SAMP8 mouse. *Mol. Biol. Rep.* **2021**, *48*, 2629–2637. [CrossRef] [PubMed]
19. Takeda, T.; Matsushita, T.; Kurozumi, M.; Takemura, K.; Higuchi, K.; Hosokawa, M. Pathobiology of the senescence-accelerated mouse (SAM). *Exp. Gerontol.* **1997**, *32*, 117–127. [CrossRef]
20. Akiguchi, I.; Pallàs, M.; Budka, H.; Akiyama, H.; Ueno, M.; Han, J.; Yagi, H.; Nishikawa, T.; Chiba, Y.; Sugiyama, H. SAMP8 mice as a neuropathological model of accelerated brain aging and dementia: Toshio Takeda's legacy and future directions. *Neuropathology* **2017**, *37*, 293–305. [CrossRef]
21. Lagartos-Donate, M.; Gonzáles-Fuentes, J.; Marcos-Rabal, P.; Insausti, R.; Arroyo-Jiménez, M. Pathological and non-pathological aging, SAMP8 and SAMR1. What do hippocampal neuronal populations tell us? *bioRxiv* **2019**, bioRxiv:598599.
22. Chen, Y.-J.; Baskaran, R.; Shibu, M.A.; Lin, W.-T. Anti-Fatigue and Exercise Performance Improvement Effect of Glossogyne tenuifolia Extract in Mice. *Nutrients* **2022**, *14*, 1011. [CrossRef]
23. Liu, J.; Luo, S.; Yang, J.; Ren, F.; Zhao, Y.; Luo, H.; Ge, K.; Zhang, H. The protective effect of sheep placental extract on concanavalin A-induced liver injury in mice. *Molecules* **2018**, *24*, 28. [CrossRef] [PubMed]
24. Bolduc, J.A.; Collins, J.A.; Loeser, R.F. Reactive oxygen species, aging and articular cartilage homeostasis. *Free Radic. Biol. Med.* **2019**, *132*, 73–82. [CrossRef]
25. Padma, V.V.; Sowmya, P.; Felix, T.A.; Baskaran, R.; Poornima, P. Protective effect of gallic acid against lindane induced toxicity in experimental rats. *Food Chem. Toxicol.* **2011**, *49*, 991–998. [CrossRef] [PubMed]
26. Baskaran, R.; Poornima, P.; Huang, C.Y.; Padma, V.V. Neferine prevents NF-κB translocation and protects muscle cells from oxidative stress and apoptosis induced by hypoxia. *Biofactors* **2016**, *42*, 407–417. [CrossRef] [PubMed]
27. Ataie, Z.; Mehrani, H.; Ghasemi, A.; Farrokhfall, K. Cinnamaldehyde has beneficial effects against oxidative stress and nitric oxide metabolites in the brain of aged rats fed with long-term, high-fat diet. *J. Funct. Foods* **2019**, *52*, 545–551. [CrossRef]
28. Song, H.; Zang, C.; Liu, X.; Hou, H.; LI, D. Anti-aging function of placenta freeze-dried powder on mice. *Chin. J. Biochem. Pharm.* **2014**, 65–67.
29. Xinghuai, Z.; Guiqin, Y.; Guangjin, H. Effect of preparations from deer fetus and placenta on anti-aging of old male rats. *J. Shenyang Agric. Univ.* **2005**, *36*, 233–235.
30. Mitsui, Y.; Bagchi, M.; Marone, P.A.; Moriyama, H.; Bagchi, D. Safety and toxicological evaluation of a novel, fermented, peptide-enriched, hydrolyzed swine placenta extract powder. *Toxicol. Mech. Methods* **2015**, *25*, 13–20. [CrossRef]
31. Park, S.; Phark, S.; Lee, M.; Lim, J.; Sul, D. Anti-oxidative and anti-inflammatory activities of placental extracts in benzo [a] pyrene-exposed rats. *Placenta* **2010**, *31*, 873–879. [CrossRef] [PubMed]
32. Sur, T.K.; Biswas, T.K.; Ali, L.; Mukherjee, B. Anti-inflammatory and anti-platelet aggregation activity of human placental extract. *Acta Pharmacol. Sin.* **2003**, *24*, 187–192. [PubMed]
33. Togashi, S.-I.; Takahashi, N.; Kubo, Y.; Shigihara, A.; Higashiyama, K.; Watanabe, S.; Fukui, T. Purification and identification of antioxidant substances in human-placenta extracts. *J. Health Sci.* **2000**, *46*, 117–125. [CrossRef]
34. Lephart, E.D. Skin aging and oxidative stress: Equol's anti-aging effects via biochemical and molecular mechanisms. *Ageing Res. Rev.* **2016**, *31*, 36–54. [CrossRef] [PubMed]
35. Zhang, R.; Chen, J.; Mao, X.; Qi, P.; Zhang, X. Anti-inflammatory and anti-aging evaluation of pigment–protein complex extracted from Chlorella pyrenoidosa. *Mar. Drugs* **2019**, *17*, 586. [CrossRef] [PubMed]
36. Roubenoff, R.; Harris, T.B.; Abad, L.W.; Wilson, P.W.; Dallal, G.E.; Dinarello, C.A. Monocyte cytokine production in an elderly population: Effect of age and inflammation. *J. Gerontol. Ser. A Biol. Sci. Med. Sci.* **1998**, *53*, M20–M26. [CrossRef]

Article

Synthesis, In Silico, and In Vitro Biological Evaluation of New Furan Hybrid Molecules

Stanimir Manolov [1], Iliyan Ivanov [1,*], Dimitar Bojilov [1] and Paraskev Nedialkov [2]

1 Department of Organic Chemistry, Faculty of Chemistry, University of Plovdiv, 24 Tzar Assen Str., 4000 Plovdiv, Bulgaria
2 Department of Pharmacognosy, Faculty of Pharmacy, Medical University of Sofia, 2 Dunav Str., 1000 Sofia, Bulgaria
* Correspondence: iiiliyan@abv.bg; Tel./Fax: +359-32-261-349

Abstract: Herein, we report the synthesis of new hybrid molecules between furan and N-containing heterocyclic compounds such as pyrrolidine, 1,2,3,4-tetrahydroquinoline, 1,2,3,4-tetrahydroisoquinoline, and piperidine. The obtained compounds were fully characterized using ^{1}H- and ^{13}C-NMR, UV-Vis, and HRMS spectra. All compounds were assessed for their anti-inflammatory, anti-arthritic, antioxidant, reducing power ability, and chelating activity. The less lipophilic molecules **H2** (60.1 ± 8.16) and **H4** (62.23 ± 0.83) had almost 12 times higher ATA compared with the used ketoprofen (720.57 ± 19.78) standard. The inhibition of albumin denaturation results makes the newly obtained hybrids potential anti-inflammatory drugs, as the expressed values are higher than the ketoprofen standard (126.58 ± 5.00), except **H3** (150.99 ± 1.16). All four compounds show significant activity regarding the in vitro biological activities, which makes them great candidates for potential future drugs.

Keywords: furan; 1,2,3,4-tetrahidroquinoline; 1,2,3,4-tetrahidroisoquinoline; piperidine; pyrrolidine; in silico; in vitro biological activity

Citation: Manolov, S.; Ivanov, I.; Bojilov, D.; Nedialkov, P. Synthesis, In Silico, and In Vitro Biological Evaluation of New Furan Hybrid Molecules. *Processes* **2022**, *10*, 1997. https://doi.org/10.3390/pr10101997

Academic Editors: Elzbieta Klewicka and Davide Dionisi

Received: 24 August 2022
Accepted: 29 September 2022
Published: 2 October 2022

Publisher's Note: MDPI stays neutral with regard to jurisdictional claims in published maps and institutional affiliations.

Copyright: © 2022 by the authors. Licensee MDPI, Basel, Switzerland. This article is an open access article distributed under the terms and conditions of the Creative Commons Attribution (CC BY) license (https://creativecommons.org/licenses/by/4.0/).

1. Introduction

Furan is a five-membered heterocyclic organic compound containing one oxygen and four carbon atoms. Its derivatives are a wide group of heterocyclic molecules owing to a vast array of biological activities. Organic compounds in which the furan structure is a building block have been described in the literature as possessing antibacterial, antiviral, pesticidal, cytotoxic, antitumorigenic, psychotropic, and anti-inflammatory properties. They also have an effect on the cardiovascular system, as well as hypoglycemic, antifertility, and ulcer-healing properties [1].

There are many different kinds of bacteria that are susceptible to Nitrofurantoin **1**. It is used for treating urinary tract infections that are caused by Escherichia coli, enterococci, Staphylococcus aureus, and certain strains of Klebsiella and Enterobacter species that are susceptible to Nitrofurantoin **1** [2] (Figure 1).

Some well-known furan derivatives are ranbezolid **2** [3], nifurzide **3** [4], and ranitidine **4** [5] widely used as anti-bacterial, anti-infective, and H2 receptor antagonists, respectively (Figure 2).

On other hand, the N-containing heterocycles are of great importance, not only because of their affluence, but above all because of their chemical, biological, and technical significance. Together, they create a large group of compounds that play a powerful role in biological investigations such as antibacterial, anticancer, anti-inflammatory, antiviral, anti-tumor, and anti-diabetic applications [6].

In the search for new biofunctional compounds, we were interested in obtaining new hybrid molecules containing both a furan fragment and a nitrogen-containing core in their structure.

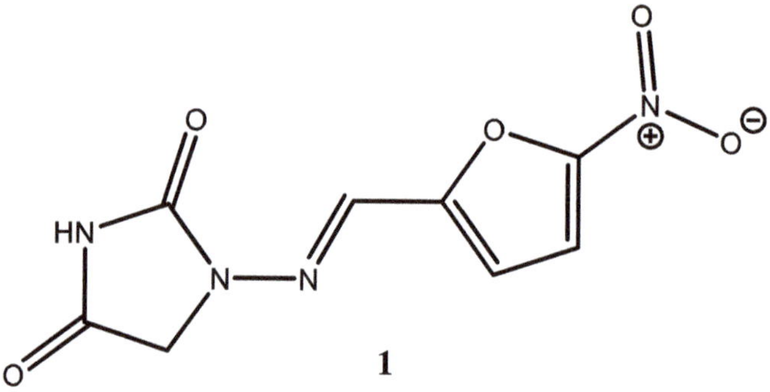

Figure 1. Structural formula of nitrofurantoin 1.

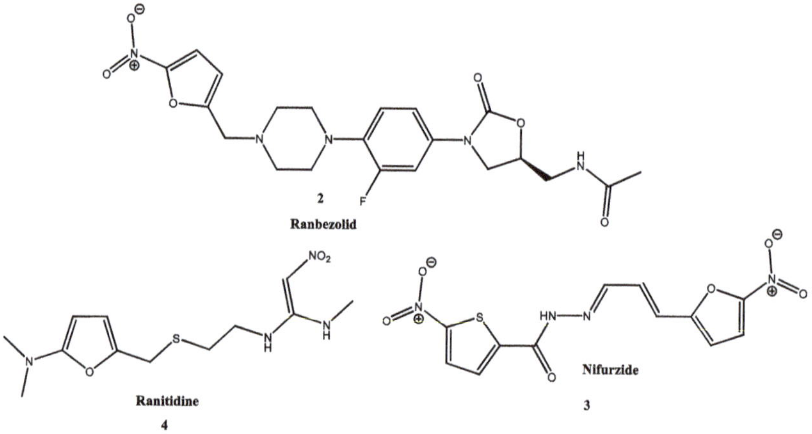

Figure 2. Structural formulas of ranbezolid, nifurzide, and ranitidine.

In a recent study, Li and co-workers synthesized new tetrahydroquinoline and tetrahydroisoquinolines containing 2-phenyl-5-furan moiety as PDE4 inhibitors [7].

Using Pd-catalyzed coupling reactions of 2-(alkyltelluro)furan with terminal alkynes, Zeni et al. managed to synthesize a series of acetylenic furan hybrids [8]. The obtained compounds were tested for anti-inflammatory activity and showed very promising results.

The significance of furan-containing amides continues to fascinate scientists. Publications in recent years have focused on the discovery and application of new synthetic methods, as well as the preparation of their biologically active analogs [9–11].

2. Materials and Methods

2.1. Synthesis

The reagents were purchased from commercial suppliers (Sigma-Aldrich S.A. and Riedel-de Haën, Sofia, Bulgaria) and used as received. A Bruker Advance II+600 spectrometer was used for the recording of the NMR spectral data (BAS-IOCCP—Sofia, Bruker, Billerica, MA, USA). All compounds were analyzed in CDCl$_3$ at 600 MHz and 150.9 MHz for ^{1}H-NMR and ^{13}C-NMR, respectively. Chemical shifts were determined to tetramethylsilane (TMS) (δ = 0.00 ppm) as an internal standard; the coupling constants are given in Hz. Recorded NMR spectra were taken at room temperature (approx. 295 K). The MS analysis was carried out on a Q Exactive Plus high-resolution mass spectrometer (HRMS) with a heated electrospray ionization source (HESI-II) (Thermo Fisher Scientific, Inc., Bremen, Germany) equipped with a Dionex Ultimate 3000RSLC ultrahigh-performance liquid chromatography (UHPLC) system (Thermo Fisher Scientific, Inc.). For the TLC analysis, precoated 0.2 mm Fluka silica gel 60 plates (Merck KGaA, Darmstadt, Germany) were used.

Synthesis of hybrid compounds 7.

Furan-2-carbonyl chloride **6** (1 mmol) was added to a solution of the corresponding amine **5a–d** (1 mmol) in dichloromethane (15 mL). Triethylamine (1.2 mmol) was added after 10 min. After 30 min, the reaction mixture was washed with diluted hydrochloric acid (H_2O:HCl = 4:1 v/v), a saturated solution of Na_2CO_3, and water. The organic layer was dried using anhydrous Na_2SO_4, concentrated, and filtered on a short column with neutral Al_2O_3.

*Furan-2-yl(pyrrolidin-1-yl)methanone **7a***: light yellow oil, yield 94% (0.155 g), ^{1}H NMR (600 MHz, CDCl$_3$) δ 7.51 (dd, J = 1.7, 0.8 Hz, 1H), 7.06 (dd, J = 3.5, 0.8 Hz, 1H), 6.49 (dd, J = 3.5, 1.8 Hz, 1H), 3.84 (t, J = 6.8 Hz, 2H), 3.66 (t, J = 6.9 Hz, 2H), 2.00 (p, J = 6.7 Hz, 2H), 1.91 (p, J = 6.7 Hz, 2H). ^{13}C NMR (151 MHz, CDCl$_3$) δ 158.1 (C=O), 148.8 (O=C-C-O), 144.0 (O-CH=CH), 115.7 (C=CH-CH), 111.3 (C-CH=CH), 47.8 (CH2-CH$_2$-N-), 47.0 (CH$_2$-CH$_2$-N-), 26.6 (CH$_2$), 23.8 (CH$_2$). UV λ_{max}, MeOH: 272 (ε = 12,200) nm. HRMS Electrospray ionization (ESI) m/z calcd for [M+H]$^+$ C$_9$H$_{12}$NO$_2$$^+$ = 166.0863, found 166.0860 (mass error Δ_m = −1.82 ppm), calcd for [M+Na]$^+$ C$_9$H$_{11}$NO$_2$Na$^+$ = 188.0682, found 188.0679 (mass error Δ_m = −1.60 ppm).

*Furan-2-yl(piperidin-1-yl)methanone **7b***: light yellow oil, yield 92% (0.165 g), ^{1}H NMR (600 MHz, CDCl$_3$) δ 7.40 (dd, J = 1.7, 0.8 Hz, 1H), 6.85 (dd, J = 3.4, 0.8 Hz, 1H), 6.39 (dd, J = 3.4, 1.8 Hz, 1H), 3.63 (s, 4H), 1.65–1.60 (m, 2H), 1.59–1.54 (m, 4H). ^{13}C NMR (151 MHz, CDCl$_3$) δ 159.3 (C=O), 148.2 (O=C-C-O), 143.4 (O-CH=CH), 115.5 (C=CH-CH=CH-O), 111.1 (C=CH-CH=CH-O), 47.7 (N-CH$_2$-CH$_2$), 26.2 (N-CH$_2$-CH$_2$), 24.7 (CH$_2$-CH$_2$-CH$_2$). UV λ_{max}, MeOH: 272 (ε = 14,800) nm. HRMS Electrospray ionization (ESI) m/z calcd for [M+H]$^+$ C$_{10}$H$_{14}$NO$_2$$^+$ = 180.1019, found 180.1017 (mass error Δ_m = −1.06 ppm), calcd for [M+Na]$^+$ C$_{10}$H$_{13}$NO$_2$Na$^+$ = 202.0838, found 202.0836 (mass error Δ_m = −0.99 ppm).

*(3,4-dihydroquinolin-1(2H)-yl)(furan-2-yl)methanone **7c***: light yellow oil, yield 95% (0.216 g), ^{1}H NMR (600 MHz, CDCl$_3$) δ 7.27 (dd, J = 1.7, 0.8 Hz, 1H), 7.11 (dd, J = 7.5, 0.8 Hz, 1H), 7.01 (td, J = 7.4, 1.2 Hz, 1H), 6.95 (ddd, J = 8.0, 4.5, 1.1 Hz, 1H), 6.84 (d, J = 8.0 Hz, 1H), 6.60 (d, J = 3.5 Hz, 1H), 6.30 (dd, J = 3.5, 1.7 Hz, 1H), 3.84 (t, J = 6.6 Hz, 2H), 2.73 (t, J = 6.6 Hz, 2H), 1.96 (p, J = 6.6 Hz, 2H). ^{13}C NMR (151 MHz, CDCl$_3$) δ 159.8, 147.9, 143.9, 138.9, 132.0, 128.4, 125.9, 124.9, 124.3, 116.4, 111.2, 44.2, 26.8, 24.0. UV λ_{max}, MeOH: 291 (ε = 10,800) nm. HRMS Electrospray ionization (ESI) m/z calcd for [M+H]$^+$ C$_{14}$H$_{14}$NO$_2$$^+$ = 228.1019, found 228.1015 (mass error Δ_m = −1.75 ppm), calcd for [M+Na]$^+$ C$_{14}$H$_{13}$NO$_2$Na$^+$ = 250.0838, found 250.0834 (mass error Δ_m = 2.40 ppm).

*(3,4-dihydroisoquinolin-2(1H)-yl)(furan-2-yl)methanone **7d***: light yellow oil, yield 90% (0.205 g), ^{1}H NMR (600 MHz, CDCl$_3$) δ 7.55 (s, 1H), 7.24–7.17 (m, 4H), 7.07 (dd, J = 3.5, 0.7 Hz, 1H), 6.53 (dd, J = 3.5, 1.8 Hz, 1H), 4.88 (s, 2H), 4.01 (s, 2H), 3.00 (s, 2H). ^{13}C NMR (151 MHz, CDCl$_3$) δ 159.6 (C=O), 148.1 (O-C-C-O), 143.9 (O-CH=CH), 133.1 (C, Ar), 129.9 (C, Ar), 127.4 (C, Ar), 126.8 (C, Ar), 126.5 (C, Ar), 125.3 (C, Ar), 116.3 (C=CH-CH), 111.3 (CH-CH=CH), 48.3 (Ar-CH$_2$-N), 44.5 (Ar-CH$_2$-CH$_2$-N), 28.5 (Ar-CH$_2$-CH$_2$-N). UV λ_{max}, MeOH: 232 (ε = 13500) nm, 272 (ε = 12,000) nm. HRMS Electrospray ionization (ESI) m/z calcd for [M+H]$^+$ C$_{14}$H$_{14}$NO$_2$$^+$ = 228.1019, found 228.1015 (mass error Δ_m = −1.75 ppm), calcd for [M+Na]$^+$ C$_{14}$H$_{13}$NO$_2$Na$^+$ = 250.0838, found 250.0835 (mass error Δ_m = 2.80 ppm).

2.2. Biological Evaluation

Chemicals and Reagents

For HPLC analysis, chromatographic-grade methanol was used (VWR, Vienna, Austria). A Millipore purifier (Millipore, USA) was used to obtain water for HPLC. Potassium dihydrogen phosphate, potassium chloride, ascorbic acid, ibuprofen, ketoprofen, dipotassium hydrogen phosphate, sodium chloride, hydrogen peroxide, trypsin, Tris-HCl buffer, and perchloric acid were purchased from Sigma-Aldrich, Taufkirchen, Germany. Human albumin 20%—BB, 200 g/L was ordered from BB-NCIPD Ltd., Sofia, Bulgaria. Chromatographic plates, Kieselgel 60 F$_{254}$, were purchased from Merck (Darmstadt, Germany).

2.3. Biological Experiments

2.3.1. Hydrogen Peroxide Scavenging Activity (HPSA)

The Manolov et al. approach was used to evaluate the furan hybrid molecules' capacity to scavenge hydrogen peroxide [12].

A 43 mM solution of H_2O_2 was prepared in potassium phosphate buffer solution (0.2 M, pH 7.4). The analysis of the samples was carried out as follows: in test tubes, 0.6 mL H_2O_2 (43 mM), 1 mL sample/standard with different concentrations (20–1000 μg/mL), and 2.4 mL potassium phosphate buffer solution were mixed. The mixture was stirred and incubated in the dark for 10 min at 37 °C. Absorbance was measured at 230 nm with a spectrophotometer (Camspec M508, Leeds, UK) against a blank solution containing phosphate buffer and H_2O_2 without the sample. Ascorbic acid and quercetin were used as standards. The percentage HPSA of the samples was evaluated by comparing with a blank sample and calculated using the following formula:

$$I, \%(HPSA) = \left[\frac{A_{blank} - (A_{TS} - A_{CS})}{A_{blank}}\right] * 100 \quad (1)$$

where A_{blank} is the absorbance of the blank sample, A_{CS} is the absorbance of the control sample, and A_{TS} is the absorbance of the test sample. The mean IC_{50} value was estimated based on three replicates by means of interpolating the graphical dependence of scavenging hydrogen peroxide on concentration.

2.3.2. Metal-Chelating Activity (MChA) on Ferrous Ions

The iron-chelating ability of the furan derivatives was studied by the method in [13].

The ability of the different fractions to chelate Fe^{2+} was evaluated. The reaction mixture contained 1 mL of sample/standard, 3.7 mL of methanol, and 0.1 mL of 2 mmol/L ferric dichloride. With the addition of 0.2 mL of 5 mmol/L ferrozine, the reaction started. The reaction mixture was vortexed and left for 10 min at room temperature. After reaching equilibrium in the mixture, the absorbance of the sample was measured at 562 nm on a UV-Vis spectrophotometer (Camspec M508, Leeds, UK). The blank was prepared the same way as the test sample, but methanol was added instead of the sample/standard. The results were reported as percent inhibition of ferrozine Fe^{2+} complex formation with the sample. The mean IC_{50} value was estimated based on three replicates. Ethylenediaminetetraacetic acid (EDTA)-sodium (Na) (0.5 mg/mL) served as a positive control.

2.3.3. Inhibition of Albumin Denaturation (IAD)

The anti-inflammatory efficacy was evaluated using an inhibition of albumin denaturation (IAD) in vitro study. Sakat method [14] with a modest modification [15] was used to finish the analysis.

Human albumin was used in the experiment. In distilled water, an albumin (1%) solution was prepared (pH 7.4). The tested compounds/standard were first dissolved in 1.2 mL DMF and PBS up to 25 mL, resulting in a stock solution with a final concentration of 1000 μg/mL. Then, in PBS, a series of working solutions with varying concentrations (20–500 μg/mL) were prepared. The reaction mixture contained 2 mL of different concentrations of test sample/standard and 1 mL of albumin (1%). The mixture was incubated at 37 °C for 15 min before being heated in a water bath at 70 °C for 15 min. After cooling, the turbidity was measured with a spectrophotometer at 660 nm (Camspec M508, Leeds, UK). The experiment was carried out three times. The inhibition of albumin denaturation (IAD) was calculated as a percentage of the control. The control sample was albumin dissolved in distilled water at the same concentration.

2.3.4. Antitryptic Activity (ATA)

This method is also known to have anti-arthritic activity in vitro. The analysis was carried out using the method described by Oyedapo and Femurewa [16], with minor modifications as described by Manolov et al. [15].

Trypsin, 2 mL 0.06 mg/mL, 1 mL Tris-HCl buffer (20 mM, pH 7.4), and 1 mL test sample/standard of various concentrations (20–1000 μg/mL) were added to the reaction mixture. For 5 min, the mixture was incubated at 37 °C. Following that, 1 mL of human albumin (4% v/v) was added. The mixture was then incubated for an additional 20 min. To stop the reaction, 2 mL of 70% perchloric acid was added to the mixture. The cloudy suspension was cooled and centrifuged for 20 min at 5000 rpm. A spectrophotometer (Camspec M508, Leeds, UK) was used to measure the absorbance of the supernatant at 280 nm in comparison to the control solution. The control solution consisted of various concentrations of sample/standard in methanol. As a baseline, ibuprofen was used. The analysis was carried out three times. By comparing the samples to a blank sample, the percentage of antitryptic activity (ATA) was determined. The blank sample was prepared in the same way as the test sample, with the exception that perchloric acid was added before the albumin.

$$\%ATA = \left[\frac{A_{blank} - (A_{TS} - A_{CS})}{A_{blank}} \right] * 100 \qquad (2)$$

where A_{blank} is the absorbance of the blank sample, A_{CS} is the absorbance of the control solution (test sample in different concentrations), and A_{TS} is the absorbance of the test samples. The mean IC_{50} values were estimated by means of interpolating the graphical dependence of ATA on concentration.

2.4. Physicochemical Characterization

2.4.1. Determination of Lipophilicity as R_M Values

The lipophilicity of the furan derivatives was estimated using the method described by Hadjipavlou-Litina and Pontiki [17].

RPTLC (reverse-phase TLC) was performed on silica gel plates impregnated with 55% (v/v) liquid paraffin in light petroleum ether. The mobile phase was a methanol/water mixture (77:23, v/v) containing 0.1 acetic acid. The plates were developed in closed chromatography tanks saturated with the mobile phase at 24 °C. Spots were detected under UV light or by iodine vapors. R_M values were determined from the corresponding R_f values (from 10 individual measurements) using the equation $R_M = \log[(1/R_f) - 1]$.

2.4.2. Prediction of Anti-Inflammatory and Anti-Arthritic Activity

A computational prediction of biological activity (anti-inflammatory and anti-arthritic) for the identified compounds was carried out using the PASS Online tool [18,19].

2.5. Statistical Analysis

All analyses were performed in triplicate. The data were presented as mean ± SD. The significance level was set at $p < 0.05$.

3. Results

3.1. Synthesis

In the present article, we report the successful synthesis of new furan hybrid molecules with different N-containing compounds, such as pyrrolidine **5a**, piperidine **5b**, 1,2,3,4-tetrahydroquinoline **5c**, and 1,2,3,4-tetrahydroisoquinoline **5d** (Figure 3), as shown in Scheme 1. In Scheme 1, we report the successful synthesis of new furan hybrid molecules with various N-containing compounds, such as pyrrolidine **5a**, piperidine **5b**, 1,2,3,4-tetrahydroquinoline **5c**, and 1,2,3,4-tetrahydroisoquinoline **5d**.

cycle
NH
Cl
O
O
$(Et)_3N$
CH_2Cl_2
r.t.
60 min.
cycle
N
O
O
5a-d
6
7a-d

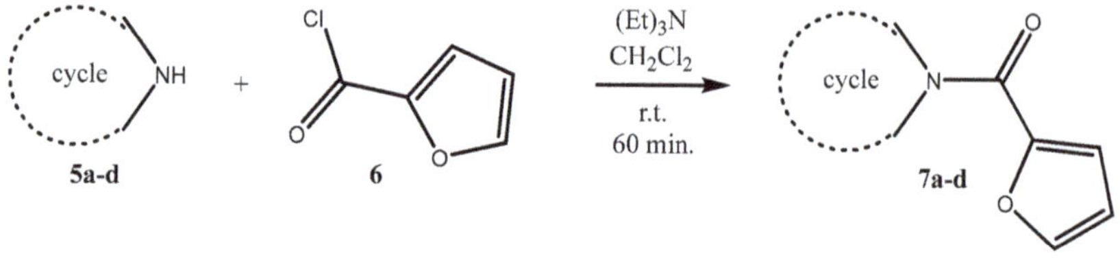

Scheme 1. Synthesis of hybrid molecules **7a–d**.

For this purpose, furan-2-carbonyl chloride 6 is added to a solution of the corresponding amine 5 (1 mmol) in dichloromethane. The reaction mixture is stirred for a further 10 min and then an excess of triethylamine (1.5 mmol) is carefully added dropwise. The secondary amines **5a–d** were fully acylated for the next 30 min (TLC).

The reaction generally works in high yields. All yields are above 90%, as can be seen in the experimental section. All compounds are fully characterized by UV, ^{1}H-, and ^{13}C-NMR and HRMS spectra.

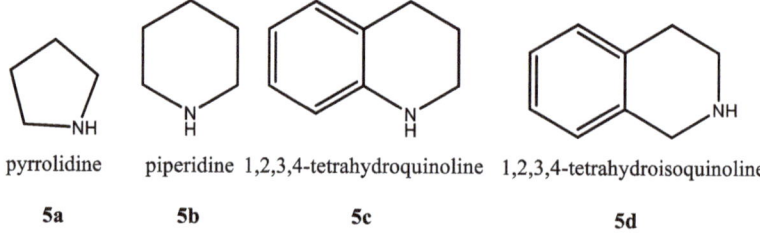

NH
N
H
N
H
NH
pyrrolidine
piperidine
1,2,3,4-tetrahydroquinoline
1,2,3,4-tetrahydroisoquinoline
5a
5b
5c
5d

Figure 3. Secondary amines took place in the acylation reaction.

All obtained compounds **7a**, **7c**, and **7d** are new compounds. Compound **7b**, the hybrid molecule of furan **6** and piperidine **5b**, has been previously reported [20–22].

Comparing the ^{1}H-NMR data for compound **7b** obtained by us with the same reported by Ramkumar and Chandrasekaran can be seen the match in the spectra [23]. ^{1}H NMR ($CDCl_3$, 400 MHz): δ = 7.34 (d, J = 0.8 Hz, 1 H, ArH), 6.79 (dd, J_1 = 3.4 Hz, J_2 = 0.8 Hz, 1 H, ArH), 6.31–6.33 (m, 1 H, ArH), 3.55 (s, 4 H, CH_2), 1.48–1.55 (m, 6H, CH_2).

A broadened singlet is observed for the methylene groups on both sides of the nitrogen atom at 3.55 ppm by them and at 3.66 ppm by us. The protons from the furan core also match, being slightly shifted (from 7.34 ppm to 7.40 ppm, from 6.79 ppm to 6.85 ppm, and from 6.31–6.33 ppm to 6.39 ppm). The rest of the signals from the hydrogenated nitrogen-containing core appear as one multiplet for six protons in the interval from 1.48 to 1.55 (m, 6H, CH_2) in the Ramkumar ^{1}H-NMR spectra. In the ^{1}H-NMR spectrum obtained by us, two multiplets stand out from 1.65–1.60 for two protons and from 1.59–1.54 for four protons, respectively. Based on the coincidence of all of these data, it can be asserted without doubt that we have succeed in obtaining the same compound.

The four compounds have been given the names hybrid 1 (**H1**)- **7a**, hybrid 2 (**H2**)- **7b**, hybrid 3 (**H3**)- **7c**, and hybrid 4 (**H4**)- **7d** for biological evaluation research.

The newly synthesized compounds (**H1–4**) differ mainly in the N-containing heterocycles. In order to investigate the structure of the new hybrid molecules, we used mass spectrometry. We used ESI in positive ionization mode. The fragmentation of the furan derivatives **H1–4** proceeds according to the mechanism presented in Scheme 2.

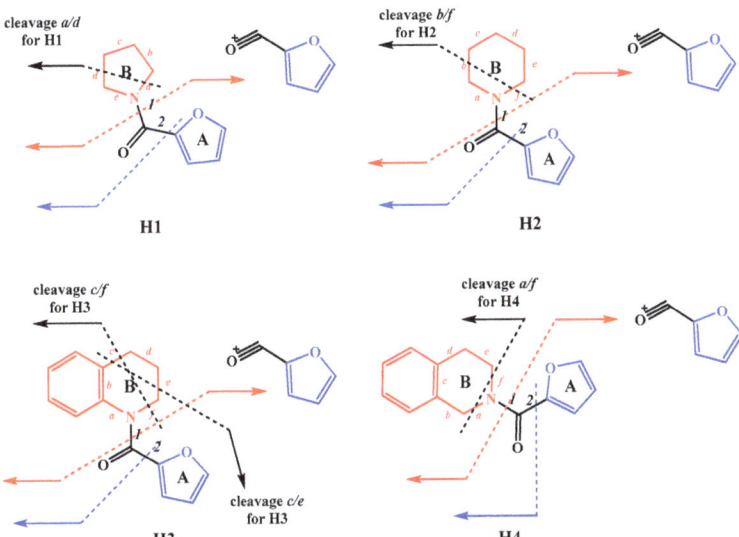

Scheme 2. Possible fragmentation pathways of compounds **H1–4** under ESI-MS/MS conditions. **1-H1**—fragmentation of pyrrolidine-furan **H1**. **1-H2**—fragmentation of **H2** compounds. **1-H3**—fragmentation of compounds **H3**. **1-H4**—fragmentation of compounds **H4**. **H3** and **H4** can be considered derivatives of piperidine, since its structure contains a piperidine ring (ring B)—benzo[b]piperidine (1,2,3,4 tetrahydroquinoline skeleton) for **H3** and benzo[c]piperidine (1,2,3,4 tetrahydroisoquinoline core) for **H4**.

The structural fragment that connects the furan (ring A) and heterocyclic (ring B) rings in furan derivatives is C-C(O)-N. Several fragmentation pathways originate from here. The main fragmentation pathways of compounds **H1–4** involve the cleavage of C-N (path 1), C-C (path 2) bonds, and retro cyclization of heteroring B with a cleavage of bonds at position a/d for **H1**, b/f for **H2**, c/f, c/e for **H3**, and a/f for **H4** (Scheme 2). The cleavage of the C-N bond (pathway 1) provides important information about the structure of the heterocyclic ring (m/z 70, 84, 132) and the furan ion at m/z 95. The structure of compounds **H1** and **H2** does not contain fused rings. Ring B, in their structure, is respectively pyrrolidine and piperidine. The retro cyclization of ring B is associated with the cleavage of the a/d bonds for **H1**, and b/f for **H4**, resulting in the same ion with m/z 124 (Figures S14 and S16).

The structure of compounds **H3** and **H4** contain fused rings: benzo[b]piperidine (1,2,3,4 tetrahydroquinoline nuclei) and benzo[c]piperidine (1,2,3,4 tetrahydroisoquinoline nuclei), respectively.

In the presence of an additional benzene nucleus during fragmentation, the formation of additional characteristic ions is observed, by which the isomers **H3** and **H4** are clearly distinguished. For both isomers, the ions m/z 132 (path 1) and m/z 160 (path 2) were obtained. Additional fragmentation pathways depend on the position of the nitrogen atom in the structure of compounds **H3** and **H4** (Scheme 2, Figures S18 and S20). For compound **H3**, which contains benzo[b]piperidine in its structure, the fragmentation proceeds in other ways, different from its isomer. Under MS conditions, a neutral CO molecule (28 au) is lost from the structure of compound **H3**, where a retro cyclization follows and leads to the m/z 172 ion. A molecule of H_2O (18 au) is lost from the same ion and an ion of m/z 154 is obtained (Scheme 2, Figure S18). These two ions are not produced in the fragmentation of compound **H4**. Additionally, the fragmentation of the m/z 160 ions produced an m/z 118 ion. In fact, ions m/z 118, m/z 154 and m/z 172 are characteristic of compound **H3** (Figure S18), while compound **H4** is ion m/z 117, which is the result of the retro cyclization of benzo[c]piperidine in position a/f (Scheme 2, Figure S20).

3.2. Biological Evaluation

All synthesized furan hybrids were analyzed for their in vitro hydrogen peroxide-scavenging activity (HPSA), inhibition albumin denaturation (IAD), metal-chelating activity (MChA), and antitryptic activity (ATA). The results observed in vitro were contrasted with those predicted in silico. The results are presented in Table 1.

Table 1. Biological evaluation findings from in vitro and in silico. The IC_{50} values are expressed for the following activities—hydrogen peroxide-scavenging activity (*HPSA*), inhibition of albumin denaturation (*IAD*), metal-chelating activity (*MChA*), and anti-tryptic activity (*ATA*). As reference points, ascorbic acid (AA), quercetin (Qrc), ibuprofen (Ibu), and ketoprofen (Ket) were used. Lipophilicity (R_M) is a non-dimensional quantity because it is acquired via thin-layer chromatography and is a function of *Rf*. The terms "calculated anti-inflammatory activity (*cAnti-I*)" and "calculated anti-arthritic activity (*cAnti-A*)" are expressed as P_a (probability "to be active"), and assess the likelihood that the investigated molecule is a member of the sub-class of active compounds (resembles the most typical molecular structures in a subset of "actives" in the PASS training set). The molecule with the highest level of activity has a value of 1. **H1–4**—furan derivatives.

Compounds	$IC_{50} \pm SD$, µg/mL			$IC_{50} \pm SD$, mg/mL	$R_M \pm SD$	P_a	
	HPSA	IAD	ATA	MChA		cAnti-I	cAnti-A
AA	24.84 ± 0.35	-	-	-	-	-	-
Qrc	69.25 ± 1.82	-	-	-	-	-	-
Ibu	-	81.50 ± 4.95	259.82 ± 9.14	-	1.11 ± 0.010	0.903	0.573
Ket	-	126.58 ± 5.00	720.57 ± 19.78	-	1.54 ± 0.015	0.925	0.469
H1	123.33 ± 7.41	114.31 ± 2.88	83.25 ± 1.69	4.92 ± 0.89	1.04 ± 0.023	0.224	-
H2	105.52 ± 10.33	116.76 ± 1.61	60.21 ± 8.16	4.26 ± 0.79	0.94 ± 0.016	0.226	-
H3	77.75 ± 0.67	150.99 ± 1.16	85.33 ± 7.26	1.02 ± 0.12	1.01 ± 0.030	-	-
H4	71.72 ± 4.63	119.08 ± 0.92	62.23 ± 0.83	1.46 ± 0.06	0.87 ± 0.015	0.201	-

3.2.1. Hydrogen Peroxide Scavenging Activity

Copper and iron are vitally essential trace elements for humans, which participate as cofactors in numerous enzymes and various physiological processes. However, in their free form, they are toxic because Cu(II) catalyzes the oxidation of ascorbic acid, producing reactive oxygen species (ROS) such as superoxide radicals ($O_2^{\bullet-}$) and H_2O_2. In this catalytic process, Cu(II) and Fe(II) react with H_2O_2, and hydroxyl radicals ($^{\bullet}OH$) are generated via the Fenton reaction [24]. It has been established that the harm they produce plays a role in the onset of a number of diseases, including Alzheimer's disease, atherosclerosis, cancer, and cardiovascular diseases [25]. Therefore, in the human body, the sulfur-containing molecules glutathione, cysteine, and ergothioneine play an important role as endogenous antioxidants [24]. For this reason, the study of compounds with high oxygen-free radical scavenging activity is a current and significant area of research.

In the current study, the goal of the research was to stop hydrogen peroxide's damaging effects. Hydrogen peroxide is an oxidant that is continuously produced in living tissues as a result of several metabolic processes. However, in order to avoid entering hazardous reactions like the Fenton reaction, its detoxification is crucial [26].

The inflammatory process also causes and accelerates the formation of ROS. Most significantly, the formation of other ROS species, such as H_2O_2, is linked to the production of superoxide anion radicals at the site of inflammation. Assuming that at least some of the oxygen produced in these processes is in the singlet state, it is also engaged in the reductive breakdown reactions of organic hydroperoxides ROOH and hydrogen peroxide (the so-called Haber–Weiss reaction) [27,28]. Therefore, it is crucial to inhibit H_2O_2 in order to stop the generation of $^{\bullet}OH$.

We compared the results obtained for the antioxidant activity of the synthetic analogs of furan with the standards of ascorbic acid and quercetin. They are natural compounds with demonstrable antioxidant activity [15].

The discovered furan hybrid molecules exhibit reduced in vitro antioxidant activity when compared to ascorbic acid (24.84 µg/mL) and quercetin (69.25 µg/mL). Compounds **H3** (77.75 µg/mL) and **H4** (71.72 µg/mL) demonstrate higher antioxidant activity compared to the rest of the synthesized compounds (**H1**, **H2**) (Table 1, Figure 4A).

The inclusion of benzene nuclei in the structure of the discovered compounds greatly boosts the antioxidant action. As a result, compounds **H3** and **H4** show strong activity.

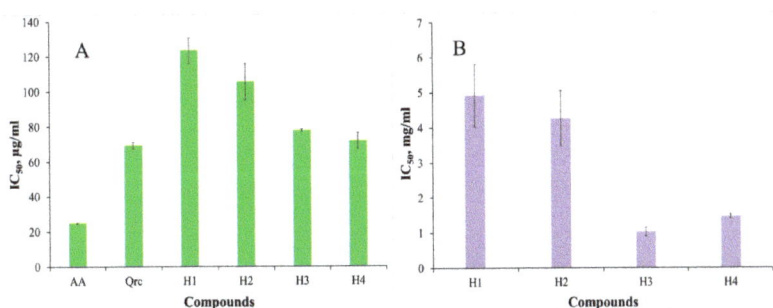

Figure 4. Ability of furan derivatives to scavenge hydrogen peroxide (HPSA) (**A**) and form chelate complexes with Fe(II) ions (MChA) (**B**). Standards employed in this study were ascorbic acid (AA) and quercetin (Qrc). IC_{50} values were used to assess the HPSA and MChA outcomes.

3.2.2. Metal-Chelating Activity (MChA) on Ferrous Ions

Fe(II) is a transition metal well known for its increased propensity for the Fenton reaction, but it also makes it one of the most important pro-oxidants involved in lipid peroxidation, i.e., the oxidation of lipids leading to cell membrane damage [13,26]. This, in turn, accelerates the aging process. Chelation is one of the antioxidant methods for reducing the catalytic action of the transition metals Fe(II) and Cu(II). Chelating agents create sigma bonds with the metal and are regarded as strong secondary antioxidants [13,29] because they can lower the redox potential and stabilize the oxidized form of the iron ion.

The metal-chelating properties of the newly synthesized furan derivatives (**H1–4**) on ferrous iron are shown in Table 1 and Figure 4B. The results for the metal-chelating activity are presented as IC_{50}. Chelating activity on ferrous iron is an important step in preventing lipid peroxidation. Therefore, we investigated the ability of compounds **H1–4** to form chelate complexes with Fe(II) ions. The concentration gradient of **H1–4** affects the chelating activity. As the concentration of compounds **H1–4** increases, their ability to chelate Fe(II) increases. At low concentrations, **H1–4** show low chelating activity. From the analysis, we found that compounds **H3** (1.02 mg/mL) and **H4** (1.46 mg/mL) showed significant metal-chelating activity four times higher than compounds **H1** and **H2**.

Here, we must clarify that there is a connection between the two methods (HPSA and MChA). In living tissues, competing reactions are most likely taking place—both hydrogen peroxide deactivation and Fe(II) chelation. The research shows that there is a good correlation dependence of 0.9672 between the two methods. This relationship is beneficial because the chelation of Fe(II) will prevent the Fenton reaction from occurring. This approach provides more information about the antioxidant activity of the compounds, proving them to be reliable exogenous antioxidants.

3.2.3. Inhibition of Albumin Denaturation (IAD)

Living tissues' response to harm is inflammation. Cell migration, tissue breakdown, mediator release, enzyme activation, fluid extravasation, and tissue healing are only a few of the numerous complicated processes involved [30]. Inflammation in rheumatoid arthritis is well known to be brought on by the denaturation of proteins. The ability of certain anti-inflammatory medications to prevent thermally induced protein denaturation has been demonstrated to depend on dose [31]. The obtained furan hybrids were examined for their ability to inhibit albumin denaturation. This method determines the extent to which albumin is protected from denaturation when heated. We used human albumin for this purpose. Figure 5A shows the percentages of inhibition of synthetic furan derivatives. The obtained results of the analysis are presented as IC_{50}. Because ibuprofen and ketoprofen have well-established properties, we decided to utilize them as a standard to compare the activities of the newly synthesized furan derivatives. The IC_{50} values of ibuprofen and ketoprofen estimated as IAD are 81.50 µg/mL and 126.58 µg/mL, respectively (Table 1, Figure 5A). All of the obtained results show that the IC_{50} values of furan hybrid molecules are in the range of 114.31 to 150.99 µg/mL (Table 1, Figure 5A).

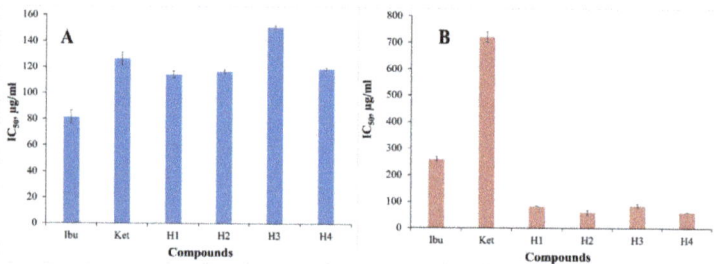

Figure 5. In vitro biological activity was assessed as inhibition of albumin denaturation (IAD) (**A**) and antitryptic activity (ATA) (**B**). As benchmarks, anti-inflammatory medications such as ibuprofen and ketoprofen were utilized. The results of both methods are presented as IC_{50}.

In general, the obtained compounds (**H1–4**) exhibited high IAD activity as profens. The in silico anti-inflammatory activity (cAnti-A) results show that the standards (ibuprofen and ketoprofen) have higher activity than the synthesized derivatives (**H1–4**), indicating that there is a directly proportional dependence between in vitro and in silico for ibuprofen and a reverse dependence for ketoprofen (Table 1).

In addition, IAD analysis reveals that lipophilicity is a significant physicochemical parameter. The lipophilicity (RM) of the synthetic furan derivatives studied ranges from 0.87 to 1.04, which influences albumin protection to some extent (Table 1).

For the stability of albumin, the hydrophobic pocket of subdomain IIA and IIIA plays an important key role, popularly known as Sudlow's sites I and II, respectively. Due to the hydrophobic nature of the interior of Sudlow's sites I pocket, the drug primarily formed hydrophobic interactions with Phe211, Trp214, Leu219, Phe223, Leu234, Leu238, Leu260, Ile264, and Ile290 [32].

As the interior of Sudlow's sites I pocket is hydrophobic in nature, the drug predominantly formed hydrophobic interactions with Phe211, Trp214, Leu219, Phe223, Leu234, Leu238, Leu260, Ile264, and Ile290 [32]. The stabilization of the albumin molecule in this study is due to hydrophobic interactions between Sudlow's sites I and the furan derivatives.

3.2.4. Antitryptic Activity (ATA)

Proteinases have been linked to the development of arthritic symptoms. Neutrophils are known to be a good source of proteinase because their lysosomal granules contain many serine proteinases. It has previously been reported that leukocyte proteinase is important in the development of tissue damage during inflammatory reactions and that proteinase inhibitors provide significant protection [16,31]. In vitro anti-arthritic activity was assessed as antitryptic activity [16]. The IC_{50} results for the ATA range from 60.21 to 85.33 µg/mL.

The results reveal that the furan derivatives **H1–4** show better antitryptic activity compared to ibuprofen and ketoprofen (Table 1, Figure 5B).

3.2.5. Lipophilicity

The most commonly used parameter in SAR drug discovery is lipophilicity. It can be determined experimentally or mathematically.

Increased permeability, solubility, target potency, and toxicity have all been linked to lipophilicity. The lipophilicity was determined as R_M values using reverse-phase thin layer chromatography (RPTLC). This is regarded as a dependable, quick, and convenient method of expressing lipophilicity [33]. Apart from the importance of lipophilicity in biologically active compound kinetics, both hydrophilic and lipophilic antioxidants are required to act as radical scavengers in the aqueous phase or as chain-breaking antioxidants in biological membranes [17].

In the present work, we have examined the antioxidant, metal-chelating, and in vitro biological activity of the newly synthesized furan derivatives. Lipophilicity is confirmed to be an important factor in their activity. The results show that compounds **H1–4** are lipophilic with good antioxidant and metal-chelating activity. This is what makes them reliable lipophilic exogenous antioxidants that are necessary to neutralize harmful radicals in the cell membrane.

In general, the in vitro studies results show that the compounds **H1–4** exhibit IAD and ATA. We learned a lot about the properties of the potential new drugs from both experiments, and both experiments are related to keeping the albumin molecule intact. Albumin is protected from denaturation (IAD) in the first case. Compounds **H1–4** allosterically bind to albumin. In the second case, it is protected from the action of the enzyme trypsin (ATA), which is inhibited by furan derivatives.

4. Conclusions

In conclusion, we have successfully obtained three new hybrid molecules containing a furan skeleton and N-heterocycle. The new hybrid molecules were fully characterized using UV, ^{1}H- and ^{13}C-NMR, and HRMS spectrometry. All four of the obtained compounds were in vitro biologically evaluated for their antioxidant, anti-inflammatory, anti-arthritic, and metal-chelating activity. Detailed HRMS analysis showed that the isomers **H3** and **H4** can be easily distinguished from each other. All compounds showed activity in all in vitro biological tests performed. As expected, hybrid molecules **H3** and **H4** show greater metal-chelating activity as well as antioxidant activity than hybrids **H1** and **H2**. The anti-inflammatory activity confirmed by the in silico analysis is also proven by the in vitro biological evaluation, and the hybrids are ranked in the following descending order: **H1 > H2 > H4 > H3**.

Supplementary Materials: The following supporting information can be downloaded at: https://www.mdpi.com/article/10.3390/pr10101997/s1, Figures S1–S4: ^{1}H-NMR spectrum of hybrids 1–4; Figures S5–S8: ^{13}C-NMR spectrum of hybrids 1–4; Figures S9–S12: UV spectrum of hybrids 1–4; Figures S13–S20: HRMS chromatograms of hybrids 1–4.

Author Contributions: Conceptualization, I.I. and S.M.; methodology, S.M. and D.B.; software, S.M.; validation, D.B., S.M. and I.I.; formal analysis, S.M., D.B. and P.N.; investigation, S.M.; resources, I.I.; data curation, I.I.; writing—original draft preparation, S.M. and D.B.; writing—review and editing, S.M., D.B. and I.I.; visualization, D.B.; supervision, I.I.; project administration, S.M.; funding acquisition, I.I. All authors have read and agreed to the published version of the manuscript.

Funding: This research was funded by the National Science Fund of the Bulgarian Ministry of Education and Science, grant number КП 06 М29/1.

Institutional Review Board Statement: Not applicable.

Informed Consent Statement: Not applicable.

Data Availability Statement: The data presented in this study are available in this article and in the supporting Supplementary Material.

Conflicts of Interest: The authors declare no conflict of interest.

References

1. Demicheva, L. Biological activity of furan derivatives (review). *Chem. Heterocycl. Compd.* **1993**, *29*, 243–267. [CrossRef]
2. Duffy, L.; Smith, A.D. Nitrofurantoin macrocrystals prevent bacteriuria in intermittment self-catheterization. *Urology* **1982**, *20*, 47–49. [CrossRef]
3. Naruganahalli, K.S.; Shirumalla, R.K.; Bansal, V.; Gupta, J.B.; Das, B.; Ray, A. Ranbezolid, a novel oxazolidinone antibacterial: In vivo characterisation of monoamine oxidase inhibitory potential in conscious rats. *Eur. J. Pharmacol.* **2006**, *545*, 167–172. [CrossRef]
4. Delsarte, A.; Faway, M.; Frère, J.M.; Coyette, J.; Calberg-Bacq, C.M.; Heinen, E. Nifurzide, a nitrofuran antiinfectious agent: Interaction with Escherichia coli cells. *Antimicro. Agents Chemother.* **1981**, *19*, 477–486. [CrossRef]
5. White, C.M.; Hernandez, A.V. Ranitidine and Risk of N-Nitrosodimethylamine (NDMA) Formation. *JAMA* **2021**, *326*, 225–227. [CrossRef]
6. Mermer, A.; Keles, T.; Sirin, Y. Recent studies of nitrogen containing heterocyclic compounds as novel antiviral agents: A review. *Bioorg. Chem.* **2021**, *114*, 105076. [CrossRef]
7. Li, Y.-S.; Liu, X.-Y.; Zhao, D.-S.; Liao, Y.-X.; Zhang, L.-H.; Zhang, F.-Z.; Song, G.-P.; Cui, Z.-N. Tetrahydroquinoline and tetrahydroisoquinoline derivatives as potential selective PDE4B inhibitors. *Bioorg. Med. Chem. Lett.* **2018**, *28*, 3271–3275. [CrossRef]
8. Zeni, G.; Lüdtke, D.; Nogueira, C.; Panatieri, R.; Braga, A.; Silveira, C.; Stefani, H.; Rocha, J. New acetylenic furan derivatives: Synthesis and anti-inflammatory activity. *Tetrahedron Lett.* **2001**, *42*, 8927–8930. [CrossRef]
9. Janczewski, Ł.; Zieliński, D.; Kolesińska, B. Synthesis of amides and esters containing furan rings under microwave-assisted conditions. *Open Chem.* **2021**, *19*, 265–280. [CrossRef]
10. Qian, Z.; Li, Q.; Wang, L.; Fu, F.; Liu, X. The chemical effect of furfural amide on the enhanced performance of the diphenolic acid derived bio-polybenzoxazine resin. *J. Poly. Sci.* **2021**, *59*, 2057–2068. [CrossRef]
11. Malladi, S.; Venkata Nadh, R.; Suresh Babu, K.; Suri Babu, P. Synthesis and antibacterial activity studies of 2,4-di substituted furan derivatives. *Beni-Suef Univ. J. Basic App. Sci.* **2017**, *6*, 345–353. [CrossRef]
12. Manolov, S.; Ivanov, I.; Bojilov, D. Synthesis of New 1,2,3,4-Tetrahydroquinoline Hybrid of Ibuprofen and Its Biological Evaluation. *Molbank* **2022**, *1*, M1350. [CrossRef]
13. Sirin, S.; Duyar, H.; Aslım, B.; Seferoğlu, Z. Synthesis and biological activity of pyrrolidine/piperidine substituted 3-amido-9-ethylcarbazole derivatives. *J. Mol. Struct.* **2021**, *1242*, 130687. [CrossRef]
14. Sakat, S.S.; Juvekar, A.R.; Gambhire, M.N. In-vitro antioxidant and anti-inflammatory activity of methanol extract of Oxalis corniculata linn. *Int. J. Pharm. Pharm. Sci.* **2010**, *2*, 146–155.
15. Manolov, S.; Ivanov, I.; Bojilov, D. Microwave-assisted synthesis of 1,2,3,4-tetrahydroisoquinoline sulfonamide derivatives and their biological evaluation. *J. Serb. Chem. Soc.* **2021**, *86*, 139–151. [CrossRef]
16. Oyedapo, O.O.; Famurewa, A.J. Antiprotease and membrane stabilizing activities of extracts of fagara zanthoxyloides, olax subscorpioides and tetrapleura tetraptera. *Int. J. Pharmacogn.* **1995**, *33*, 65–69. [CrossRef]
17. Pontiki, E.; Hadjipavlou-Litina, D. Synthesis and pharmachemical evaluation of novel aryl-acetic acid inhibitors of lipoxygenase, antioxidants, and anti-inflammatory agents. *Bioorg. Med. Chem.* **2007**, *15*, 5819–5827. [CrossRef]
18. Sadym, A.; Lagunin, A.; Filimonov, D.; Poroikov, V. Prediction of biological activity spectra via the Internet. *SAR QSAR Environ. Res.* **2003**, *14*, 339–347. [CrossRef] [PubMed]
19. Filimonov, D.; Lagunin, A.; Gloriozova, T.; Rudik, A.; Druzhilovskii, D.; Pogodin, P.; Poroikov, V. Prediction of the Biological Activity Spectra of Organic Compounds Using the Pass Online Web Resource. *Chem. Heterocycl. Compd.* **2014**, *50*, 444–457. [CrossRef]
20. Das, S.; Addis, D.; Zhou, S.; Junge, K.; Beller, M. Zinc-catalyzed reduction of amides: Unprecedented selectivity and functional group tolerance. *J. Am. Chem. Soc.* **2010**, *132*, 1770–1771. [CrossRef] [PubMed]
21. Pathak, U.; Bhattacharyya, S.; Pandey, L.; Mathur, S.; Jain, R. An easy access to tertiary amides from aldehydes and N, N-dialkylchlorothiophosphoramidates. *RSC Adv.* **2014**, *4*, 3900. [CrossRef]
22. Khosravi, K.; Naserifar, S. 1,1,2,2-Tetrahydroperoxy-1,2-diphenylethane: An efficient and high oxygen content oxidant in various oxidative reactions. *Tetrahedron* **2018**, *74*, 6584–6592. [CrossRef]
23. Ramkumar, R.; Chandrasekaran, S. Catalyst-free, metal-free, and chemoselective transformation of activated secondary amides. *Synthesis* **2019**, *51*, 921–932. [CrossRef]
24. Chalana, A.; Kumar, R.; Karri, R.; Kumar, K.; Kumar, B.; Roy, G. Interplay of the intermolecular and intramolecular interactions in stabilizing the thione-based copper(I) complexes and their significance in protecting the biomolecules against metal-mediated oxidative damage. *Polyhedron* **2022**, *215*, 115647. [CrossRef]

25. Galano, A.; Macías-Ruvalcaba, A.; Campos, M.; Pedraza-Chaverri, J. Mechanism of the OH radical scavenging activity of nordihydroguaiaretic acid: A combined theoretical and experimental study. *J. Phys. Chem. B* **2010**, *114*, 6625–6635. [CrossRef] [PubMed]
26. Halliwell, B.; Gutterdge, J.M.C. *Free Radicals in Biology and Medicine*, 5th ed.; Oxford Academic: Oxford, UK, 2015; Available online: https://doi.org/10.1093/acprof:oso/9780198717478.001.0001 (accessed on 5 August 2022).
27. Khan, A.U. Singlet molecular oxygen from superoxide anion and sensitized fluorescence of organic molecules. *Science* **1970**, *168*, 467–477. [CrossRef]
28. Kellog, E.W.; Fridovich, I. Superoxide, hydrogen peroxide, and singlet oxygen in lipid peroxidation by a xanthine oxidase system. *J. Biol. Chem.* **1975**, *250*, 8812–8817. Available online: https://www.jbc.org/article/S0021-9258(19)40745-X/pdf (accessed on 15 August 2022). [CrossRef]
29. Bandgar, B.P.; Adsul, L.K.; Lonikar, S.V.; Chavan, H.V.; Shringare, S.N.; Patil, S.A.; Jalde, S.S.; Koti, B.A.; Dhole, N.A.; Gacche, R.N.; et al. Synthesis of novel carbazole chalcones as radical scavenger, antimicrobial and cancer chemopreventive agents. *J. Enzyme Inhib. Med. Chem.* **2013**, *28*, 593–600. [CrossRef]
30. Vane, J.R.; Botting, R.M. New insights into the mode of action of anti-inflammatory drug. *Inflamm. Res.* **1995**, *44*, 1–10. [CrossRef]
31. Jayashree, V.; Bagyalakshmi, S.; Manjula Devi, K.; Richard Daniel, D. In-vitro anti-inflamatory activity of 4-benzylpiperidine. *Asian J. Pharm. Clin. Res.* **2016**, *9*, 108–110. [CrossRef]
32. Mondal, M.; Lakshmi, P.; Krishna, R.; Sakthivel, N. Molecular interaction between human serum albumin (HSA) and phloroglucinol derivative that shows selective anti-proliferative potential. *J. Lumin.* **2017**, *192*, 990–998. [CrossRef]
33. Hansch, C.; Leo, D.; Hoekman, D.H. *Exploring QSAR: Hydrophobic, Electronic, and Steric Constants*; American Chemical Society: Washington, DC, USA, 1995; Available online: https://www.amazon.com/Exploring-QSAR-Hydrophobic-Electronic-Professional/dp/0841229910 (accessed on 2 August 2022).

Article

Synthesis, Characterization, and Antibacterial Potential of Poly(*o*-anisidine)/BaSO$_4$ Nanocomposites with Enhanced Electrical Conductivity

Mirza Nadeem Ahmad [1], Sohail Nadeem [2,*], Raya Soltane [3,4], Mohsin Javed [2], Shahid Iqbal [5,*], Zunaira Kanwal [1], Muhammad Fayyaz Farid [1], Sameh Rabea [6], Eslam B. Elkaeed [6], Samar O. Aljazzar [7], Hamad Alrbyawi [8] and Walid F. Elkhatib [9,10]

[1] Department of Applied Chemistry, Government College University, Faisalabad 38030, Pakistan; pioneeravian@hotmail.com (M.N.A.); zunairakanwal995@gmail.com (Z.K.); fiazawan050@gmail.com (M.F.F.)
[2] Department of Chemistry, School of Science, University of Management and Technology, Lahore 54770, Pakistan; mohsin.javed@umt.edu.pk
[3] Department of Basic Sciences, Adham University College, Umm Al-Qura University, Makkah 21955, Saudi Arabia; rasoltan@uqu.edu.sa
[4] Department of Biology, Faculty of Sciences, Tunis El Manar University, Tunis 1068, Tunisia
[5] Department of Chemistry, School of Natural Sciences (SNS), National University of Science and Technology (NUST), H-12, Islamabad 46000, Pakistan
[6] Department of Pharmaceutical Sciences, College of Pharmacy, AlMaarefa University, Riyadh 13713, Saudi Arabia; srabea@mcst.edu.sa (S.R.); ikaeed@mcst.edu.sa (E.B.E.)
[7] Department of Chemistry, College of Science, Princess Nourah bint Abdulrahman University, P.O. Box 84428, Riyadh 11671, Saudi Arabia; soaljazar@pnu.edu.sa
[8] Pharmaceutics and Pharmaceutical Technology Department, College of Pharmacy, Taibah University, Medina 42353, Saudi Arabia; hrbyawi@taibahu.edu.sa
[9] Microbiology and Immunology Department, Faculty of Pharmacy, Ain Shams University, African Union Organization St., Abbassia, Cairo 11566, Egypt; walid-elkhatib@pharma.asu.edu.eg
[10] Department of Microbiology & Immunology, Faculty of Pharmacy, Galala University, New Galala City, Suez 43511, Egypt
* Correspondence: sohail.nadeem@umt.edu.pk (S.N.); shahidgcs10@yahoo.com (S.I.)

Abstract: The poly(o-anisidine)/BaSO$_4$ nanocomposites were prepared by oxidative polymerization of *o*-anisidine monomer with BaSO$_4$ filler for the potential antibacterial properties of the composite materials. To achieve the optimal and tunable properties of the nanocomposites, the ratio of BaSO$_4$ filler was changed at the rates of 1%, 3%, 5%, 7%, and 10% with respect to matrix. Different analytical techniques, i.e., FTIR and UV-visible spectroscopy, were employed for functional identification and optical absorption of the poly(*o*-anisidine)/BaSO$_4$ nanocomposites. The FTIR data revealed the significant interaction between POA and BaSO$_4$, as well as the good absorption behavior of the UV-visible spectra. The conducting properties were controllable by varying the load percentage of the BaSO$_4$ filler. Furthermore, different bacterial strains, i.e., *Pseudomonas aeruginosa* (Gram-negative) and *Staphylococcus aureus* (Gram-positive), were used to evaluate the antibacterial activity of the POA/BaSO$_4$ nanocomposites. The largest zones of inhibition 0.8 and 0.9 mm were reached using 7% and 10% for *Staphylococcus aureus* and *Pseudomonas aeruginosa*, respectively.

Keywords: nanocomposites; antibacterial; electrical conductivity; BaSO$_4$; polymerization

1. Introduction

The development of nanomaterials has been a useful and important task during the last fifteen years. These nanomaterials have original multifunctionalities, which make them very helpful products in a variety of industries. The nanocomposites are classified into structural and functional materials. Structural nanocomposites focus on the improvement of mechanical properties due to the addition of nanoparticles, while for the functional ones, the presence of nanoparticles adds extra functionalities to the polymer, such as electrical conductivity, special optical properties, antimicrobial features, and many more. The hybrid nanomaterials consist of organic–inorganic components, e.g., organic polymer matrix and inorganic filler, and have potential applications as antimicrobial agents, biocompatible materials, sensors, and electrical devices. Moreover, they are utilized as high-value coatings in gas separation, ultra- and nanofiltration, and corrosion prevention, as well as in artificial membranes [1–3]. In addition, they can be used as catalysts, toxic-compound adsorbents, biomaterials, and information display materials with specialized optical, magnetic, and electrical capabilities [4–6].

These nanomaterials offer unique properties due to the fact that their components have a different nature to prepare some tailor-made novel nanomaterials [7–10]. In particular, the composites made by adding multi-walled carbon nanotubes to thin layers of poly(3,4-ethylenedioxythophene) and poly(4-styrenesulfonate) exhibited intriguing characteristics regarding the interdependence of electrical properties towards the large temperature range [11]. Thus, the inorganic and organic species interact at a molecular level at the nanoscopic range. Similarly, Mt-PS-BZO-PANI hybrid composites showed valuable luminescence properties [12]. Moreover, polyurethane (PU) and poly(hydroxyethyl methacrylate) composited with poly(titanium oxide) showed excellent viscoelastic and thermophysical properties [13]. The ability of these organic–inorganic hybrid materials to blend the heterogeneous properties of inorganic–organic components in a single material is its key benefit. In recent years, nanomaterials have offered one of the most interesting developments in the subject of materials chemistry [14–16].

The conducting polymer is one of a new class of polymers that are being used for a variety of applications. The merging of different unique properties, such as the chemical, electrical, and electrochemical properties of these polymers, may cause them to be used in several scientific applications [17–20]. Polyaniline has the unique property to coordinate with metallic ions, which provides the multi-metallic system and also prepares the nanocomposite materials with some other species. Polyaniline and its derivatives have achieved considerable attention over the last few years due to their conducting properties. The electro-active characteristics of polyaniline films, in particular, have been determined useful for the development of batteries, electrochromic displays, and microelectronics devices. Poly(o-anisidine) exhibits some interesting and unique properties [21–24]. Poly(o-anisidine) was found to have greater solubility compared with polyaniline in some organic solvents as it has a low boiling point while maintaining its crystalline property. This is very important because this polyaniline derivative can be used in technological applications [25–29].

Polymeric nanocomposites are modern and superior composites that are obtained from nanoparticles and the polymeric matrix, in which NPs are covered by polymers; through this, a core-shell arrangement can be developed [30]. Due to the particular shape, chemical nature, and exceptional structure of polymers, nanoparticles can be spread in a polymer matrix in the best and unique shapes. As a result of covering nanoparticles and their functionalization, the Van der Waals forces among NPs become reduced and the distribution of NPs in the matrix is improved and amplified [31,32]. The importance of polymer nanocomposites lies in the fact that polymers are always preferred for the covering of nanoparticles. On the other hand, appropriate functional groups in polymer structures can be used as reaction sites to organize the one-pot synthesis of nanocomposites [33,31].

Polymeric materials can be extensively employed in industries because of their lightweight and their simplicity of preparation, in addition to their elastic properties. Nevertheless, these materials have several drawbacks, such as their small modulus as well as potency compared to ceramics and metals [35]. Thus, to improve the mechanical properties, a very useful and valuable approach can be the addition of fibers, platelets, particles, or whiskers into such polymeric materials as reinforcements. The polymeric materials are composited with other filler materials to tune the properties of the materials, such as the temperature resistance, impact opposition, mechanical strength, fire or flame resistance, gas permeability, conductivity, and microbial resistance [36].

Due to the addition of reinforcements or filler materials to a wide range of polymer resins, their unique properties have greatly been improved. This is a great example of polymer-matrix nanocomposites acting as eco-friendly systems. Furthermore, the latest information on nano-industries' nanomaterials showed complete and effective applications in the form of polymer nanocomposites and has been successfully reported [31,37]. Moreover, the barium sulfate has versatile applications and can be utilized as an antibacterial agent, a paint, a film, a fiber, and a luminescence material, and for photocatalysis [38–41]. Consequently, the versatile properties of barium-sulfate-nanocomposite materials proved their widespread applications in every field. Thus, there exists a big and wonderful demand for such hybrid materials. Generally, these hybrid materials, based on organic polymers, have many advantages, such as good processability, long-term stability, and wonderful optical, electronic, and catalytic as well as magnetic properties [42,43]. Thus, the resulting nanocomposites could offer numerous applications in different fields, for example, antibacterial, optoelectronics, electrical devices, sensors, etc. [44]. Moreover, barium-sulfate nanoparticles show antibacterial properties due to its small size, greater surface area, and high penetration power to inhibit the growth of bacteria [43]. The barium-sulfate nanoparticles not only exhibit the antimicrobial properties towards the nanocomposite but also control the electrical conductivity of the materials. Therefore, the antibacterial properties of $BaSO_4$ nanoparticles as filler and conducting properties of the poly(o-anisidine) matrix were combined to obtain novel composite materials with bactericidal and conducting properties, and for biomedical and sensor applications.

2. Material and Methods

For the present research work, all the chemicals, such as ammonium persulfate [$(NH_4)_2S_2O_8$], o-anisidine monomer as an oxidant, diaminodiphenylamine (DDPA), HCl/H_2SO_4, DI water, barium chloride, and ammonium sulfate, were procured from the Sigma Aldrich company (St. Louis, MI, USA). They were chemicals of the analytical and research grade, and they were utilized directly.

2.1. Preparation of Poly(o-anisidine)

Using a chemical oxidative approach, the o-anisidine was polymerized into poly(o-anisidine) (Figure 1). Two solutions were prepared and named solution A and solution B using the conventional process. By combining 3 g of monomer with 10 mL of 1 M HCl, solution A was created. Then, with magnetic stirring, solution A was added to an ice bath. The next step was to create solution B, which was made by combining 4 g of ammonium persulfate with 10 mL of 1 M HCl and keeping it in an ice bath. Solution B was added dropwise in solution A, which was already constantly stirred. The reaction contents were continuously stirred for 2 h. After that, it was left undisturbed overnight. The bluish-green suspension was prepared, which was subjected to centrifugation for isolation and purification. The material was isolated and washed with deionized water three times for better purification. Finally, the material was dried for 6 h at 80 °C in an electric oven and we proceeded with the characterization [17].

2.2. Preparation of Barium-Sulfate Nanoparticles

$BaSO_4$ nanoparticles were prepared by using the method already reported by Ahmad et al. [44]. The 20 mL of the $BaCl_2$ (0.5 M) solution, along with 20 mL absolute ethanol, was taken in a conical flask. Then, the $(NH_4)_2SO_4$ solution (0.1 M) was prepared in distilled water. Then, the ammonium-sulfate solution was added dropwise to the conical flask containing the barium-chloride solution for 20 min with constant stirring at 25 °C. Afterwards, the gelatinous white barium-sulfate nanoparticles were formed, which were centrifuged to separate them from each other. Distilled water and ethanol were used to wash the item multiple times. Finally, the material was dried in an electric oven at 120 °C for 12 h. The dried material was collected to obtain the barium-sulfate nanoparticles.

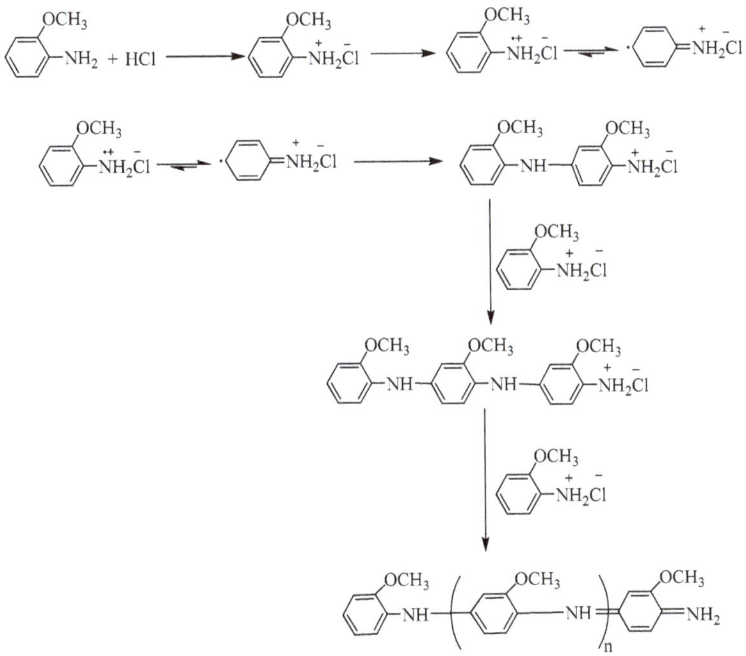

Figure 1. Proposed scheme for polymerization of o-anisidine.

2.3. Synthesis of Poly(o-anisidine)/BaSO₄ Nanocomposite

The preparation of the poly(o-anisidine)/BaSO$_4$ nanocomposites was carried out by following the method of poly(o-anisidine) polymerization. Two solutions were prepared, which were designated as solution A and solution B. For solution A, 0.246 g of o-anisidine monomer was added in 20 mL of HCl (1 M), along with 0.026 g of the DDPA and BaSO$_4$ nanoparticles. The load percentage of the BaSO$_4$ nanoparticles was changed for different composites (1, 3, 5, 7, and 10), as shown in Table 1. Then, solution A was sonicated for 10 min and placed to stir in an ice bath. Additionally, 20 mL of HCl and 0.246 g of APS were combined to create solution B (1 M). After that, solution B was continuously stirred into solution A while being introduced dropwise in an ice bath. The mixture was stirred for 2 h and kept overnight for stabilization. The greenish suspension was obtained and we proceeded to the centrifugation. The material was separated and washed with distilled water three times. Then, the material was dried in an electric oven at 80 °C for 6 h [17]. Similarly, other composites were prepared by varying the percentage of the BaSO$_4$ nanofiller according to the above method. Finally, the materials were collected and subjected to the characterization.

Table 1. Scheme for nanocomposite preparation.

Sr. No.	o-anisidine Monomer	BaSO₄ NPs	% of BaSO₄ NPs
1	0.246 g	0 g	0%
2	0.246 g	0.0024 g	1%
3	0.246 g	0.0073 g	3%
4	0.246 g	0.0123 g	5%
5	0.246 g	0.0172 g	7%
6	0.246 g	0.0246 g	10%

2.4. Characterization

The samples were characterized by various analytical techniques, as mentioned below.

2.4.1. FTIR Spectroscopic Studies

For the FTIR spectroscopic studies, the Agilent carrying 630 spectrometers was used to conduct the FTIR analysis of the poly(o-anisidine)/BaSO$_4$ nanocomposites. The sample discs were prepared by using KBr powder and dried to remove any moisture at 80 °C in an electric oven. Then, the samples were analyzed by the FTIR spectrometer in the wavenumber range of 400–4000 cm^{-1}.

2.4.2. UV-Visible Spectroscopy

The UV-visible spectroscopic study of the poly(o-anisidine)/BaSO4 nanocomposites was carried out using the Shimadzu double-bean spectrophotometer. The samples were prepared by dispersing 1 mg of the composite in 5 mL of distilled water. Then, the samples were scanned in the wavelength range of 200–800 nm.

2.4.3. TEM Analysis

On the H-7650 (Hitachi, Tokyo, Japan), running at an acceleration voltage of 100 kV, the TEM examination was performed.

2.4.4. Conductivity Measurement

The CyberScan PC 510 conductivity meter was used to conduct the conductivity measurement of the poly(o-anisidine)/BaSO$_4$ nanocomposites. The samples were prepared by dispersing 1 mg of the composite in 5 mL of distilled water. Then, the dispersion was used for the measurement of electrical conductivity by the conductivity meter.

2.4.5. Antibacterial Activity

The antibacterial capability of the poly(o-anisidine)/BaSO4 nanocomposites was assessed using the disc-diffusion technique. Antibacterial testing was carried out by using the disc-diffusion method. Bacterial growth was carried out by using the nutrient-growth agar method. Only Gram-negative bacteria (*Pseudomonas aeruginosa*) were used in the MacConkey agar. *Pseudomonas aeruginosa* can grow on MacConkey but *Staphylococcus aureus* cannot grow on MacConkey. The media were autoclaved, chilled, and deposited in petri plates for 41 min. Forty-one minutes later, hardened agar plates were covered with a fresh inoculum (20 mL) of bacteria cultures. The UVAS diagnostic lab provided bacterial cultures. Every plate contained sterile paper discs dipped in various suspensions of the POA/BaSO$_4$ nanocomposites. For 24 h, the agar plates were incubated at 37 °C (310 K). The zones of inhibition were studied following a 24-hour incubation period. The magnitude of the presence or absence of growth inhibition zones was used to determine the sensitivity and resistance of bacteria in sensitivity tests.

3. Results and Discussion

The synthesized poly(o-anisidine)/BaSO$_4$ nanocomposites were characterized by various techniques, as mentioned above. Six samples were prepared, in which one was the pure polymer and other five were nanocomposites with BaSO$_4$ nanoparticles (1%, 3%, 5%, 7%, and 10%). The barium-sulfate nanoparticles were characterized by powder XRD and the average particle size was 25.26 nm, as reported by Ahmad et al. [44]. The composites were further analyzed by using the following analytical techniques.

3.1. FTIR Spectroscopic Studies

The FTIR analysis of the poly(o-anisidine)/BaSO$_4$ nanocomposites was carried out to determine the structural changes in the material. The C=C stretching mode of the quinoid ring appeared at 1591 cm^{-1}, which was attributed to the stretching vibration of the o-methoxy group. The poly(2-methoxy aniline) signal appeared at 3379 cm^{-1}, which was attributed to the N-H stretching mode. The C=C stretching modes of the benzenoid ring appeared at 1469, 1206, and 1032 cm^{-1}, which were attributed to the stretching vibration of the o-methoxy group. The methoxy band clearly indicated the existence of poly(o-methoxy aniline) in the spectra. The spectrum in Figure 2b shows the addition of 1% of the BaSO$_4$ nanoparticles in the poly(o-anisidine) matrix at 2837 cm^{-1}, demonstrating the interaction of the poly(o-anisidine) with the BaSO$_4$ that appeared as the peak was deeper and wider.

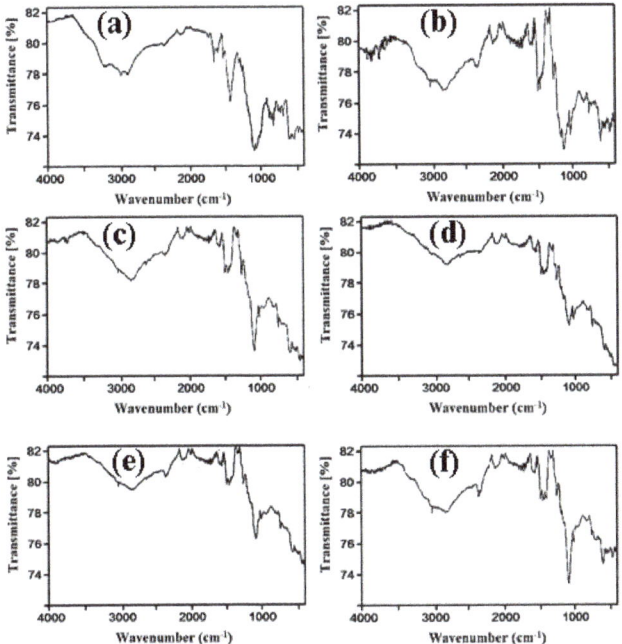

Figure 2. FTIR spectra of (**a**) pure POA, (**b**) 1% of POA/BaSO$_4$, (**c**) 3% of POA/BaSO$_4$, (**d**) 5% of POA/BaSO$_4$, (**e**) 7% of POA/BaSO$_4$, and (**f**) 10% of POA/BaSO$_4$.

The peak that appeared at 2837 cm^{-1} with the addition of 3 percent of the BaSO$_4$ nanoparticles in the poly(o-anisidine) matrix was practically the same as in the 1 percent, indicating that the 3 percent addition of the BaSO$_4$ nanoparticles improved the properties of the polymer. The peak at 2116 cm^{-1} indicated the interaction of the BaSO$_4$ nanoparticles with the polymer matrix when 3 percent of the BaSO$_4$ nanoparticles was added to the polymer matrix. The peak at 1568 cm^{-1} was linked to the quinoid ring's (C=C) stretching modes.

The characteristic peak was observed at 1266 cm^{-1} after the addition of 5% of the BaSO$_4$ nanoparticles in the polymer matrix, as shown in Figure 2d, indicating that the 5 percent addition of the BaSO$_4$ nanoparticles increased the features of the polymer. The peak at 1628 cm^{-1} indicated the interaction of the BaSO$_4$ nanoparticles with the polymer matrix when 5 percent of the BaSO$_4$ nanoparticles was added to the polymer matrix. The peak at 1568 cm^{-1} was linked to the quinoid ring's (C=C) stretching modes.

The peak at 3511 cm^{-1} was, again, found by adding 7% of the BaSO$_4$ nanoparticles to the polymer matrix, as shown in Figure 2e, which was deep but not as wide as in the prior cases, indicating that the 7% addition of the BaSO$_4$ nanoparticles improved the polymer properties. Furthermore, the spectrum revealed that when 7% of the BaSO$_4$ nanoparticles are added to the polymer matrix, a peak at 1628 cm^{-1} indicates that the BaSO$_4$ nanoparticles are bound to the polymer matrix. The quinoid ring's (C=C) stretching modes were attributed to the peak at 1589 cm^{-1}.

Figure 2f shows the spectra of 10% of the BaSO$_4$ nanoparticles. It can be seen that due to the addition of 10% of the BaSO$_4$ nanoparticles in the poly(o-anisidine), a sharp and comparatively deep peak appeared at 3511 cm^{-1}, which showed the addition of 10% of BaSO$_4$ NPs, which increased the features of the poly(o-anisidine) differently. The peak that formed at 1626 cm^{-1} after the addition of 10% of the BaSO$_4$ nanoparticles to the poly(o-anisidine) matrix illustrated the binding of BaSO$_4$ nanoparticles to the polymer matrix. The quinoid-ring stretching modes (C=C) were exhibited at 1589 cm^{-1} [30].

3.2. UV-Visible Spectroscopy

The POA/BaSO$_4$ nanocomposite was characterized to understand the absorption behavior of the POA/BaSO$_4$ nanocomposite. To analyze the absorption behavior of composite materials, the materials were scanned in the wavelength range of 200–800 nm at the rate of 400 nm/min under a UV-visible spectrophotometer. The spectrophotometer was furnished with UV Winlab programming to record and analyze the information. A blank reference was used to conduct baseline correction on the spectrophotometer. The absorption spectra of the pure POA and POA/BaSO$_4$ nanocomposites are shown in Figure 3. The significant absorption peak appeared at 273 nm, related to π-π* aromatic C=C band transitions and n-π* aromatic C=N band transitions. The o-anisidine monomer-to-monomer attachment was responsible for conjugation and polarity. The interaction of the BaSO$_4$ nanofiller and POA matrix resulted in a change in the absorption behavior of the composites. The absorption behavior of the composites was significantly different than that of the pure polymer, as shown in Figure 3a. There was a notable shifting of the absorption peak towards higher wavelengths, resulting in the marvel of the red shift at ~520 nm. Moreover, the absorption intensity of the composites was much higher than that of the pure polymer, indicating the role of barium-sulfate filler in improving the absorption properties of the POA/BaSO$_4$ composites. The other composites also showed similar absorption behaviors, as depicted in Figure 3b–f. The absorption intensity was enhanced with the increase in the barium-sulfate-load percentage in the composites [17].

3.3. TEM Analysis

By using TEM micrograms at various magnifications to better examine the morphology of the BaSO$_4$ nanoparticles, Figure 4 was created. Due to their strong magnetic interaction and large surface area due to their nanoscale size, the produced BaSO$_4$ nanoparticles were found to be crystalline and include agglomerated irregular particles with an average size of 30–60 nm (Figure 4).

3.4. Conductivity Test

The POA/BaSO$_4$ nanocomposites were subjected to the measurement of electrical conductivity by the conductivity meter. The electrical conductivity of the nanocomposites depends upon the properties of the individual components and also depends on the temperature factor.

Table 2 shows the electrical conductivity of POA at 1080 S/m. The electrical conductivities of the POA/BaSO$_4$ nanocomposites determined were lower than the pure POA. The nanocomposite with 1% of loading of the BaSO$_4$ nanoparticles showed conductivity at 992 S/m. Other nanocomposites containing 3%, 5%, 7%, and 10% of loading of BaSO$_4$ NPs exhibited conductivity at 919, 843, 784, and 578 S/m respectively (Figure 5). These results not only confirm the presence of nanoparticles but also show that the increase in the percentage of the BaSO$_4$ nanoparticles in the composites results in a decrease in the conductivity of the materials [17]. The addition of barium sulfate in poly(o-anisidine) allowed the composite properties to be tunable, which can be used for designing the sensor materials for future use.

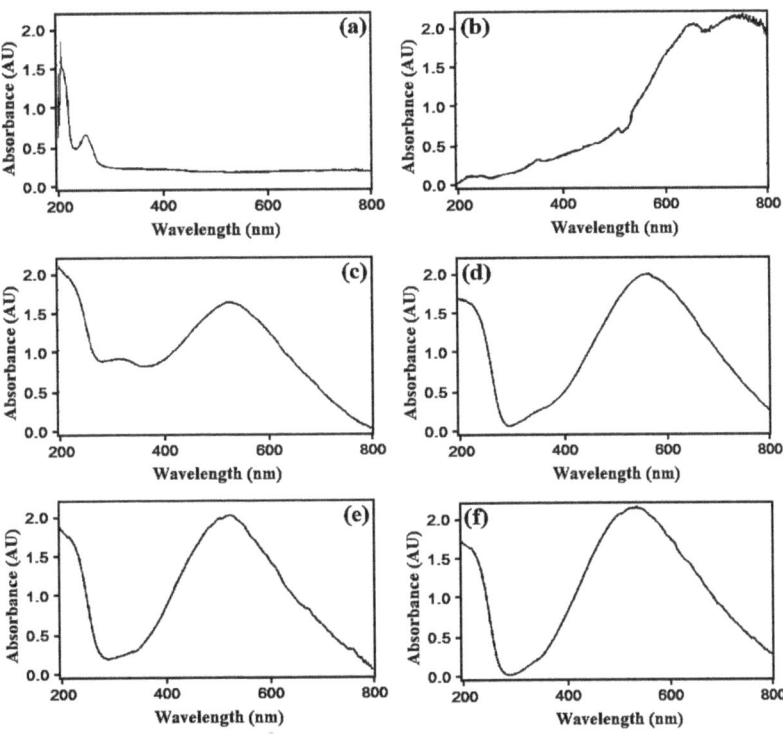

Figure 3. UV-visible spectra of (**a**) pure POA, (**b**) 1% of POA/BaSO$_4$, (**c**) 3% of POA/BaSO$_4$, (**d**) 5% of POA/BaSO$_4$, (**e**) 7% of POA/BaSO$_4$, and (**f**) 10% of POA/BaSO$_4$.

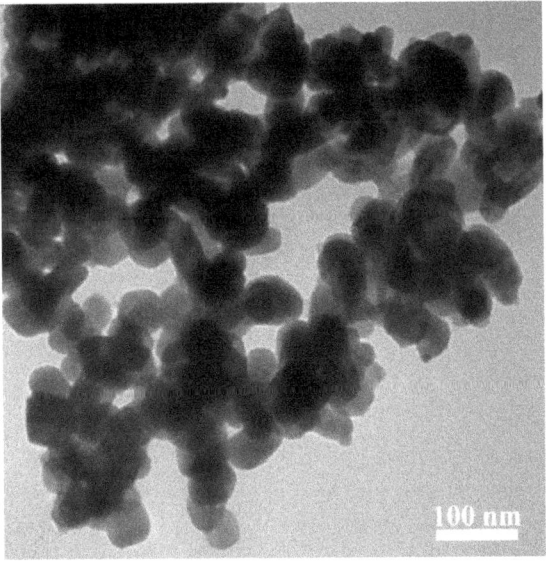

Figure 4. TEM micrograph of BaSO$_4$ nanoparticles.

Table 2. Scheme of conductivity test.

Sr. No.	Sample Name	% BaSO₄ NPs	Temperature (°C)	Conductivity (S/cm)
1	Poly(o-anisidine)	0%	25 °C	1080
2	POA/BaSO₄	1%	25 °C	992
3	POA/BaSO₄	3%	25 °C	919
4	POA/BaSO₄	5%	25 °C	843
5	POA/BaSO₄	7%	25 °C	784
6	POA/BaSO₄	10%	25 °C	578

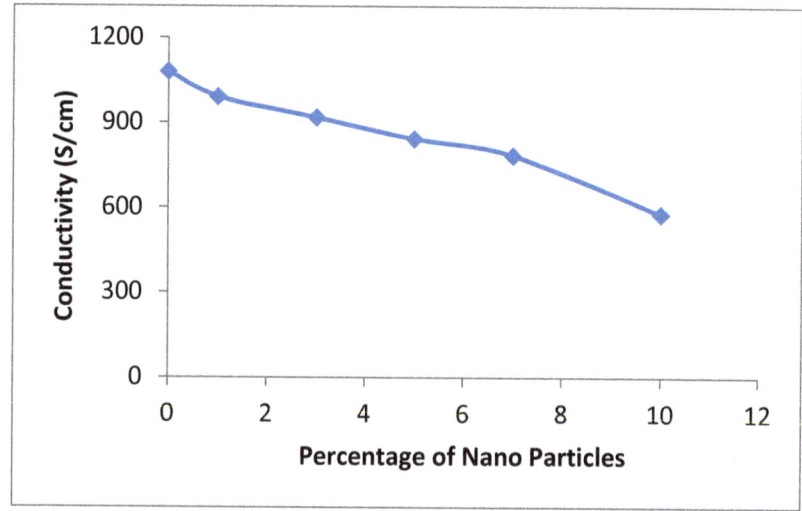

Figure 5. The plot of conductivity and nanoparticles' percentage relationship.

3.5. Disc-Diffusion Method

The synthesized poly(o-anisidine)/BaSO₄ nanocomposites were evaluated for their antibacterial activity against different strains of bacteria using the disc-diffusion technique. It was found that poly(o-anisidine)/BaSO₄ nanocomposites are effective for inhibiting the bacterial strains. The composites showed significant efficacy against *Pseudomonas aeruginosa* (Gram-negative) and *Staphylococcus aureus* (Gram-positive). The antibacterial activity of the BaSO₄ nanocomposites is noteworthy and reported in Table 3.

Table 3. Inhibition zone of antibacterial test of the BaSO₄ nanocomposite.

Sample No.	Nanocomposite Concentrations (% w/w)	Inhibition Zone Diameter (mm) of *Staphylococcus aureus*	Inhibition Zone Diameter (mm) of *Pseudomonas aeruginosa*
1	1	-	-
2	3	-	-
3	5	0.6	0.5
4	7	0.8	0.8
5	10	0.9	0.7

The POA/BaSO₄ nanocomposite that inhibited *Pseudomonas aeruginosa* growth was found to be optimal at 7% (Table 3 and Figure 6). Similarly, *Staphylococcus aureus* growth was inhibited by the POA/BaSO$_4$ nanocomposites. The maximum inhibition was associated with 10% of the POA/BaSO$_4$ nanocomposites against Gram-positive bacteria (*Staphylococcus aureus*). Nanoparticles have a small size, a larger surface area than bulky particles, and better penetration; therefore, they have a greater bactericidal effect. The specific process by which nanoparticles enter bacteria is unclear; however, studies have shown that treating bacterial cultures with nanoparticles alters the shape of the membranes and significantly increases membrane permeability, impairing normal transport through the plasma membrane. Bacterial cells die when they are unable to manage the transport across the plasma membrane. It was believed that due to their small size, barium-sulphate nanoparticles enter the bacterial cell membrane and link to functional groups of proteins, causing denaturation. They are also hypothesized to cause bacterial cell death by interacting with phosphorus and sulphur compounds, such as DNA. Total bacterial inhibition was proportional to BaSO$_4$-nanoparticle concentrations [30]. It was found that the BaSO$_4$ nanoparticles have a significant antibacterial activity, which may be utilized for practical uses.

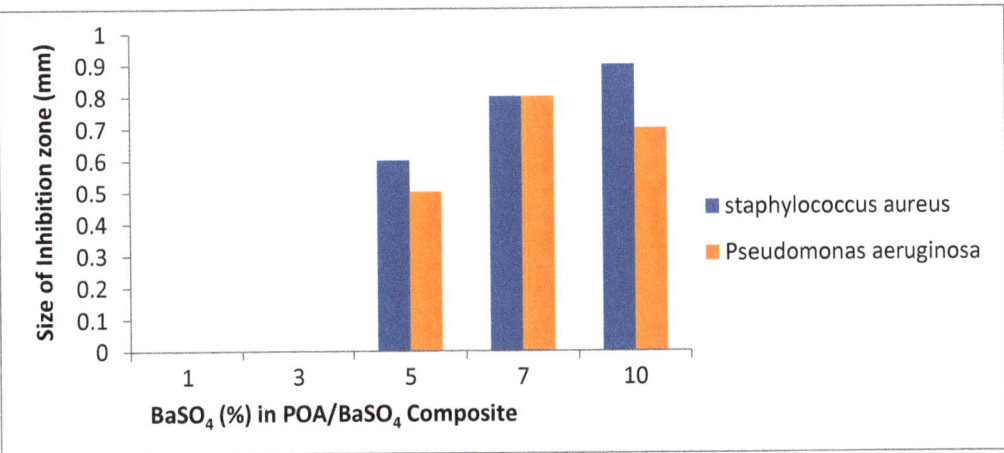

Figure 6. The graph shows the size inhibition zone of BaSO$_4$ nanoparticles in nanocomposites against load percentage.

4. Conclusions

The current study provides significant results regarding the antibacterial and conductivity properties of POA/BaSO$_4$ nanocomposites. Poly(*o*-anisidine)/BaSO$_4$ nanocomposites were synthesized by the oxidative polymerization of *o*-anisidine and variable percentages of BaSO$_4$ nanoparticles. The nanocomposites were tested by using different characterization techniques, i.e., FTIR, UV, and conductivity and disc-diffusion methods. The FTIR results show the strong communication among poly(*o*-anisidine) and BaSO$_4$ nanoparticles, and the UV-visible results indicate good absorption behavior. Moreover, the POA/BaSO$_4$ nanocomposites exhibited significant bactericidal potential against *Pseudomonas aeruginosa* and *Staphylococcus aureus*, and a tunable electrical conductivity feature. Therefore, the versatile properties of poly(*o*-anisidine)/BaSO$_4$ nanocomposites identify these materials as future candidates for biomedical, electrical device, and sensor applications.

Author Contributions: The manuscript was written with the contributions of all authors. All authors have approved the final version of the manuscript.

Funding: This research was funded by Princess Nourah bint Abdulrahman University Researchers Supporting Project number (PNURSP2022R134), Princess Nourah bint Abdulrahman University, Riyadh, Saudi Arabia. The authors would like to thank the Deanship of Scientific Research at Umm Al-Qura University for supporting this work by grant code (22UQU4331312DSR06). The authors extend their appreciation to the Research Center at AlMaarefa University for funding this work under TUMA project agreement number (TUMA-2021-22).

Institutional Review Board Statement: Not applicable.

Informed Consent Statement: Not applicable.

Data Availability Statement: The data will be available on request.

Acknowledgments: This research was funded by Princess Nourah bint Abdulrahman University Researchers Supporting Project number (PNURSP2022R134), Princess Nourah bint Abdulrahman University, Riyadh, Saudi Arabia. The authors would like to thank the Deanship of Scientific Research at Umm Al-Qura University for supporting this work by grant code (22UQU4331312DSR06). The authors extend their appreciation to the Research Center at AlMaarefa University for funding this work under TUMA project agreement number (TUMA-2021-22).

Conflicts of Interest: The authors declare no conflict of interest.

References

1. Sebastian, J.; Samuel, J.M. Recent advances in the applications of substituted polyanilines and their blends and composites. *Polym. Bull.* **2019**, *77*, 6641–6669. [CrossRef]
2. Thakur, S.; Patil, P. Enhanced LPG sensing-performance at room temperature of poly(o-anisidine)–CeO2 nanocomposites. *RSC Adv.* **2016**, *6*, 45768–45782. [CrossRef]
3. Bonakdar, M.; Hosseini, S.R.; Ghasemi, S. Poly(o-Anisidine)/carbon nanotubes/graphene nanocomposite as a novel and cost-effective supercapacitor material. *Mater. Sci. Eng. B* **2021**, *267*, 115099. [CrossRef]
4. Alijani, H.; Abdouss, M.; Khataei, H. Efficient photocatalytic degradation of toxic dyes over BiFeO3/CdS/rGO nanocomposite under visible light irradiation. *Diam. Relat. Mater.* **2022**, *122*, 108817. [CrossRef]
5. Ahmad, M.N.; Nadeem, S.; Javed, M.; Iqbal, S.; Khan, M.; Alsaab, H.O.; Awwad, N.S.; Ibrahim, H.A.; Mohyuddin, A. Photocatalytic Degradation of Yellow-50 Using Zno/Polyorthoethylaniline Nanocomposites. *JOM* **2022**, *74*, 2106–2112. [CrossRef]
6. Rajkumar, R.; Vedhi, C. Preparation, characterization and anticorrosion behavior of Poly(o-anisidine)-SiO2 nanocomposites on mild steel. *Mater. Today Proc.* **2022**, *48*, 169–173. [CrossRef]
7. Sangjan, S.; Wisasa, K.; Deddeaw, N. Enhanced photodegradation of reactive blue dye using Ga and Gd as catalyst in reduced graphene oxide-based TiO2 composites. *Mater. Today Proc.* **2019**, *6*, 19–23. [CrossRef]
8. De Alvarenga, G.; Hryniewicz, B.M.; Jasper, I.; Silva, R.J.; Klobukoski, V.; Costa, F.S.; Cervantes, T.N.M.; Amaral, C.D.B.; Schneider, J.T.; Bach-Toledo, L.; et al. Recent trends of micro and nanostructured conducting polymers in health and environmental applications. *J. Electroanal. Chem.* **2020**, *879*, 114754. [CrossRef]
9. Pandian, P.; Kalimuthu, R.; Arumugam, S.; Kannaiyan, P. Solid phase mechanochemical synthesis of Poly(o-anisidine) protected Silver nanoparticles for electrochemical dopamine sensor. *Mater. Today Commun.* **2021**, *26*, 102191. [CrossRef]
10. Jadoun, S.; Riaz, U.; Yáñez, J.; Pal Singh Chauhan, N. Synthesis, characterization and potential applications of Poly(o-phenylenediamine) based copolymers and Nanocomposites: A comprehensive review. *Eur. Polym. J.* **2021**, *156*, 110600. [CrossRef]
11. Karbovnyk, I.; Klym, H.; Piskunov, S.; Popov, A.A.; Chalyy, D.; Zhydenko, I.; Lukashevych, D. The impact of temperature on electrical properties of polymer-based nanocomposites. *Low Temp. Phys.* **2020**, *46*, 1231–1234. [CrossRef]
12. Aksimentyeva, O.I.; Savchyn, V.P.; Dyakonov, V.P.; Piechota, S.; Horbenko, Y.Y.; Opainych, I.Y.; Demchenko, P.Y.; Popov, A.; Szymczak, H. Modification of Polymer-Magnetic Nanoparticles by Luminescent and Conducting Substances. *Mol. Cryst. Liq. Cryst.* **2014**, *590*, 35–42. [CrossRef]
13. Tsebriienko, T.; Popov, A.I. Effect of Poly(Titanium Oxide) on the Viscoelastic and Thermophysical Properties of Interpenetrating Polymer Networks. *Crystals* **2021**, *11*, 794. [CrossRef]
14. Ahmad, M.N.; Hussain, A.; Anjum, M.N.; Hussain, T.; Mujahid, A.; Khan, M.H.; Ahmed, T. Synthesis and characterization of a novel chitosan-grafted-polyorthoethylaniline biocomposite and utilization for dye removal from water. *Open Chem.* **2020**, *18*, 843–849. [CrossRef]
15. Akyüz, D. rGO-TiO2-CdO-ZnO-Ag photocatalyst for enhancing photocatalytic degradation of methylene blue. *Opt. Mater.* **2021**, *116*, 111090. [CrossRef]
16. Landa, R.A.; Calvino, J.J.; López-Haro, M.; Antonel, P.S. Nanostructure, compositional and magnetic studies of Poly(aniline)–CoFe2O4 nanocomposites. *Nano-Struct. Nano-Objects* **2021**, *28*, 100808. [CrossRef]

17. Ahmad, M.N.; Rafique, F.; Nawaz, F.; Farooq, T.; Anjum, M.N.; Hussain, T.; Hassan, S.; Batool, M.; Khalid, H.; Shehzad, K. Synthesis of antibacterial poly (o-chloroaniline)/chromium hybrid composites with enhanced electrical conductivity. *Chem. Cent. J.* **2018**, *12*, 46. [CrossRef]
18. Nadeem, S.; Iqbal, S.; Javed, M.; Ahmad, M.N.; Alsaab, H.O.; Awwad, N.S.; Ibrahim, H.A.; Ibrar, A.; Mohyuddin, A.; Haroon, S.M. Kinetic and Isothermal Studies on the Adsorptive Removal of Direct Yellow 12 Dye from Wastewater Using Propionic Acid Treated Bagasse. *Chem. Sel.* **2021**, *6*, 12146–12152. [CrossRef]
19. Iqbal, S.; Bahadur, A.; Saeed, A.; Zhou, K.; Shoaib, M.; Waqas, M. Electrochemical performance of 2D polyaniline anchored CuS/Graphene nano-active composite as anode material for lithium-ion battery. *J. Colloid Interface Sci.* **2017**, *502*, 16–23. [CrossRef]
20. Bahadur, A.; Iqbal, S.; Shoaib, M.; Saeed, A.J.D.T. Electrochemical study of specially designed graphene-Fe$_3$O$_4$-polyaniline nanocomposite as a high-performance anode for lithium-ion battery. *Dalton Trans.* **2018**, *47*, 15031–15037. [CrossRef]
21. Iqbal, S.; Nadeem, S.; Bahadur, A.; Javed, M.; Ahmad, Z.; Ahmad, M.N.; Shoaib, M.; Liu, G.; Mohyuddin, A.; Raheel, M. The effect of Ni-doped ZnO NPs on the antibacterial activity and degradation rate of polyacrylic acid-modified starch nanocomposite. *Jom* **2021**, *73*, 380–386. [CrossRef]
22. Lv, X.; Li, J.; Xu, L.; Zhu, X.; Tameev, A.; Nekrasov, A.; Kim, G.; Xu, H.; Zhang, C. Colorless to Multicolored, Fast Switching, and Highly Stable Electrochromic Devices Based on Thermally Cross-Linking Copolymer. *ACS Appl. Mater. Interfaces* **2021**, *13*, 41826–41835. [CrossRef] [PubMed]
23. Zaera, F. Designing Sites in Heterogeneous Catalysis: Are We Reaching Selectivities Competitive With Those of Homogeneous Catalysts? *Chem. Rev.* **2022**, *122*, 8594–8757. [CrossRef]
24. Wu, H.; Lin, S.; Chen, C.; Liang, W.; Liu, X.; Yang, H. A new ZnO/rGO/polyaniline ternary nanocomposite as photocatalyst with improved photocatalytic activity. *Mater. Res. Bull.* **2016**, *83*, 434–441. [CrossRef]
25. Boutaleb, N.; Chouli, F.; Benyoucef, A.; Zeggai, F.Z.; Bachari, K. A comparative study on surfactant cetyltrimethylammoniumbromide modified clay-based poly(p-anisidine) nanocomposites: Synthesis, characterization, optical and electrochemical properties. *Polym. Compos.* **2021**, *42*, 1648–1658. [CrossRef]
26. Peng, T.; Xiao, R.; Rong, Z.; Liu, H.; Hu, Q.; Wang, S.; Li, X.; Zhang, J. Polymer Nanocomposite-based Coatings for Corrosion Protection. *Chem. Asian J.* **2020**, *15*, 3915–3941. [CrossRef] [PubMed]
27. Elugoke, S.E.; Adekunle, A.S.; Fayemi, O.E.; Mamba, B.B.; Nkambule, T.T.I.; Sherif, E.-S.M.; Ebenso, E.E. Progress in electrochemical detection of neurotransmitters using carbon nanotubes/nanocomposite based materials: A chronological review. *Nano Sel.* **2020**, *1*, 561–611. [CrossRef]
28. Bahadur, A.; Saeed, A.; Shoaib, M.; Iqbal, S.; Anwer, S.J.J.o.A.P.S. Modulating the burst drug release effect of waterborne polyurethane matrix by modifying with polymethylmethacrylate. *J. Appl. Polym. Sci.* **2019**, *136*, 47253. [CrossRef]
29. Nadeem Ahmad, M.; Anjum, M.N.; Nawaz, F.; Iqbal, S.; Saif, M.J.; Hussain, T.; Mujahid, A.; Farooq, M.U.; Nadeem, M.; Rahman, A. Synthesis and antibacterial potential of hybrid nanocomposites based on polyorthochloroaniline/copper nanofiller. *Polym. Compos.* **2018**, *39*, 4524–4531. [CrossRef]
30. Hussain, T.; Jabeen, S.; Shehzad, K.; Mujahid, A.; Ahmad, M.N.; Farooqi, Z.H.; Raza, M.H. Polyaniline/silver decorated-MWCNT composites with enhanced electrical and thermal properties. *Polym. Compos.* **2018**, *39*, E1346–E1353. [CrossRef]
31. Merangmenla; Nayak, B.; Baruah, S.; Puzari, A. 1D copper (II) based coordination polymer/PANI composite fabrication for enhanced photocatalytic activity. *J. Photochem. Photobiol. A Chem.* **2022**, *427*, 113803. [CrossRef]
32. Potle, V.D.; Shirsath, S.R.; Bhanvase, B.A.; Saharan, V.K. Sonochemical preparation of ternary rGO-ZnO-TiO$_2$ nanocomposite photocatalyst for efficient degradation of crystal violet dye. *Optik* **2020**, *208*, 164555. [CrossRef]
33. Hussain, A.; Ahmad, M.N.; Jalal, F.; Yameen, M.; Falak, S.; Noreen, S.; Naz, S.; Nazir, A.; Iftikhar, S.; Soomro, G.A. Investigating the Antibacterial Activity of POMA Nanocomposites. *Pol. J. Environ. Stud.* **2019**, *28*, 4191–4198. [CrossRef]
34. Hussain, T.; Ahmad, M.N.; Nawaz, A.; Mujahid, A.; Bashir, F.; Mustafa, G. Surfactant incorporated Co nanoparticles polymer composites with uniform dispersion and double percolation. *J. Chem.* **2017**, *2017*, 7191590. [CrossRef]
35. Monga, D.; Basu, S. Enhanced photocatalytic degradation of industrial dye by g-C3N4/TiO$_2$ nanocomposite: Role of shape of TiO$_2$. *Adv. Powder Technol.* **2019**, *30*, 1089–1098. [CrossRef]
36. Li, X.; Peoples, J.; Yao, P.; Ruan, X. Ultrawhite BaSO$_4$ Paints and Films for Remarkable Daytime Subambient Radiative Cooling. *ACS Appl. Mater. Interfaces* **2021**, *13*, 21733–21739. [CrossRef]
37. Li, M.; Mann, S. Emergence of Morphological Complexity in BaSO$_4$ Fibers Synthesized in AOT Microemulsions. *Langmuir* **2000**, *16*, 7088–7094. [CrossRef]
38. Atone, M.S.; Dhoble, S.J.; Moharil, S.V.; Dhopte, S.M.; Muthal, P.L.; Kondawar, V.K. Luminescence in BaSO$_4$. *Eu Radiat. Eff. Defects Solids* **1993**, *127*, 225–230. [CrossRef]
39. Cui, W.; Chen, L.; Li, J.; Zhou, Y.; Sun, Y.; Jiang, G.; Lee, S.; Dong, F. Ba-vacancy induces semiconductor-like photocatalysis on insulator BaSO$_4$. *Appl. Catal. B: Environ.* **2019**, *253*, 293–299. [CrossRef]
40. Ugur, N.; Bilici, Z.; Ocakoglu, K.; Dizge, N. Synthesis and characterization of composite catalysts comprised of ZnO/MoS2/rGO for photocatalytic decolorization of BR 18 dye. *Colloids Surf. A Physicochem. Eng. Asp.* **2021**, *626*, 126945. [CrossRef]
41. Wu, G.; Zhou, H.; Zhu, S. Precipitation of barium sulfate nanoparticles via impinging streams. *Mater. Lett.* **2007**, *61*, 168–170. [CrossRef]
42. Birhan, D.; Tekin, D.; Kiziltas, H. Thermal, photocatalytic, and antibacterial properties of rGO/TiO$_2$/PVA and rGO/TiO$_2$/PEG composites. *Polym. Bull.* **2021**, *79*, 2585–2602. [CrossRef]

43. Aninwene, G.E.; Stout, D.; Yang, Z.; Webster, J.T. Nano-BaSO$_4$: A novel antimicrobial additive to pellethane. *Int. J. Nanomed.* **2013**, *8*, 1197–1205.
44. Ahmad, M.N.; Nadeem, S.; Hassan, S.U.; Jamil, S.; Javed, M.; Mohyuddin, A.; Raza, H. UV/VIS absorption properties of metal sulphate polymer nanocomposites. *Dig. J. Nanomater. Biostructures* **2021**, *16*, 1557–1563.

Article

Synthesis, Characterization, and Biological Evaluation of Novel *N*-{4-[(4-Bromophenyl)sulfonyl]benzoyl}-*L*-valine Derivatives

Theodora-Venera Apostol [1,*], Mariana Carmen Chifiriuc [2], Laura-Ileana Socea [1], Constantin Draghici [3], Octavian Tudorel Olaru [1,*], George Mihai Nitulescu [1], Diana-Carolina Visan [1], Luminita Gabriela Marutescu [2], Elena Mihaela Pahontu [1], Gabriel Saramet [1] and Stefania-Felicia Barbuceanu [1]

[1] Faculty of Pharmacy, "Carol Davila" University of Medicine and Pharmacy, 6 Traian Vuia Street, District 2, 020956 Bucharest, Romania
[2] Department of Botany and Microbiology, Faculty of Biology, University of Bucharest, 1-3 Intrarea Portocalelor, District 6, 60101 Bucharest, Romania
[3] "Costin D. Nenițescu" Centre of Organic Chemistry, Romanian Academy, 202 B Splaiul Independenței, District 6, 060023 Bucharest, Romania
* Correspondence: theodora.apostol@umfcd.ro (T.-V.A.); octavian.olaru@umfcd.ro (O.T.O.)

Citation: Apostol, T.-V.; Chifiriuc, M.C.; Socea, L.-I.; Draghici, C.; Olaru, O.T.; Nitulescu, G.M.; Visan, D.-C.; Marutescu, L.G.; Pahontu, E.M.; Saramet, G.; et al. Synthesis, Characterization, and Biological Evaluation of Novel *N*-{4-[(4-Bromophenyl)sulfonyl]benzoyl}-*L*-valine Derivatives. *Processes* **2022**, *10*, 1800. https://doi.org/10.3390/pr10091800

Academic Editors: Iliyan Ivanov and Stanimir Manolov

Received: 17 August 2022
Accepted: 3 September 2022
Published: 7 September 2022

Publisher's Note: MDPI stays neutral with regard to jurisdictional claims in published maps and institutional affiliations.

Copyright: © 2022 by the authors. Licensee MDPI, Basel, Switzerland. This article is an open access article distributed under the terms and conditions of the Creative Commons Attribution (CC BY) license (https://creativecommons.org/licenses/by/4.0/).

Abstract: In this article, we present the design and synthesis of novel compounds, containing in their molecules an *L*-valine residue and a 4-[(4-bromophenyl)sulfonyl]phenyl moiety, which belong to *N*-acyl-α-amino acids, 4*H*-1,3-oxazol-5-ones, 2-acylamino ketones, and 1,3-oxazoles chemotypes. The synthesized compounds were characterized through elemental analysis, MS, NMR, UV/VIS, and FTIR spectroscopic techniques, the data obtained being in accordance with the assigned structures. Their purities were verified by reversed-phase HPLC. The new compounds were tested for antimicrobial action against bacterial and fungal strains, for antioxidant activity by DPPH, ABTS, and ferric reducing power assays, and for toxicity on freshwater cladoceran *Daphnia magna* Straus. Furthermore, in silico studies were performed concerning the potential antimicrobial effect and toxicity. The results of antimicrobial activity, antioxidant effect, and toxicity assays, as well as of in silico analysis revealed a promising potential of *N*-{4-[(4-bromophenyl)sulfonyl]benzoyl}-*L*-valine and 2-{4-[(4-bromophenyl)sulfonyl]phenyl}-4-isopropyl-4*H*-1,3-oxazol-5-one for developing novel antimicrobial agents to fight Gram-positive pathogens, and particularly *Enterococcus faecium* biofilm-associated infections.

Keywords: *N*-acyl-*L*-valine; 4*H*-1,3-oxazol-5-one; 2-acylamino ketone; 1,3-oxazole; diphenyl sulfone scaffold; drug design; in silico studies; antimicrobial and antibiofilm actions; antioxidant effect; alternative toxicity testing

1. Introduction

Antimicrobial resistance represents a significant threat to global health and development today. Without effective antimicrobials (e.g., antibiotics, antifungals, and antiparasitics) and antiviral drugs, the success of modern medicine in preventing and treating infections would be at increased risk [1]. The discovery and development of new antimicrobial agents are therefore of considerable importance and one of the main focuses of today's scientific community is the synthesis and identification of potent active substances.

An exhaustive survey of the literature on natural and synthetic oxazole-based molecules shows that they have numerous biological properties, which include antibacterial, antifungal, anti-inflammatory, antioxidant, and cytotoxic activities [2–11]. For example, a physiologically important active cyclic peptide with a 1,3-oxazole ring is dalfopristin, a semi-synthetic streptogramin A antibiotic analog marketed under the trade name Synercid in the combination with quinupristin, a streptogramin B derivative. The structures of some representative antimicrobial agents containing the 1,3-oxazole skeleton are shown in Figure 1.

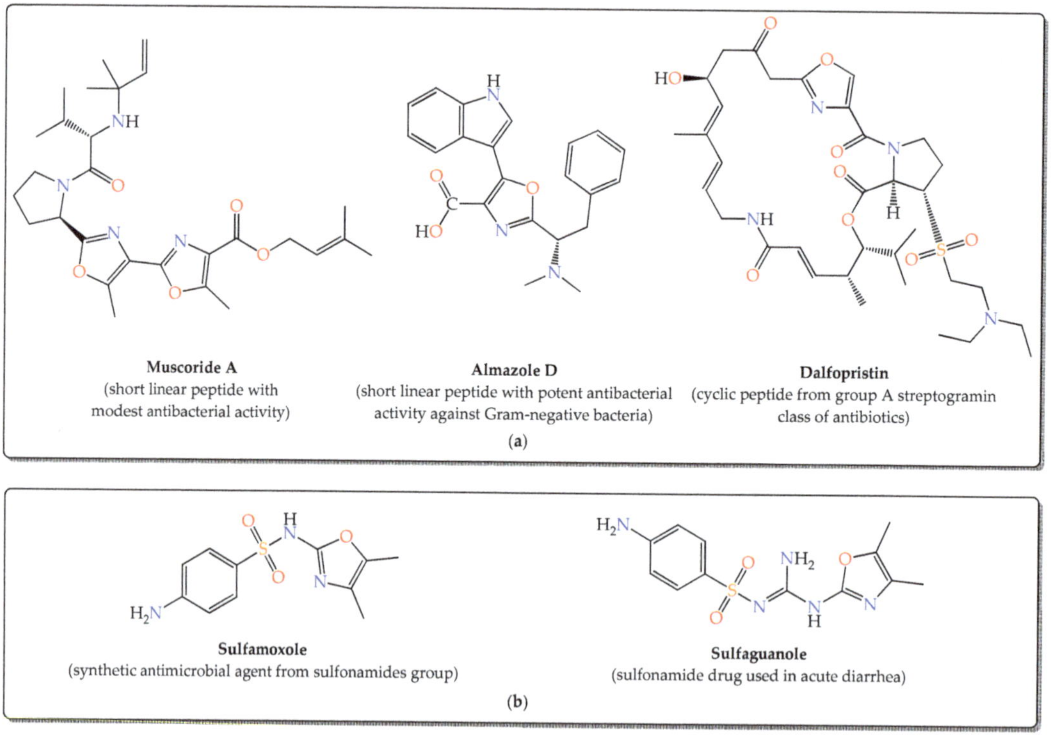

Figure 1. Structures of some representative bioactive compounds sharing the 1,3-oxazole scaffold as antimicrobial agents: (**a**) Natural or semi-synthetic products; (**b**) Synthetic compounds.

The isolable 5-keto-tautomers of 1,3-oxazol-5-ols are 5-oxo-4,5-dihydro-1,3-oxazoles, known as 4H-1,3-oxazol-5-ones or 2-oxazolin-5-ones, which are also mentioned for their antimicrobial, cytotoxic, antiprotozoal, and antiviral properties [12–15]. Moreover, the N-acylated α-amino acids are reported to display many pharmaceutical activities, such as antimicrobial, antiviral, anticancer, mucolytic, antioxidant, and antihypertensive effects [16–22], and the 2-acylamino ketones present antiviral, anti-inflammatory, antithrombotic, and antihypertensive actions [23–28].

Furthermore, diaryl sulfones represent an important class of bioactive compounds with diverse biological properties, including antimicrobial and antioxidant actions [29–37]. The structural prototype of this class is dapsone (4,4′-sulfonyldianiline, 4,4′-diaminodiphenyl sulfone, DDS), which is used alone or as part of a multi-drug regimen in the treatment or prophylaxis of certain infectious diseases, such as leprosy (also known as Hansen's disease). Currently, the therapeutic potential of organic compounds with a sulfonyl group in their molecules is being studied extensively [38,39]. In this regard, a new drug candidate from the organosulfones class is masupirdine (SUVN-502), a 1-[(2-bromophenyl)sulfonyl]-1H-indole derivative with a selective 5-HT$_6$ receptor antagonist effect, developed for the symptomatic treatment of Alzheimer's disease [40].

Based on the promising therapeutic potential of these scaffolds and in the continuation of our research [41–46], we designed, obtained, and characterized novel L-valine-derived analogs with a 4-[(4-bromophenyl)sulfonyl]phenyl fragment and also evaluated their in silico and in vitro antimicrobial, antioxidant, and toxicity properties with the aim of identifying new biologically active compounds. The used synthesis methodology enables synthetic chemical diversification with the potential to prepare drug-like compounds.

The in silico approach to the biological activity of the new bromine compounds allowed the assessment of their toxicity and potential antimicrobial effect.

The in vitro antimicrobial testing of the new compounds included both qualitative screening (measuring the diameters of the zones of inhibition of microbial growth) and quantitative analysis (determining the values of minimal inhibitory concentration), as well as the assay of the antibiofilm action (determining the values of minimal biofilm inhibitory concentration). Particular emphasis was placed on evaluating the inhibition of microbial adhesion to surfaces as a possible means of reducing the burden of biofilm-associated infections.

The in vitro antioxidant activity of newly obtained compounds was investigated colorimetrically according to three electron transfer-based methods, namely 2,2-diphenyl-1-picrylhydrazyl (DPPH) assay, 2,2′-azino-bis(3-ethylbenzothiazoline-6-sulfonic acid) (ABTS) radical cation test, and ferric reducing power assay.

The compounds' toxicity was assessed on *Daphnia magna*. This bioassay is a widely used method for the toxicity evaluation of natural and synthetic compounds, and it is in accordance with the "3 R's" (reduction, refinement, and replacement) concept regarding experiments on vertebrates. It can be used as a prescreening method for rats and other mammals' acute toxicity tests. The main advantage of this method is the crustacean reproduction mode by parthenogenesis, which leads to populations with lower variability than other invertebrate species [47,48].

2. Results

2.1. Drug Design Strategy

2.1.1. Structure-Based Similarity Analysis

Based on our previous research on *N*-{4-[(4-chlorophenyl)sulfonyl]benzoyl}-*L*-valine derivatives as antimicrobial and antibiofilm agents, we made modifications in the structures of the target molecules by replacing the chlorine atom with the more lipophilic bromine atom. The objective of this study was to increase the lipophilic character of the newly synthesized compounds while maintaining the substituent's electronic effects. The proposed structures **5–8** were virtually explored on the ChEMBL database to demonstrate their original character and to assess their antimicrobial potential using similar compounds.

The search for similar compounds of *N*-acyl-*L*-valine **5** and α-acylamino ketones **7a** and **7b** returned 34 results and no similar structure for 4*H*-1,3-oxazol-5-one **6** and 1,3-oxazoles **8a** and **8b**, highlighting the originality of the proposed structures. No antimicrobial results were found for the 34 structurally similar compounds recorded in ChEMBL.

The substructure search based on the common 1-bromo-4-(phenylsulfonyl)benzene scaffold returned 32 compounds with 237 registered minimal inhibitory concentration (MIC) values, ranging from 2.5 up to 1024 μg/mL. All 32 compounds share an R-{4-[(4-bromophenyl)sulfonyl]phenyl} structure (Figure 2). The wide range of the MIC values indicates that the nature of the R group is very important for the potency of the antimicrobial effect. The lipophilic character of the compounds, expressed by the clogP value, seems to confirm the hypothesis of a better antimicrobial effect for most tested bacterial strains (Figure 2). The newly designed bromo derivatives have clogP values above the average of their similar compounds, indicating a potentially improved antimicrobial effect.

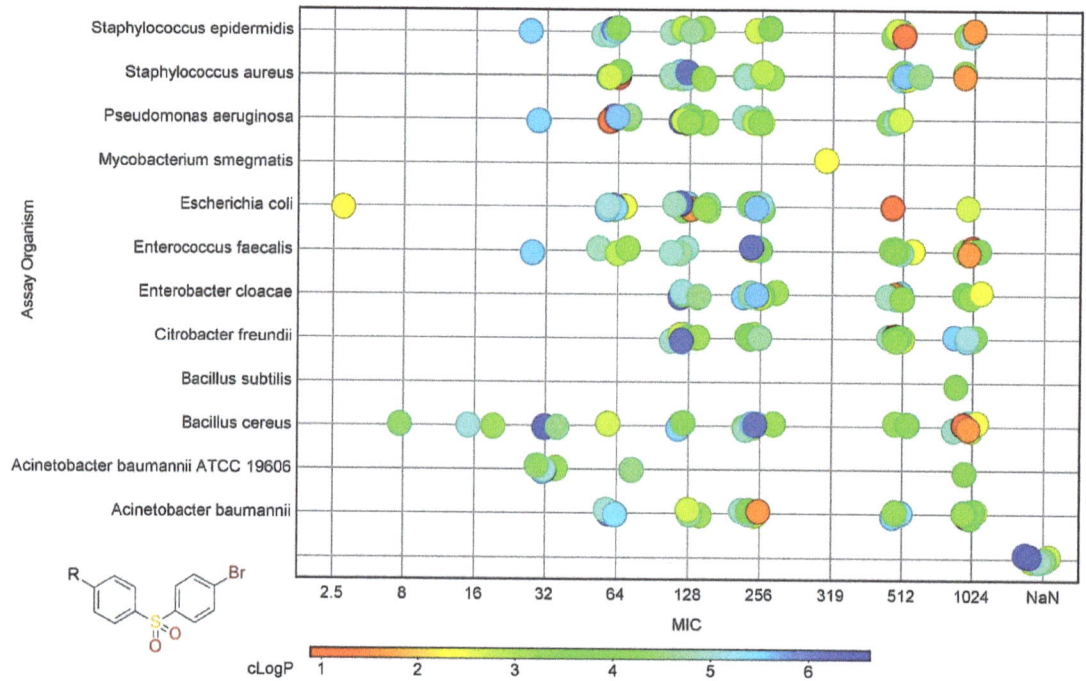

Figure 2. MIC values for the compounds resulted after the substructure search based on the 1-bromo-4-(phenylsulfonyl)benzene scaffold. NaN presents the new compounds **5**, **6**, **7a,b**, and **8a,b**.

2.1.2. PASS Prediction

The 2D chemical structure of each target compound was input into the PASS application to obtain the probability of the corresponding compound to produce (Pa) or not (Pi) a certain biological effect. The application returns an array of Pa and Pi values for a very wide variety of pharmacological actions [49]. In the case of newly synthesized compounds **5**, **6**, **7a**, **7b**, **8a**, and **8b**, the 896 listed biological activities were filtered in relation to the antibacterial effect. The predicted probabilities Pa and Pi are presented in Table 1.

Table 1. PASS predicted the probabilities that the new compounds **5**, **6**, **7a,b**, and **8a,b** to be active (Pa) as antimicrobial agents.

Target	5	6	7a	7b	8a	8b
Antibiotic glycopeptide-like	-	0.40	-	-	-	-
Antimycobacterial	0.49	0.58	0.51	0.51	0.36	0.37
Antituberculosis	0.48	0.41	0.49	0.49	0.30	0.30
Anti-infective	0.58	0.37	0.36	0.41	0.30	0.34

The two new α-acylamino ketones **7a** and **7b** were predicted to have very similar chances to produce antimycobacterial, antituberculosis, or anti-infective effects, and the corresponding values are significantly higher than those of the 1,3-oxazoles derivatives **8a** and **8b**. The Pa values reflect the likelihood of a certain effect to be produced, and not of its potency. Based on the Pa values, the addition of an aromatic fragment to N-acylated L-valine **5** has little impact on its predicted antimycobacterial and antituberculosis activities.

2.2. Chemistry

2.2.1. Synthesis of the New Compounds

The synthesis of new 4-[(4-bromophenyl)sulfonyl]benzoic acid derivatives **5–8** was performed according to the synthetic route, as outlined in Scheme 1.

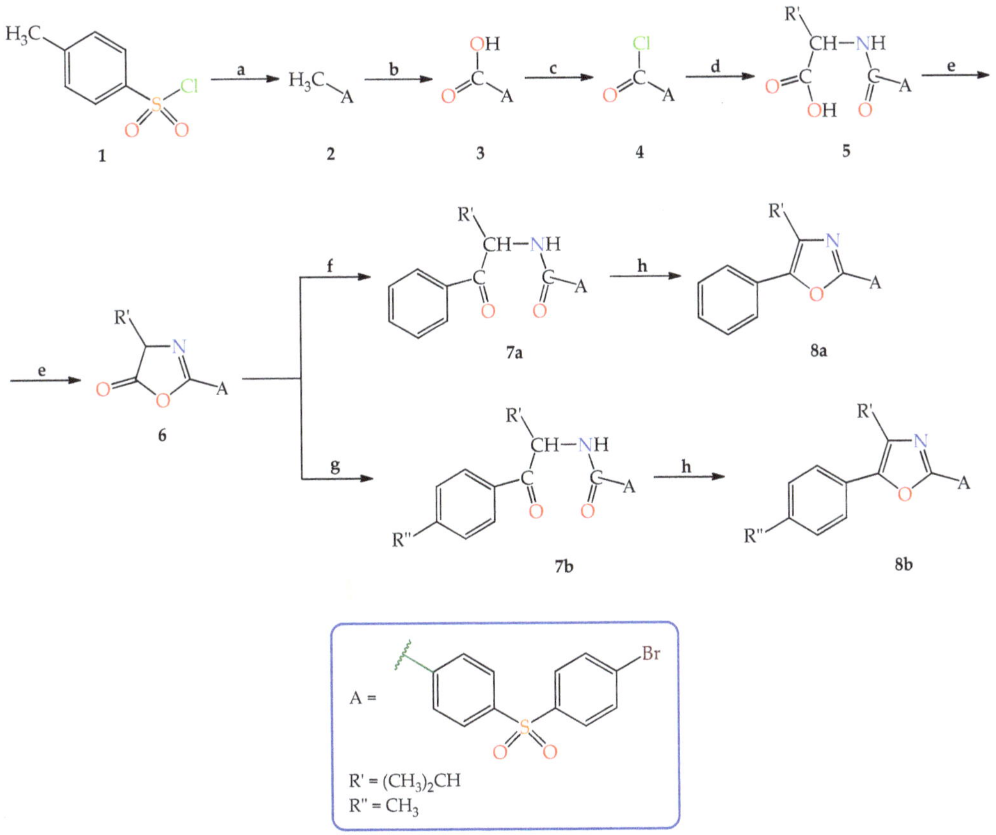

Scheme 1. Synthesis pathway for the obtainment of new N-{4-[(4-bromophenyl)sulfonyl]benzoyl}-L-valine analogs. Reagents and reaction conditions: (**a**) $C_6H_5Br/AlCl_3$, reflux; (**b**) CrO_3/CH_3COOH, reflux [50]; (**c**) $SOCl_2$, reflux, 30 h, 99% yield [46]; (**d**) (i) L-valine/NaOH, CH_2Cl_2, 0–5 °C, 30 min; (ii) rt (room temperature), 1 h; (iii) 2 N hydrochloric acid; 94% yield; (**e**) ethyl chloroformate/4-methylmorfoline, CH_2Cl_2, rt, 30 min, 90% yield; (**f**) $C_6H_6/AlCl_3$, rt, 20 h, 80% yield; (**g**) $C_6H_5CH_3/AlCl_3$, rt, 20 h, 86% yield; (**h**) $POCl_3$, reflux, 4 h, 91% (**8b**) and 94% (**8a**) yields.

The key starting material, 4-[(4-bromophenyl)sulfonyl]benzoic acid **3** was prepared by Friedel–Crafts sulfonylation of bromobenzene with commercially available tosyl chloride **1** in presence of aluminum trichloride, followed by oxidation of the sulfonation product (1-bromo-4-tosylbenzene) **2** with chromium trioxide in acetic acid [50]. The previously obtained compound **3** was then transformed by reaction with chlorinating reagent $SOCl_2$ to the corresponding acyl chloride **4** [46,51], which was used in the next step in the raw state. Subsequently, the 2-{4-[(4-bromophenyl)sulfonyl]benzamido}-3-methylbutanoic acid **5** was synthesized via Schotten–Baumann-type N-acylation of L-valine with 4-[(4-bromophenyl)sulfonyl]benzoyl chloride **4** in dichloromethane at room temperature with a reaction yield of 94%. The 2-{4-[(4-bromophenyl)sulfonyl]phenyl}-4-isopropyl-4H-1,3-oxazol-5-one **6** was produced by intramolecular cyclodehydration of compound **5** in presence of ethyl chloroformate and 4-methylmorpholine at room temperature in 90% yield.

Then, the *N*-(1-aryl-3-methyl-1-oxobutan-2-yl)-4-[(4-bromophenyl)sulfonyl]benzamides **7a,b** were obtained via Friedel–Crafts acylation of aromatic hydrocarbons (benzene, toluene) with 2-oxazolin-5-one **6** using anhydrous $AlCl_3$ as catalyst at room temperature in about 83% yield. Finally, the 5-aryl-2-{4-[(4-bromophenyl)sulfonyl]phenyl}-4-isopropyl-1,3-oxazoles **8a,b** were prepared via Robinson–Gabriel-type intramolecular cyclization reaction of the 2-acylamino ketones **7a,b** with phosphoryl trichloride at reflux in yields of 91% (**8b**) and 94% (**8a**). Structure elucidation of all newly synthesized *L*-valine derivatives was performed by means of spectral (UV/VIS, FTIR, NMR, MS) and elemental analyses.

2.2.2. Spectroscopic Characterization of the New Compounds

Various spectral analysis methods were applied to perform the detailed structural characterization of the newly obtained 4-[(4-bromophenyl)sulfonyl]benzoic acid derivatives **5–8**, the resulting data being consistent with the depicted molecular structures.

Ultraviolet and Visible Absorption Spectroscopy Data

The UV/VIS absorption spectra of novel compounds **5–8** dissolved in methanol presented bands at 202.6 nm (**5–8**), at 227.3 nm (**6**), in the 249.3–255.5 nm region (**5–8**), and at 333.9 (**8a**) or 337.4 nm (**8b**). When acetonitrile was used as solvent, the first peak was shifted in the interval of 195.3–198.0 nm (**5–8**), while the rest of the bands were recorded at approximately the same wavelength values: at 228.2 nm (for **5**, this peak appearing as a "shoulder" when methanol was used as solvent) or 229.1 nm (**6**), in the range of 248.6–254.8 nm (**5–8**), and at 335.6 (**8a**) or 340.1 nm (**8b**). The last absorption maximum was present only in the electronic spectra of compounds **8a** and **8b** due to the extension of the π-electron conjugation by the formation of the 1,3-oxazole chromophore.

Fourier-Transform Infrared Absorption Spectroscopy Data

The FTIR absorption spectra of acyclic precursors **5** and **7** showed a characteristic peak in the range of 3347–3281 cm^{-1} due to the valence vibration of the N–H bond. In addition, the absorption maximum at 1746 cm^{-1} due to carbonyl valence vibration, and another at 1635 cm^{-1} due to stretching vibration of amidic carbonyl were remarked in the FTIR spectrum of **5**. For α-acylamino ketones **7a** and **7b**, these two carbonyl absorption bands are overlapped, as suggested by the very strong single peak recorded in their FTIR spectra at 1655 cm^{-1}. A broad absorption band in the spectral region from 3300 to 2500 cm^{-1}, centered at ≈ 3000 cm^{-1}, due to stretching vibration of the O–H bond and two noticeably weak satellite peaks at 2676 and 2599 cm^{-1} are also characteristic of the hydrogen-bonded *N*-acylated *L*-valine **5**.

As evidence that the intramolecular cyclocondensations occurred, significant changes were observed in the FTIR absorption spectra of five-membered *O,N*-heterocyclic compounds **6**, and **8a,b** compared with the corresponding spectra of open-chain intermediates **5**, and **7a,b**, respectively. The FTIR spectrum of 4*H*-1,3-oxazol-5-one **6** presented a peak at 1825 cm^{-1} due to carbonyl stretching vibration, which is shifted at a higher wavenumber compared with the C=O absorption maximum from the spectrum of **5**. The FTIR absorption spectra of 5-membered heterocycles **6** and **8** showed an absorption band at 1650 (**6**), 1602 (**8a**), or 1601 cm^{-1} (**8b**) due to the valence vibration of the C=N bond. The absorption maxima at 1099 (**8a**), 1097 (**8b**), or 1040 cm^{-1} (**6**) due to the C–O–C symmetric stretching vibration and at 1280 (**8a** and **8b**) or 1243 cm^{-1} (**6**) due to the asymmetric stretching vibration of the C–O–C group were also observed.

Nuclear Magnetic Resonance Spectroscopy Data

The nuclear magnetic resonance spectroscopic data also confirmed the structures of the new *L*-valine analogs. Complete assignments of the signals from the ^1H and ^{13}C NMR spectra of the new compounds **5–8** were performed using combinations of standard NMR spectroscopic techniques, namely 2D COSY and HETCOR experiments.

The molecular structures of the new bromine-containing derivatives **5–8** with the numbering of the atoms, used for the assignment of NMR signals, are presented in Figure 3.

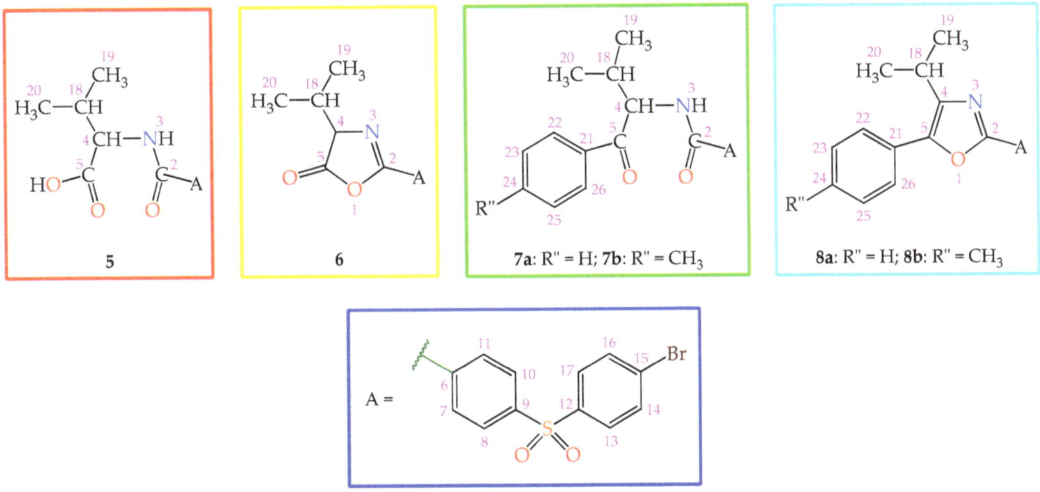

Figure 3. Molecular structures of the new 4-[(4-bromophenyl)sulfonyl]benzoic acid derivatives **5–8** with the atom numbering (for assigning NMR signals).

Proton NMR Spectroscopy Data

The proton NMR spectra of **5** and **7** showed a doublet signal in the region between 8.73 and 8.99 ppm, attributed to deshielded H-3 proton. The ^{1}H NMR spectrum of **5** revealed a signal assigned to the H-4 proton at 4.29 ppm as a doublet of doublets due to the vicinal coupling to H-3 and H-18 protons. The proton NMR spectra of α-acylamino ketones **7a** and **7b** showed the H-4 signal at 5.36 (**7b**) or 5.38 ppm (**7a**) as a triplet, due to the coupling to H-3 and H-18. The ^{1}H NMR spectra of **5**, **7a**, and **7b** also highlighted characteristic signals for the isopropyl substituent, i.e., an octet (**5**) or a multiplet (**7a,b**) signal registered in the range of 2.17–2.28 ppm assigned to the methine proton (H-18) and two strongly shielded doublet signals, the first in the interval of 0.89–0.94 ppm and the second at 0.92 or 0.95 ppm, due to the non-equivalent protons (H-19, H-20) of the two methyl groups. The signal of proton of the carboxyl group of N-acyl-α-amino acid **5** was not recorded in the ^{1}H NMR spectrum probably due to hydrogen-deuterium exchange, but the presence of the COOH group was confirmed both by IR and ^{13}C NMR spectral data.

The ^{1}H NMR spectrum of **6** displayed the H-4 doublet signal at 4.32 ppm due to the coupling only to the H-18 proton (with ^3J = 4.7 Hz). The methine proton of the 1-methylethyl group was revealed by the presence of a septet of doublets signal at 2.40 ppm due to coupling to protons H-19, H-20 (with ^3J = 6.9 Hz), and to H-4 proton (with ^3J = 4.7 Hz). The ^{1}H NMR spectra of **8a** and **8b** showed for the isopropyl substituent a signal as septet at 3.26 (**8b**) or 3.29 ppm (**8a**) attributed to the H-18 proton, and a shielded doublet at 1.35 (**8b**) or 1.36 ppm (**8a**), due to the two CH$_3$ groups' protons.

Carbon-13 NMR Spectroscopy Data

The ^{13}C NMR spectrum of N-acylated L-valine **5** presented a signal at 58.43 ppm attributed to the C-4, for the 1-methylethyl group revealed a signal at 29.48 ppm due to the methine carbon (C-18) and two signals at 18.61 and 19.25 ppm due to the C-19 and C-20 non-equivalent carbon atoms. The ^{13}C NMR spectrum of **6** highlighted the C-4 signal at 71.04 ppm, which is deshielded by 12.61 ppm as a result of cyclization of the open-chain precursor **5**. Further, the C-2 atom of 4H-1,3-oxazol-5-one **6** resonated at 160.36 ppm, being more shielded with 5.42 ppm compared with the C-2 of **5**, and the C-5 at 176.98 ppm, being more deshielded with 4.19 ppm than the corresponding carbon atom of its precursor (**5**).

The carbon-13 NMR spectra of aromatic O,N-heterocycles **8a,b** showed a signal assigned to the C-4 atom at 143.63 (**8b**) or 144.19 ppm (**8a**), which is more deshielded by ≈ 84.77 ppm than the corresponding signal of the compounds from which they were obtained by cyclization, from 59.05 (**7b**) or 59.23 ppm (**7a**). The cyclization of **7a,b** to compounds **8a,b** induced a shielding effect for the C-2 of the 1,3-oxazole ring, resulting in an about 7.87 ppm lower chemical shift. The C-5 signal of 1,3-oxazoles **8a,b** was observed at 145.47 (**8a**) or 145.69 ppm (**8b**), whereas the corresponding signal of acyclic precursors **7a,b** appeared at 198.57 (**7b**) or 199.19 ppm (**7a**), revealing a shift of this carbon signal at a smaller δ_C of approximately 53.30 ppm.

Gas Chromatography Coupled to Electron Ionization Mass Spectrometry Data

For saturated azlactone **6**, supplementary evidence was obtained by its mass spectrum obtained by GC/EI–MS analysis. The two molecular ions of compound **6** corresponding to bromine isotopes (^{79}Br/^{81}Br) were energetically unstable upon interaction with high-energy (70 eV) electrons and fragmented with the elimination of a molecule of propene. The base peak with m/z 381 and the corresponding radical cation with m/z 379 (with a relative abundance of 71.61%) were resulted according to the ^{79}Br/^{81}Br isotopic ratio of ≈1:1. Other characteristic structural fragments of **6** are reported in the Materials and Methods section.

2.3. Evaluation of Antimicrobial Activity

2.3.1. Qualitative Assessment of Antimicrobial Activity

The agar diffusion assay showed that compounds **7a**, **8a**, and **8b** did not interfere with microbial growth (Table 2). *Enterococcus faecium* E5 was the most susceptible species to compounds **5** and **6** with a growth inhibition zone diameter of 17 and 15 mm, respectively. Compounds **6** and **7b** also inhibited the growth of *Staphylococcus aureus* ATCC 6538 and *Bacillus subtilis* ATCC 6683, respectively with diameters of growth inhibition zones of 8 mm and of 10 mm.

Table 2. Results of the qualitative analysis of the antimicrobial effect of compounds **5–8** tested at 5000 µg/mL, using an adapted disk diffusion method (diameters of growth inhibition zones were measured in mm).

Tested Compound	Gram-Positive Bacteria			Gram-Negative Bacteria		Fungus
	Bacillus subtilis ATCC 6683	*Enterococcus faecium* E5	*Staphylococcus aureus* ATCC 6538	*Escherichia coli* ATCC 8739	*Pseudomonas aeruginosa* ATCC 27857	*Candida albicans* 393
5	0	17	0	0	0	0
6	0	15	8	0	0	0
7a	0	0	0	0	0	0
7b	10	0	0	0	0	0
8a	0	0	0	0	0	0
8b	0	0	0	0	0	0
Ciprofloxacin	28	30	26	34	30	-*
Fluconazole	-	-	-	-	-	30

* -: not tested.

2.3.2. Effects of Compounds on Antibiotic Susceptibility Profile

For compounds **5** and **6** only, their influence on the *E. faecium* E5 and *S. aureus* ATCC 6538 strains' susceptibility to different antibiotics was evaluated. Although compound **7b** was active against *B. subtilis* ATCC 6683, the influence on this strain's susceptibility to antibiotics was not investigated, as there are no specific guidelines regarding the susceptibility breakpoints of this species.

No significant changes were noticed in the *E. faecium* E5 strain's susceptibility to current antibiotics after culture in the presence of compound **5** tested at subinhibitory concentration (Table 3), suggesting both a low selective pressure for resistance occurrence and a different mechanism of action. Regarding the *E. faecium* E5 strain cultured in the presence of compound **6**, it was observed that it determined an increase in the antimicrobial effect of the tested antibiotics (ampicillin, penicillin, linezolid, and vancomycin). The diameters of growth inhibition zones (in mm) are shown in Table 3.

Table 3. Antibiotic susceptibility testing of *E. faecium* E5 strain cultured in the presence of compounds **5** and **6** tested at a subinhibitory concentration of 250 µg/mL and of dimethyl sulfoxide (DMSO) (growth inhibition zone diameters were measured in mm).

Microbial Culture	Growth Inhibition Zone Diameter (In mm)			
	Ampicillin	Linezolid	Penicillin	Vancomycin
Control *	24	27	14	19
5	22	25	13	17
6	29	39	15	23
DMSO	22	23	0	18

* liquid culture medium inoculated with standardized suspension of *E. faecium* E5.

With respect to the *S. aureus* ATCC 6538 strain cultured in the presence of compound **6**, a slight increase in the diameter of growth inhibition zone was observed in the case of cefoxitin (Table 4). Regarding other studied antibiotics, namely vancomycin, linezolid, clindamycin, and rifampicin, the susceptibility was reduced, suggesting a phenotypic change in the tested strain exposed to compound **6** at subinhibitory concentration. There were no changes in the growth inhibition zone diameters compared with those of the microbial growth control in the case of the studied antibiotics: azithromycin, penicillin, and trimethoprim-sulfamethoxazole.

Table 4. Antibiotic susceptibility testing of *S. aureus* ATCC 6538 strain cultured in the presence of compound **6** tested at 250 µg/mL and of DMSO control (growth inhibition zone diameters were measured in mm).

Microbial Culture	Growth Inhibition Zone Diameter (In mm)							
	Azithromycin	Cefoxitin	Clindamycin	Linezolid	Penicillin	Rifampicin	Trimethoprim-Sulfamethoxazole	Vancomycin
Control *	15	20	26	32	0	21	0	23
6	15	21	23	29	0	19	0	21
DMSO	13	15	20	28	0	19	0	18

* liquid culture medium inoculated with standardized suspension of *S. aureus* ATCC 6538.

2.3.3. Quantitative Testing of Antimicrobial Activity

The results of the quantitative antimicrobial testing by the standard broth microdilution method are presented in Table 5. The majority of the tested compounds exhibited MIC values equal to or higher than 500 µg/mL. The most active proved to be compound **6** which was found to have a moderate antimicrobial activity (MIC value of 250 µg/mL) against the *S. aureus* ATCC 6538 reference strain.

Table 5. MIC (minimal inhibitory concentration) and MBIC (minimal biofilm inhibitory concentration) values in µg/mL determined for compounds **5–8** tested at concentrations between 500 and 0.97 µg/mL.

Tested Compound	Gram-Positive Bacteria						Gram-Negative Bacteria				Fungus	
	Bacillus subtilis ATCC 6683		*Enterococcus faecium* E5		*Staphylococcus aureus* ATCC 6538		*Escherichia coli* ATCC 8739		*Pseudomonas aeruginosa* ATCC 27857		*Candida albicans* 393	
	MIC	MBIC	MIC	MBIC	MIC	MBIC	MIC	MBIC	MIC	MBIC	MIC	MBIC
5	>500	>500	500	62.5	>500	>500	>500	>500	>500	>500	>500	>500
6	>500	>500	500	1.95	250	250	>500	>500	>500	>500	>500	>500
7a	>500	>500	>500	>500	>500	>500	>500	>500	>500	>500	>500	>500
7b	>500	>500	>500	>500	>500	>500	>500	>500	>500	>500	>500	>500
8a	>500	>500	>500	>500	>500	>500	>500	>500	>500	>500	>500	>500
8b	>500	>500	>500	>500	>500	>500	>500	>500	>500	>500	>500	>500
Ciprofloxacin	<0.03	<0.03	0.62	0.62	0.15	0.15	0.012	0.012	0.15	0.15	-*	-
Fluconazole	-	-	-	-	-	-	-	-	-	-	<0.12	<0.12

* -: not tested.

2.3.4. Evaluation of Antibiofilm Activity

The results obtained in this study showed that compounds **7a**, **7b**, **8a**, and **8b** did not interfere with microbial adherence on surfaces. Exceptional antibiofilm activity was observed for compounds **5** and **6** against *E. faecium* E5 strain, with MBIC values of 62.5 and 1.95 µg/mL, respectively (Table 5). Compound **6** also displayed a moderate antibiofilm effect in the case of the Gram-positive *S. aureus* ATCC 6538 strain, with an MBIC value of 250 µg/mL.

2.4. Evaluation of Antioxidant Activity

The antioxidant effect of newly synthesized derivatives **5–8** was studied using three methods based on electron transfer reactions, namely DPPH, ABTS, and ferric reducing power tests. The results were compared with those of the following standard antioxidant agents: ascorbic acid (AA), butylated hydroxyanisole (BHA, a mixture of 2-(*tert*-butyl)-4-methoxyphenol and 3-(*tert*-butyl)-4-methoxyphenol) and butylated hydroxytoluene (BHT, 2,6-di-*tert*-butyl-4-methylphenol), used as positive controls. The antioxidant capacity of the key starting materials L-valine and 4-[(4-bromophenyl)sulfonyl]benzoic acid **3** was also determined.

2.4.1. Antioxidant Activity Assay by DPPH Method

The results of the study concerning the antioxidant effect evaluation of the new 4-[(4-bromophenyl)sulfonyl]benzoic acid derivatives **5–8** by the DPPH method are presented in Table 6. In terms of experimental results obtained by this test, of all the newly synthesized compounds, 4H-1,3-oxazol-5-one **6** had the best antioxidant activity with a DPPH inhibition rate of $16.75 \pm 1.18\%$, its effect being better than that of key starting materials (L-valine and carboxylic acid **3**), but lower than the standard antioxidants used. This compound was followed by N-acylated α-amino acid **5** with a DPPH inhibition percentage of $4.70 \pm 1.88\%$. In contrast, compounds **7a,b**, and **8a,b** had the lowest scavenging effect values in the range of 1.35–1.82%.

Table 6. Results of the assessment of the antioxidant effect of the compounds tested by the DPPH method.

Compound	Concentration (µM)	Scavenging Effect (%)
5	250	4.70 ± 1.88
6	250	16.75 ± 1.18
7a	250	1.42 ± 0.06
7b	250	1.35 ± 1.72
8a	250	1.82 ± 0.36
8b	250	1.50 ± 0.11
L-Valine	250	7.87 ± 0.07
3	250	2.32 ± 0.22
AA	250	85.09 ± 1.67
BHA	250	77.99 ± 0.56
BHT	250	31.79 ± 1.52

2.4.2. Antioxidant Activity Assay by ABTS Method

The results of the assessment of the antioxidant potential of the compounds tested by the ABTS method are shown in Table 7. The obtained values of the percentage scavenging effect of new compounds **5–8** are small in the interval of 0.40–7.66%. Among the new compounds, 1,3-oxazole **8b** showed the best antioxidant activity ($7.66 \pm 0.71\%$), followed by α-acylamino ketone **7a** ($7.14 \pm 1.51\%$), while compound **7b** had the lowest ($0.40 \pm 0.27\%$). By this method, L-valine had a better effect ($44.71 \pm 0.66\%$) than the new products **5–8**, but key raw material **3** showed a lower activity ($0.17 \pm 0.85\%$) than these.

Table 7. Results of the evaluation of the scavenging effect of the compounds analyzed by the ABTS method.

Compound	Concentration (μM)	Scavenging Effect (%)
5	250	0.78 ± 0.91
6	250	1.49 ± 0.37
7a	250	7.14 ± 1.51
7b	250	0.40 ± 0.27
8a	250	1.77 ± 0.56
8b	250	7.66 ± 0.71
L-Valine	250	44.71 ± 0.66
3	250	0.17 ± 0.85
AA	250	99.93 ± 0.11
BHA	250	99.74 ± 0.08
BHT	250	98.46 ± 1.23

2.4.3. Ferric Reducing Power Assay

The results of the investigation of the antioxidant capacity of the compounds tested by the ferric reducing power method are presented in Table 8. All the tested compounds possessed the ability to reduce iron(III) to iron(II). The α-acylamino ketone **7a** was a better iron(III) reducer (absorbance at 700 nm, A_{700} of 0.0722 ± 0.0013) than the other new derivatives, followed by three compounds, which showed similar values of the ferric reducing power, in descending order: **6** (A_{700} = 0.0461 ± 0.0088), **8b** (A_{700} = 0.0439 ± 0.0057), and **7b** (A_{700} = 0.0437 ± 0.0105). Of the compounds tested, N-acyl-L-valine **5** was the least active tested compound (A_{700} = 0.0224 ± 0.0019) and L-valine had the best antioxidant activity (A_{700} = 0.0854 ± 0.0051). None of the activities of the analyzed compounds were comparable to those of the AA, BHA, and BHT positive controls.

Table 8. Results of the screening of the antioxidant activity of the compounds tested by the ferric reducing power method.

Compound	Concentration (μM)	Reducing Power (Absorbance at 700 nm, A_{700})
5	500	0.0224 ± 0.0019
6	500	0.0461 ± 0.0088
7a	500	0.0722 ± 0.0013
7b	500	0.0437 ± 0.0105
8a	500	0.0369 ± 0.0060
8b	500	0.0439 ± 0.0057
L-Valine	500	0.0854 ± 0.0051
3	500	0.0568 ± 0.0078
AA	500	0.8727 ± 0.0315
BHA	500	1.1363 ± 0.0096
BHT	500	0.7282 ± 0.1686

2.5. *Daphnia magna* Toxicity Bioassay

The results of the *Daphnia magna* toxicity test are presented in Table 9, and the lethality curves are shown in Figure 4. At 24 h, the highest toxicity was induced by **7a** and **7b**, and the compounds **5**, **6**, **8a**, and **8b** were non-toxic on *D. magna* at the tested concentrations. At 48 h, the most toxic compound was **7a**. At the lowest concentration, its lethality was 65%; therefore, the LC_{50} being lower than 2 μg/mL, the estimated value by extrapolation is 0.21 μg/mL. The compounds **5**, **6**, **7b**, and **8a** induced medium to high toxicity, their LC_{50} values varying from 21.07 to 54.62 μg/mL. Except for **7a**, all compounds induced lower toxicity than the key starting material **3**. The compound **8b** at the highest concentration (50 μg/mL) induced a lethality of 25%, and at all other concentrations, the lethality varies from 0 to 10%. The predicted LC_{50} values were significantly lower than those obtained experimentally, except for compound **3**.

Table 9. *Daphnia magna* toxicity test results.

Tested Compound	Predicted LC$_{50}$ [1] (48 h) (μg/mL)	L % Max (48 h) [2]	Determined LC$_{50}$ (24 h) (μg/mL)	95% CI [3] of LC$_{50}$ (24 h) (μg/mL)	Determined LC$_{50}$ (48 h) (μg/mL)	95% CI of LC$_{50}$ (48 h) (μg/mL)
5	0.51	60	ND [4] *	ND *	43.5	ND **
6	0.33	70	ND *	ND *	31.25	22.32 to 43.75
7a	0.11	100	42.93	38.14 to 48.31	ND ***	ND **
7b	0.11	90	58.83	46.28 to 74.78	21.07	12.92 to 34.35
8a	0.1	60	ND *	ND *	54.62	41.73 to 71.48
8b	0.04	30	ND *	ND *	ND *	ND *
L-Valine (Control 1)	1078.3	20	ND *	ND *	ND *	ND *
3 (Control 2)	11.8	100	31.11	ND **	1.144	0.13 to 9.93

[1] 50% lethal concentration; [2] maximum lethality induced at 48 h; [3] 95% confidence interval; [4] not determined because of the results obtained; * lethality ranged between 0 and 40%; ** 95% CI could not be determined due to the results; *** lethality was higher than 50% at all concentrations.

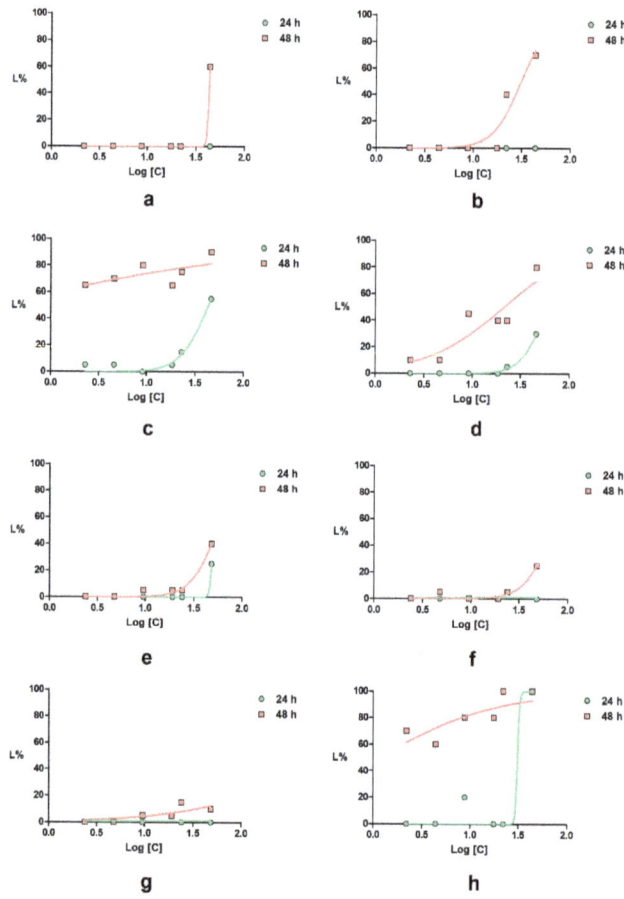

Figure 4. *Daphnia magna* lethality curves for the tested compounds: (**a**) **5**; (**b**) **6**; (**c**) **7a**; (**d**) **7b**; (**e**) **8a**; (**f**) **8b**; (**g**) *L*-valine (control 1); (**h**) **3** (control 2).

3. Discussions

The 1,3-oxazoles are five-membered heteromonocyclic scaffolds, containing two heteroatoms, O and N at positions 1 and 3, respectively, of particular importance in synthetic and medicinal chemistry, since representatives of this class have great structural diversity and are of significant biological interest.

Over time, researchers have paid particular attention to the preparation of new derivatives incorporating the 1,3-oxazole ring [52–55], as well as to the evaluation, among other pharmacological properties, of their antimicrobial and antioxidant activities, a large number of articles being published on this topic. Currently, various bioactive 1,3-oxazole-embedded heterocycles have been reported, such as 1,3-oxazole clubbed pyridyl-pyrazolines having good to excellent antimicrobial action [56], cytisine-containing 1,3-oxazoles as potential inhibitors of *Candida* spp. glutathione reductase [57], thioxo-1,3-oxazole analogs that prevent bacterial growth and with antioxidant ability [58], 1,3-oxazole-quinoxaline amine hybrids showing antibacterial activity [59], steroidal 1,3-oxazole derivatives displaying effective antimicrobial and antibiofilm properties [60], binaphthyl-based, functionalized 1,3-oxazole peptidomimetics with moderate to excellent antimicrobial activity [61], substituted 1,3-oxazole-benzamides as antibacterial inhibitors of the essential bacterial cell division protein FtsZ [62], *N*-(oxazolylmethyl)thiazolidinediones as selective inhibitors of *Candida albicans* biofilm formation [63], and some antimicrobial 2-amino-1,3-oxazoles [64]. Very recently, we investigated the in silico and in vitro antimicrobial and antibiofilm effects of a series of 1,3-oxazole-based compounds and their isosteric analogs derived from alanine and phenylalanine, some of them presenting a promising profile [65]. Quantitative studies also indicated that of all 2,5-diaryl-1,3-oxazoles tested, only some of those with a bromine atom in their molecules showed antimicrobial and/or antibiofilm action. In contrast, all other derivatives from the same class, unsubstituted or with a chlorine atom in their structures, proved to be inactive at the concentrations tested. Other researchers have also observed that the incorporation of the chlorine or bromine atom, usually grafted onto an aromatic nucleus, led to an improvement in the antimicrobial activity of compounds of different classes, as well as that brominated derivatives have shown greater efficacy than chlorinated analogs [66–69]. The antimicrobial effect due to the presence of halogen atoms appears to be related to the compounds' relative hydrophobicity. It was found that hydrophobicity of the compounds enhanced with the addition of a halogen atom in their molecules and that chlorine derivatives displayed a less pronounced effect, while the incorporation of a bromine atom led to analogs with a higher hydrophobic profile [66]. In view of these findings, it is of interest to obtain new bromine-bearing 1,3-oxazoles-type molecules and to explore their application as biologically active agents.

In the present work, we synthesized novel organic compounds, carrying a bromine atom in their molecules, based on the 1,3-oxazole and diphenyl sulfone pharmacophores. The molecular structures of the newly obtained bromine-containing compounds (which are *N*-acyl-*L*-valine, 4-isopropyl-4*H*-1,3-oxazol-5-one, 2-acylamino ketone, and 4-isopropyl-1,3-oxazole derivatives) were elucidated based on spectroscopic investigations. Moreover, the synthesized analogs were assessed for antimicrobial properties, namely qualitative (zone of inhibition) and quantitative (MIC) activities, antibiofilm action (MBIC), and for antioxidant and toxic effects. Furthermore, in silico investigations on the toxicity and potential antimicrobial action were performed.

The structure-based similarity analysis of the new *L*-valine derivatives **5–8** highlighted the originality of the proposed structures and indicated that the presence of the bromine atom in the proposed structures of the new compounds can enhance the potential antimicrobial effect. The PASS prediction showed the probability that all new compounds to exhibit anti-infective, antimycobacterial, and antituberculosis activities and, in addition, that 4*H*-1,3-oxazol-5-one **6** to have a glycopeptide-like antibiotic effect.

The antimicrobial activity assays revealed a promising antimicrobial potential against Gram-positive cocci strains, particularly in the case of *E. faecium* biofilm, when the MBIC values were as from 8 to 256 times lower than the corresponding MIC values.

Concerning the results of quantitative tests for the evaluation of antimicrobial and antibiofilm activity, it was also found that the MIC and MBIC values of the standard broad-spectrum antibacterial drug ciprofloxacin and antifungal agent fluconazole, which served as positive controls, were lower than those determined for the compounds tested. This can be explained by the fact that the novel L-valine derivatives, which belong to four classes, may have distinct mechanisms of action from those of the used control drugs and, in addition, unlike them, are not standardized active ingredients in optimized drug formulations.

It was also found that all compounds analyzed show antioxidant activity through all spectroscopic tests used (DPPH, ABTS, and ferric reducing power methods), but lower than the bioactive standards (AA, BHA, and BHT). The tested compounds thus demonstrated weak electron-donating properties.

The in vitro evaluation regarding antimicrobial, antioxidant, and toxicity features highlighted that acyclic intermediate **5** is promising due to its very good antibiofilm activity, low antioxidant effect, and moderate toxicity. It is worth noting that by cyclodehydration of compound **5** to the corresponding 4H-1,3-oxazol-5-one **6**, an increase in the antibiofilm effect and antioxidant activity was observed, but also an increase in its toxicity. Moreover, it appears that the conversion of 2-oxazolin-5-one **6** to α-acylamino ketone **7b** and the cyclization of **7b** to 1,3-oxazole **8b** leads to a decrease in antimicrobial activity, and compound **7a** and corresponding 1,3-oxazole **8a** were found to be inactive at the tested concentrations. No regular variation in the antioxidant effect of compounds **7a**, **7b**, **8a**, and **8b** was noticed, but of all the new compounds tested, **7a** and **8b** stood out, which showed the best activity by all three methods used. It was also shown that by converting saturated azlactone **6** to α-acylamino ketones **7a,b**, the toxicity increases, and by cyclization of open-chain intermediates **7a,b** to 1,3-oxazoles **8a,b**, the toxicity decreases significantly, **8b** being non-toxic. However, the *Daphnia magna* assay showed medium to high toxicity for compounds **5**, **6**, and **7b**, and significant-high toxicity for compounds **3** and **7a**. Our findings indicate a high potential for biological targets, whereas the 1,3-oxazole-containing compound **8b** showed no toxicity at concentrations lower than 50 µg/mL. The toxicity study also showed that, with the exception of the LC_{50} value obtained for key intermediate **3**, the experimental values of LC_{50} for all other compounds were significantly higher than those predicted by the GUSAR analysis.

Recently, we reported the synthesis and results of the antimicrobial and antibiofilm assessment of a series of similar N-{4-[(4-X-phenyl)sulfonyl]benzoyl}-L-valine derivatives (X = H, Cl) [70,71]. The non-substituted analog of **7b**, N-[3-methyl-1-oxo-1-(p-tolyl)butan-2-yl]-4-(phenylsulfonyl)benzamide presented a better antimicrobial effect compared with **7b** or its chloro-derivative. The increase in the lipophilic character was perhaps detrimental because of the associated lower water solubility. The introduction into the molecule of the bromine atom proved slightly beneficial in the case of the 4H-1,3-oxazol-5-one **6**.

Subsequent structural and biological investigations into these kinds of privileged scaffolds can lead to the discovery of new more potent derivatives that can act as optimized candidates for the development of new effective preventive and therapeutic agents. In this respect, we consider three critical positions on the molecular templates that may have an effect on biological action. A first possibility for the future improvement of the antimicrobial property of these analogs is the replacement of the bromine atom in the arylsulfonylphenyl fragment with different substituents, such as fluorine, iodine, trifluoromethyl, or nitro. For this purpose, other benzene analogs can be used as starting molecules in the Friedel–Crafts sulfonylation reaction. As a second optimization choice, in the N-acylation reaction instead of L-valine, we consider the use of other natural or unnatural α-amino acids (such as histidine and tryptophan) or the incorporation of several α-amino acid residues in the molecules as it is well-known that some natural polypeptides bearing the 1,3-oxazole ring (e.g., microcin B17) are inhibitors of DNA-gyrase [3]. The use of other aromatic compounds with different substituents (e.g., I, NO_2, CF_3, OCH_3) in the reaction with **6** is the third alternative route of synthesis to enhance the biological activity of these molecular scaffolds. The iodine atom proved more active than other halogens, CF_3-functionalized compounds showed higher antibacterial activity than those with a CH_3 group, hydrophobic substituents (e.g., benzyloxy, *tert*-butyl) [72], or electron-withdrawing groups (such as nitro) on aromatic cores increased the antibacterial effect [73], small substituents (e.g., hydroxy) at 4-position of the phenyl nucleus improved the antibacterial potency [62]. In addition, it was noticed that the electron-donating substituents (such as OCH$_3$) on the aromatic moiety enhanced the antioxidant activity compared with the electron-withdrawing groups and that, in general, this pharmacological action increased with the increasing electron-donating effect of the substituent [74–77].

4. Materials and Methods
4.1. Prediction of the Molecular Mechanism of Action and Toxicity
4.1.1. Structure-Based Similarity Analysis

Each target structure was inputted into the search engine of the ChEMBL database [78] and a similarity search was performed using a 50% threshold. The output structures were extracted together with any associated data of biological activity on bacteria. The DataWarrior v5.2.1 software (Actelion Pharmaceuticals Ltd., Allschwil, Switzerland, https://openmolecules.org/) [79] was used to filter the duplicate structures and to calculate the clogP values.

4.1.2. PASS Prediction

The web-based Prediction of Activity Spectra for Substances (PASS) application was used to evaluate the antimicrobial potential of the newly designed bromo derivatives based on their chemical structures. The array of results was analyzed for any effect with a predicted Pa value higher than the Pi value.

4.2. General Information

Melting points (mp) (in °C), were determined on a Boëtius hot plate microscope (VEB Wägetechnik Rapido, PHMK 81/3026, Radebeul, Germany) and are reported uncorrected. UV/VIS spectra were recorded for solutions of the new compounds in methanol at concentration of ≈0.025 mM and in acetonitrile at ≈0.015 mM on an Analytik Jena Specord 40 ultraviolet/visible spectrophotometer (Analytik Jena AG, Jena, Germany) using a quartz cuvette with a path length of 1.00 cm. Values of the wavelength (corresponding to the maximum absorbance), λ_{max} (in nm), and of the logarithm to the base 10 of the molar absorption coefficient, ε_{max} (M$^{-1}\cdot$cm^{-1}) are provided. FTIR spectra were collected with a Bruker Optics Vertex 70 Fourier-transform infrared spectrophotometer (Bruker Optics GmbH, Ettlingen, Germany) using the conventional KBr pellet technique. The position of the selected absorption bands in the FTIR spectra is reported using the wavenumber at the absorption maximum, $\tilde{\nu}_{max}$ (in cm^{-1}). The intensity of the FTIR peaks is provided as very strong (vs); strong (s); medium (m); and weak (w). Nuclear magnetic resonance spectra were registered on a Varian Gemini 300 BB NMR spectrometer (Varian, Inc., Palo Alto, CA, USA) operating at 300 MHz for ^{1}H NMR and 75 MHz for ^{13}C NMR, in deuterated solvents, i.e., dimethyl sulfoxide-d_6 (DMSO-d_6) or deuterochloroform (CDCl$_3$). Additional evidence was provided by the 2D HETCOR and COSY experiments. Chemical shifts, δ, were measured in ppm with respect to the ^{1}H or ^{13}C resonance of (CH$_3$)$_4$Si (tetramethylsilane, TMS), and the coupling constants, J, are expressed in Hz. The multiplicity of ^{1}H NMR signals is provided as per the following convention: singlet (s); doublet (d); doublet of doublets (dd); triplet (t); triplet of triplets (tt); septet (sept); septet of doublets (septd); octet (oct); multiplet (m); and broad (br). Proton NMR spectroscopic data are quoted in the following order: ^{1}H chemical shift, multiplicity, proton number, signal/hydrogen atom assignment, coupling constants, and ^{13}C NMR spectroscopic data are reported as follows: δ_C-value, signal/carbon atom attribution. GC/EI-MS data were acquired on a Fisons Instruments GC 8000 series gas chromatograph coupled to an MD 800 mass spectrometer detector equipped with an electron ionization source and with a quadrupole mass analyzer (Fisons Instruments SpA, Rodano, Milano, Italy), using an SLB-5ms capillary column (d_f 0.25 µm; L × I.D. 30 m × 0.32 mm), CH$_2$Cl$_2$ as solvent, and a flow rate of helium carrier gas of 2 mL/min. Reversed-phase high-performance liquid chromatography (HPLC) was performed on a Beckman Coulter System Gold 126 liquid chromatograph, equipped with a

System Gold 166 UV/VIS spectrophotometer detector (Beckman Coulter, Inc., Fullerton, CA, USA), a Rheodyne injection system, and a LiChrosorb RP-18 column (d_p 5 µm; L × I.D. 25 cm × 4.6 mm). The flow rate of eluent, a mixture of CH_3OH–H_2O in different volume ratios, was 1 mL/min. The compounds' purities (%) and retention times, t_R (in min), were reported. Elemental analysis was carried out on a Costech ECS 4010 instrument (Costech Analytical Technologies Inc., Valencia, CA, USA).

4.3. Chemistry

The chemicals and reagents were obtained from commercial sources of analytical grade and were used as received, except for CH_2Cl_2 which was dried over anhydrous $CaCl_2$.

4.3.1. Preparation and Characterization of 2-{4-[(4-Bromophenyl)sulfonyl]benzamido}-3-methylbutanoic Acid **5**

The solution of 2.34 g (20 mmol) of *L*-valine in 20 mL (0.80 g, 20 mmol) of 1 N NaOH solution was cooled at 0–5 °C in an ice bath and then two other solutions were added simultaneously in droplets under magnetic stirring during half an hour, namely a solution of 7.19 g (20 mmol) of crude 4-[(4-bromophenyl)sulfonyl]benzoyl chloride **4** in 45 mL of anhydrous dichloromethane, and a 2 N NaOH solution (10 mL, 0.80 g, 20 mmol). The reaction mixture was stirred for another 1 h at room temperature and then the separated aqueous phase was acidified with 2 N HCl when a white precipitate appeared. The solid was filtered off at low-pressure, washed thoroughly with distilled water on the filter to remove traces of hydrochloric acid, air-dried, and recrystallized from water when white acicular crystals resulted.

Yield 94% (8.28 g, 18.80 mmol). mp 194–196 °C.

UV/VIS (CH_3OH) λ_{max} nm ($\log_{10} \varepsilon_{max}$): 202.6 (4.48); 252.0 (4.20); (CH_3CN) λ_{max} nm ($\log_{10} \varepsilon_{max}$): 196.2 (4.83); 228.2 (4.36); 252.2 (4.57).

FTIR (KBr disc) $\tilde{\nu}_{max}$ cm^{-1}: 3347 s (νN–H); 3300–2500 vs (νO–H); 3092 s; 3069 s (νC–H, aromatic); 2966 vs ($\nu_{asym}CH_3$); 2933 s (νC–H, aliphatic); 2876 s ($\nu_{sym}CH_3$); 2676 m; 2599 m (combination bands); 1746 vs ($\nu\underline{O=C}$–O); 1635 vs ($\nu\underline{O=C}$–N, amide I); 1599 s; 1573 vs; 1487 s; 1467 s (νC=C, aromatic); 1536 vs (δN–H, amide II); 1324 vs; 1296 vs; 1281 vs ($\nu_{asym}SO_2$); 1161 vs ($\nu_{sym}SO_2$); 852 s (γC–H, aromatic); 613 vs; and 575 vs (νC–Br).

^{1}H NMR (300 MHz, DMSO-d_6) δ ppm: 8.73 (d, 1H, H-3, J = 8.0 Hz); 8.08 (d, 2H, H-7, H-11, J = 8.8 Hz); 8.04 (d, 2H, H-8, H-10, J = 8.8 Hz); 7.92 (d, 2H, H-13, H-17, J = 8.8 Hz); 7.85 (d, 2H, H-14, H-16, J = 8.8 Hz); 4.29 (dd, 1H, H-4, J = 8.1, 6.9 Hz); 2.17 (oct, 1H, H-18, J = 6.9 Hz); 0.95 (d, 3H, H-19, J = 6.9 Hz); and 0.94 (d, 3H, H-20, J = 6.9 Hz).

^{13}C NMR (75 MHz, DMSO-d_6) δ ppm: 172.79 (C-5); 165.78 (C-2); 142.66 (C-9); 139.92 (C-12); 139.03 (C-6); 132.96 (C-14, C-16); 129.48 (C-7, C-11); 129.08 (C-8, C-10); 128.22 (C-15); 127.49 (C-13, C-17); 58.43 (C-4); 29.48 (C-18); 19.25 (C-19); and 18.61 (C-20).

Reversed-phase HPLC (CH_3OH–H_2O 30:70, *V*/*V*; flow rate 1 mL/min; λ 250 nm): purity 99.99%; t_R 4.47 min.

Elem. anal. (%) found: C, 49.06; H, 4.11; N, 3.19; and S, 7.28; calcd. for $C_{18}H_{18}BrNO_5S$ (M_r 440.31): C, 49.10; H, 4.12; N, 3.18; and S, 7.28.

4.3.2. Preparation and Characterization of 2-{4-[(4-Bromophenyl)sulfonyl]phenyl}-4-isopropyl-4*H*-1,3-oxazol-5-one **6**

An amount of 4.62 g (10.5 mmol) of **5** was suspended under magnetic stirring at room temperature in 50 mL of dry CH_2Cl_2 and then an equimolar quantity (1.15 mL, 1.06 g, 10.5 mmol) of 4-methylmorpholine was added. A volume of 1 mL (1.14 g, 10.5 mmol) of ethyl chloroformate was added dropwise with continuous stirring to the solution obtained above. The reaction mixture was stirred for an additional 30 min and further poured into 100 mL of ice water. The separated organic phase was washed with 5% $NaHCO_3$ solution, then with distilled water and dried ($MgSO_4$). After concentration under low-pressure, solid product **6** was recrystallized from cyclohexane when white crystals were obtained.

Yield 90% (3.99 g, 9.45 mmol). mp 144–145 °C.

UV/VIS (CH$_3$OH) λ_{max} nm (log$_{10}$ ε_{max}): 202.6 (4.49); 227.3 (4.14); 252.0 (4.35); (CH$_3$CN) λ_{max} nm (log$_{10}$ ε_{max}): 197.1 (4.93); 229.1 (4.48); 252.2 (4.64).

FTIR (KBr disc) $\tilde{\nu}_{max}$ cm^{-1}: 3092 m; 3067 w (νC–H, aromatic); 2963 m (ν_{asym}CH$_3$); 2930 m (νC–H, aliphatic); 2875 m (ν_{sym}CH$_3$); 1825 vs (νC=O); 1650 vs (νC=N); 1599 w; 1573 s; 1469 m (νC=C, aromatic); 1329 vs; 1293 s (ν_{asym}SO$_2$); 1243 m (ν_{asym}C–O–C); 1160 vs (ν_{sym}SO$_2$); 1040 vs (ν_{sym}C–O–C); 845 m (γC–H, aromatic); 614 vs; and 574 s (νC–Br).

^{1}H NMR (300 MHz, CDCl$_3$) δ ppm: 8.16 (d, 2H, H-7, H-11, J = 8.5 Hz); 8.05 (d, 2H, H-8, H-10, J = 8.5 Hz); 7.83 (d, 2H, H-13, H-17, J = 8.5 Hz); 7.68 (d, 2H, H-14, H-16, J = 8.5 Hz); 4.32 (d, 1H, H-4, J = 4.7 Hz); 2.40 (septd, 1H, H-18, J = 6.9, 4.7 Hz); 1.14 (d, 3H, H-19, J = 6.9 Hz); and 0.99 (d, 3H, H-20, J = 6.9 Hz).

^{13}C NMR (75 MHz, CDCl$_3$) δ ppm: 176.98 (C-5); 160.36 (C-2); 144.89 (C-9); 139.92 (C-12); 132.97 (C-14, C-16); 130.58 (C-6); 129.49 (C-13, C-17); 129.23 (C-15); 128.99 (C-7, C-11); 128.19 (C-8, C-10); 71.04 (C-4); 31.42 (C-18); 18.90 (C-19); and 17.65 (C-20).

GC/EI–MS (70 eV) m/z (rel. abund. %): 379 (^{79}Br)/381 (^{81}Br) (71.61/100, BP) [M-C$_3$H$_6$]$^{·+}$; 323/325 (41.31/38.56) [^{79}BrC$_6$H$_4$SO$_2$C$_6$H$_4$CHNH]$^{·+}$/[^{81}BrC$_6$H$_4$SO$_2$C$_6$H$_4$CHNH]$^{·+}$ or [^{79}BrC$_6$H$_4$SO$_2$C$_6$H$_4$CO]$^+$/[^{81}BrC$_6$H$_4$SO$_2$C$_6$H$_4$CO]$^+$; 207 (29.87); 203 (40.89) [^{79}BrC$_6$H$_4$SO]$^+$; 76 (24.15) [C$_6$H$_4$]$^{·+}$; 43 (37.71) [C$_3$H$_7$]$^+$; and t_R 36.53 min.

Reversed-phase HPLC (CH$_3$OH–H$_2$O 60:40, V/V; flow rate 1 mL/min; λ 250 nm): purity 94.16%; t_R 4.05 min.

Elem. anal. (%) found: C, 51.24; H, 3.81; N, 3.33; and S, 7.57; calcd. for C$_{18}$H$_{16}$BrNO$_4$S (M_r 422.29): C, 51.19; H, 3.82; N, 3.32; and S, 7.59.

4.3.3. General Procedure for the Preparation of 2-Acylamino Ketones 7a,b and Their Characterization

To a solution of raw 2-{4-[(4-bromophenyl)sulfonyl]phenyl}-4-isopropyl-4H-1,3-oxazol-5-one **6** (2.11 g, 5 mmol) in 25 mL of anhydrous arene (i.e., benzene and toluene, respectively), 2.00 g (15 mmol) of anhydrous aluminum trichloride were added portion-wise with magnetic stirring at room temperature. The reaction mixture was stirred for a further 20 h until hydrogen chloride emission ceased and then poured into 100 mL of ice water acidified with 5 mL of 37% HCl. A precipitate was obtained, which was isolated by filtration, washed on the filter with cold distilled water, and further with a cold ethanol/distilled water (1:1, V/V) mixture. The aqueous filtrate was extracted with CH$_2$Cl$_2$ (2 × 15 mL), then the organic phase was washed with distilled water, dried (Na$_2$SO$_4$), and concentrated to dryness by vacuum distillation, when the second fraction of **7** was isolated. Purification of the crude product by recrystallization from ethanol afforded colorless crystals.

4-[(4-Bromophenyl)sulfonyl]-N-(3-methyl-1-oxo-1-phenylbutan-2-yl)benzamide **7a**

Compound **7a** was synthesized by the reaction of **6** with benzene (25 mL, 21.85 g, 279.7 mmol).

Yield 80% (2.00 g, 4.00 mmol). mp 164–166 °C.

UV/VIS (CH$_3$OH) λ_{max} nm (log$_{10}$ ε_{max}): 202.6 (4.49); 252.0 (4.19); (CH$_3$CN) λ_{max} nm (log$_{10}$ ε_{max}): 198.0 (5.03); 250.2 (4.77).

FTIR (KBr disc) $\tilde{\nu}_{max}$ cm^{-1}: 3301 s (νN–H); 3089 m; 3061 m; 3040 m (νC–H, aromatic); 2962 m (ν_{asym}CH$_3$); 2931 m (νC–H, aliphatic); 2872 m (ν_{sym}CH$_3$); 1655 vs (νO=C–O and νO=C–N, amide I; overlapped); 1597 m; 1574 s; 1483 m; 1469 m; 1448 m (νC=C, aromatic); 1528 s (δN–H, amide II); 1323 vs; 1289 s (ν_{asym}SO$_2$); 1161 vs (ν_{sym}SO$_2$); 854 m (γC–H, aromatic); 615 vs; and 580 s (νC–Br).

^{1}H NMR (300 MHz, DMSO-d_6) δ ppm: 8.99 (d, 1H, H-3, J = 8.0 Hz); 8.05 (m, 6H, H-7, H-8, H-10, H-11, H-22, H-26); 7.91 (d, 2H, H-13, H-17, J = 8.5 Hz); 7.84 (d, 2H, H-14, H-16, J = 8.5 Hz); 7.64 (br t, 1H, H-24, J = 7.6 Hz); 7.53 (br t, 2H, H-23, H-25, J = 7.6 Hz); 5.38 (t, 1H, H-4, J = 7.7 Hz); 2.28 (m, 1H, H-18); 0.92 (d, 3H, H-19, J = 7.1 Hz); and 0.90 (d, 3H, H-20, J = 7.1 Hz).

^{13}C NMR (75 MHz, DMSO-d_6) δ ppm: 199.19 (C-5); 165.57 (C-2); 142.76 (C-9); 139.88 (C-12); 138.78 (C-6); 136.17 (C-21); 133.44 (C-24); 132.94 (C-14, C-16); 129.48 (C-13, C-17);

129.04 (C-7, C-11); 128.82 (C-23, C-25); 128.24 (C-22, C-26); 128.22 (C-15); 127.53 (C-8, C-10); 59.23 (C-4); 29.50 (C-18); 19.72 (C-19); and 18.36 (C-20).

Reversed-phase HPLC (CH$_3$OH–H$_2$O 60:40, V/V; flow rate 1 mL/min; λ 250 nm): purity 92.02%; t_R 4.35 min.

Elem. anal. (%) found: C, 57.63; H, 4.41; N, 2.81; and S, 6.40; calcd. for C$_{24}$H$_{22}$BrNO$_4$S (M_r 500.40): C, 57.60; H, 4.43; N, 2.80; and S, 6.41.

4-[(4-Bromophenyl)sulfonyl]-N-[3-methyl-1-oxo-1-(p-tolyl)butan-2-yl]benzamide **7b**

Compound **7b** was synthesized by the reaction of **6** with toluene (25 mL, 21.63 g, 234.8 mmol).

Yield 86% (2.21 g, 4.30 mmol). mp 154–155 °C.

UV/VIS (CH$_3$OH) λ_{max} nm (log$_{10}$ ε_{max}): 202.6 (4.47); 255.5 (4.18); (CH$_3$CN) λ_{max} nm (log$_{10}$ ε_{max}): 197.8 (5.01); 254.8 (4.78).

FTIR (KBr disc) $\widetilde{\nu}_{max}$ cm^{-1}: 3281 s (νN–H); 3086 m; 3058 m; 3037 m (νC–H, aromatic); 2962 m (ν_{asym}CH$_3$); 2929 m (νC–H, aliphatic); 2868 m (ν_{sym}CH$_3$); 1655 vs ($\nu\underline{O=C}$–O and $\nu\underline{O=C}$–N, amide I; overlapped); 1606 m; 1572 s; 1483 m; 1467 m (νC=C, aromatic); 1530 s (δN–H, amide II); 1325 s; 1305 m; 1287 m (ν_{asym}SO$_2$); 1161 vs (ν_{sym}SO$_2$); 858 m (γC–H, aromatic); 617 s; and 576 s (νC–Br).

^{1}H NMR (300 MHz, DMSO-d_6) δ ppm: 8.94 (d, 1H, H-3, J = 8.0 Hz); 8.06 (d, 2H, H-7, H-11, J = 8.8 Hz); 8.03 (d, 2H, H-8, H-10, J = 8.8 Hz); 7.95 (d, 2H, H-22, H-26, J = 8.2 Hz); 7.91 (d, 2H, H-13, H-17, J = 8.8 Hz); 7.84 (d, 2H, H-14, H-16, J = 8.8 Hz); 7.33 (d, 2H, H-23, H-25, J = 8.2 Hz); 5.36 (t, 1H, H-4, J = 7.7 Hz); 2.36 (s, 3H, CH$_3$); 2.27 (m, 1H, H-18); 0.92 (d, 3H, H-19, J = 7.4 Hz); and 0.89 (d, 3H, H-20, J = 7.4 Hz).

^{13}C NMR (75 MHz, DMSO-d_6) δ ppm: 198.57 (C-5); 165.47 (C-2); 143.89 (C-24); 142.72 (C-9); 139.88 (C-12); 138.81 (C-6); 133.61 (C-21); 132.92 (C-14, C-16); 129.44 (C-13, C-17); 129.34 (C-23, C-25); 129.00 (C-7, C-11); 128.36 (C-22, C-26); 128.18 (C-15); 127.49 (C-8, C-10); 59.05 (C-4); 29.56 (C-18); 21.12 (CH$_3$); 19.71 (C-19); and 18.33 (C-20).

Reversed-phase HPLC (CH$_3$OH–H$_2$O 60:40, V/V; flow rate 1 mL/min; λ 250 nm): purity 99.99%; t_R 4.95 min.

Elem. anal. (%) found: C, 58.34; H, 4.69; N, 2.71; and S, 6.21; calcd. for C$_{25}$H$_{24}$BrNO$_4$S (M_r 514.43): C, 58.37; H, 4.70; N, 2.72; and S, 6.23.

4.3.4. General Procedure for the Preparation of 2,5-Disubstituted 4-Isopropyl-1,3-oxazoles **8a,b** and Their Characterization

A mixture of 10 mmol of crude **7** and 20 mL (33.40 g, 217.8 mmol) of phosphoryl trichloride was heated under reflux during 4 h. The excess phosphoryl trichloride was evaporated under reduced pressure. The resulting oily residue was cooled, treated with an ice water mixture, and extracted with 2 × 20 mL CH$_2$Cl$_2$. The organic phase was separated and washed with 5% NaHCO$_3$ solution, then with distilled water and dried (Na$_2$SO$_4$). The solvent was evaporated off under vacuum. Purification of the resulting solid was performed by recrystallization from ethanol when colorless crystals of **8** were obtained.

2-{4-[(4-Bromophenyl)sulfonyl]phenyl}-4-isopropyl-5-phenyl-1,3-oxazole **8a**

Compound **8a** was synthesized from 5.00 g (10 mmol) of 2-acylamino ketone **7a**. Yield 94% (4.53 g, 9.39 mmol). mp 165–167 °C.

UV/VIS (CH$_3$OH) λ_{max} nm (log$_{10}$ ε_{max}): 202.6 (4.48); 249.3 (4.10); 333.9 (4.13); (CH$_3$CN) λ_{max} nm (log$_{10}$ ε_{max}): 196.2 (4.95); 248.6 (4.56); 335.6 (4.64).

FTIR (KBr disc) $\widetilde{\nu}_{max}$ cm^{-1}: 3092 m; 3063 m; 3041 w (νC–H, aromatic); 2963 s (ν_{asym}CH$_3$); 2930 m (νC–H, aliphatic); 2869 m (ν_{sym}CH$_3$); 1602 m (νC=N); 1588 m; 1574 s; 1543 w; 1494 m; 1471 m; 1446 m (νC=C, aromatic); 1323 vs; 1293 m (ν_{asym}SO$_2$); 1280 m (ν_{asym}C–O–C); 1155 vs (ν_{sym}SO$_2$); 1099 s (ν_{sym}C–O–C); 844 m (γC–H, aromatic); 617 s; and 568 s (νC–Br).

^{1}H NMR (300 MHz, CDCl$_3$) δ ppm: 8.22 (d, 2H, H-7, H-11, J = 8.8 Hz); 8.00 (d, 2H, H-8, H-10, J = 8.8 Hz); 7.82 (d, 2H, H-13, H-17, J = 8.8 Hz); 7.65 (d, 2H, H-14, H-16, J = 8.8 Hz); 7.64

(dd, 2H, H-22, H-26, J = 7.4, 1.4 Hz); 7.48 (br t, 2H, H-23, H-25, J = 7.4 Hz); 7.37 (tt, 1H, H-24, J = 7.4, 1.4 Hz); 3.29 (sept, 1H, H-18, J = 6.9 Hz); and 1.36 (d, 6H, H-19, H-20, J = 6.9 Hz).

^{13}C NMR (75 MHz, CDCl$_3$) δ ppm: 157.78 (C-2); 145.47 (C-5); 144.19 (C-4); 141.71 (C-12); 140.56 (C-9); 132.80 (C-14, C-16); 132.33 (C-6); 129.29 (C-13, C-17); 129.01 (C-23, C-25); 128.81 (C-21); 128.78 (C-15); 128.39 (C-24); 128.26 (C-7, C-11); 127.09 (C-8, C-10); 126.25 (C-22, C-26); 26.10 (C-18); and 22.07 (C-19, C-20).

Reversed-phase HPLC (CH$_3$OH–H$_2$O 70:30, V/V; flow rate 1 mL/min; λ 335 nm): purity 98.95%; t_R 5.97 min.

Elem. anal. (%) found: C, 59.71; H, 4.18; N, 2.91; and S, 6.67; calcd. for C$_{24}$H$_{20}$BrNO$_3$S (M_r 482.39): C, 59.76; H, 4.18; N, 2.90; and S, 6.65.

2-{4-[(4-Bromophenyl)sulfonyl]phenyl}-4-isopropyl-5-(p-tolyl)-1,3-oxazole **8b**

Compound **8b** was synthesized from 5.14 g (10 mmol) of 2-acylamino ketone **7b**. Yield 91% (4.52 g, 9.10 mmol). mp 215–217 °C.

UV/VIS (CH$_3$OH) λ_{max} nm (log$_{10}$ ε_{max}): 202.6 (4.48); 249.3 (4.11); 337.4 (4.09); (CH$_3$CN) λ_{max} nm (log$_{10}$ ε_{max}): 195.3 (4.96); 249.5 (4.59); 340.1 (4.64).

FTIR (KBr disc) $\tilde{\nu}_{max}$ cm^{-1}: 3091 m; 3069 w; 3028 w (νC–H, aromatic); 2963 m (ν_{asym}CH$_3$); 2925 m (νC–H, aliphatic); 2869 m (ν_{sym}CH$_3$); 1601 m (νC=N); 1592 m; 1574 m; 1545 w; 1508 m; 1471 m (νC=C, aromatic); 1322 s; 1292 m (ν_{asym}SO$_2$); 1280 m (ν_{asym}C–O–C); 1156 vs (ν_{sym}SO$_2$); 1097 s (ν_{sym}C–O–C); 846 m (γC–H, aromatic); 617 s; and 568 s (νC–Br).

^{1}H NMR (300 MHz, CDCl$_3$) δ ppm: 8.21 (d, 2H, H-7, H-11, J = 8.5 Hz); 8.00 (d, 2H, H-8, H-10, J = 8.5 Hz); 7.82 (d, 2H, H-13, H-17, J = 8.5 Hz); 7.65 (d, 2H, H-14, H-16, J = 8.5 Hz); 7.53 (d, 2H, H-22, H-26, J = 8.2 Hz); 7.28 (d, 2H, H-23, H-25, J = 8.2 Hz); 3.26 (sept, 1H, H-18, J = 6.9 Hz); 2.41 (s, 3H, CH$_3$); and 1.35 (d, 6H, H-19, H-20, J = 6.9 Hz).

^{13}C NMR (75 MHz, CDCl$_3$) δ ppm: 157.52 (C-2); 145.69 (C-5); 143.63 (C-4); 141.57 (C-12); 140.61 (C-9); 138.48 (C-24); 132.80 (C-14, C-16); 132.43 (C-6); 129.70 (C-23, C-25); 129.29 (C-13, C-17); 128.76 (C-15); 128.25 (C-7, C-11); 127.03 (C-8, C-10); 126.22 (C-22, C-26); 126.01 (C-21); 26.08 (C-18); 22.07 (C-19, C-20); and 21.47 (CH$_3$).

Reversed-phase HPLC (CH$_3$OH–H$_2$O 70:30, V/V; flow rate 1 mL/min; λ 335 nm): purity 99.99%; t_R 7.32 min.

Elem. anal. (%) found: C, 60.52; H, 4.46; N, 2.81; and S, 6.48; calcd. for C$_{25}$H$_{22}$BrNO$_3$S (M_r 496.42): C, 60.49; H, 4.47; N, 2.82; and S, 6.46.

4.4. Assessment of Antimicrobial Activity

The evaluation of the influence of the compounds on microbial growth and biofilm formation was performed using an adapted disk diffusion assay, the standard broth microdilution method, and the microtiter plate test [80].

4.4.1. Pathogenic Microbial Strains

The pathogenic microorganisms tested included three Gram-positive bacteria, i.e., *B. subtilis* 6683, *E. faecium* E5, *S. aureus* ATCC 6538, and two Gram-negative bacteria, namely *E. coli* ATCC 8739, *P. aeruginosa* ATCC 27853, as well as a fungal *C. albicans* 393 strain.

4.4.2. Agar Diffusion Assay

The antimicrobial activity was screened using an adapted agar diffusion method [81]. Briefly, the bacterial and fungal suspensions were prepared in phosphate-buffered saline from fresh microbial cultures and their density was adjusted to 0.5 McFarland. The compounds were dissolved in DMSO at a concentration of 5 mg/mL and then a volume of 5 µL was spotted on the inoculated plates. The diameters of the growth inhibition zones were measured after incubation for 24 h at 37 °C. Standardized discs of ciprofloxacin, 5 µg (Oxoid), and fluconazole, 25 µg (Oxoid) were used as positive controls.

4.4.3. Evaluation of the Effects of the Tested Compounds on the Antibiotic Susceptibility Profile

Standardized microbial suspensions of *E. faecium* E5 and *S. aureus* ATCC 6538 were put in contact with subinhibitory concentrations (0.25 mg/mL) of tested compounds **5** and **6** prepared in broth. After incubation at 37 °C for 24 h, the bacterial cultures were used to evaluate their antibiotic susceptibility profile. The disk diffusion method was performed to assess the susceptibility to the following antibiotics: ampicillin, linezolid, penicillin, vancomycin for Gram-positive *E. faecium* E5 strain and azithromycin, cefoxitin, clindamycin, linezolid, penicillin, rifampicin, trimethoprim-sulfamethoxazole, and vancomycin (bioMérieux, Paris, France) for Gram-positive *S. aureus* ATCC 6538 strain. The reading of the results was performed following the Clinical and Laboratory Standards Institute (CLSI, Berwyn, PA, USA) guidelines.

4.4.4. Broth Microdilution Assay

The standard broth microdilution test was performed on 96-well microtiter plates. The compounds were solubilized in DMSO at a concentration of 10 mg/mL. Serial dilutions of the tested compounds were prepared in sterile broth and the obtained solutions of different concentrations were inoculated with each standardized microbial suspension. Ciprofloxacin and fluconazole (Oxoid) served as positive controls. After incubation for 24 h at 37 °C, the MIC values were determined as the lowest concentration of compound required to inhibit the microbial growth, as detected spectrophotometrically (Apollo LB 911 ELISA Reader, Berthold Technologies GmbH & Co. KG, Waltham, MA, USA) [82]. The assays were performed in triplicate.

4.4.5. Microplate Microtiter Assay

The influence of the compounds on the biofilm formation was determined using the microtiter plate assay. Briefly, after the determination of the MIC values, the microplates were emptied and washed three times with sterile distilled water for the removal of planktonic microbial cells. The biofilms adherent on the plastic wells were fixed with methanol for 5 min and stained with 1% crystal violet solution for a quarter of an hour. The excess dye was removed and thereafter the fixed dye was resuspended in 33% acetic acid. Ciprofloxacin (Oxoid) was used as a positive control for antibacterial tests and fluconazole (Oxoid) for antifungal testing. The absorbance of the colored solutions was recorded with an ELISA Reader (Apollo LB 911). The minimal biofilm inhibitory concentration (MBIC), defined as the lowest compound concentration required to inhibit biofilm formation, was determined. The results from three separate biological replicates were mediated [83].

4.5. Evaluation of Antioxidant Activity

The antioxidant activity of newly obtained derivatives **5–8** was studied spectrophotometrically by means of the DPPH, ABTS, and ferric reducing power methods [84–86].

4.5.1. Antioxidant Activity Assay by DPPH Method

The antioxidant effect of new compounds was determined according to the well-known DPPH assay based on the fact that the antioxidant samples react with stable 2,2-diphenyl-1-(2,4,6-trinitrophenyl)hydrazin-1-yl (2,2-diphenyl-1-picrylhydrazyl, DPPH), an effective free radical trap with an absorption band at 517 nm, which converts to its reduced form, i.e., 1,1-diphenyl-2-(2,4,6-trinitrophenyl)hydrazine (2,2-diphenyl-1-picryl hydrazine), with a color change from purple to yellow accompanied by a decrease in absorption at 517 nm [84].

The antioxidant potential of the tested compounds was investigated according to the method previously described by Blois [87–89], with some modifications, and compared with the free radical scavenging activity of AA, BHA, and BHT standards.

Briefly, 2 mL of a solution of DPPH in ethanol at a concentration of 400 µM was added to 2 mL of each solution of the tested compound in DMSO at a concentration of 500 µM. After maintaining the samples in the dark at room temperature for 30 min, the absorbance of each sample at the wavelength of 517 nm was measured on a Specord 40 UV/VIS spectrophotometer (Analytik Jena AG, Jena, Germany). Then, the radical scavenging activity (*RSA*) in percent (%) was calculated using the following formula:

$$RSA\ (\%) = (1 - \frac{A_\text{sample} - A_\text{blank sample}}{A_\text{DPPH control}}) \times 100$$

where A_sample is the absorbance of the tested compound solution with the DPPH solution, $A_\text{blank sample}$ is the absorbance of the tested compound solution (without the DPPH solution), and $A_\text{DPPH control}$ is the absorbance of the DPPH solution (without the tested compound solution) [90]. Each analysis was performed on three replicates and the results were averaged.

4.5.2. Antioxidant Activity Assay by ABTS Method

The ABTS discoloration test is based on the ability of the antioxidant agents to scavenge the long-life ABTS radical cation, $ABTS^{\bullet+}$ [91]. The stable green-blue radical cationic chromophore, $ABTS^{\bullet+}$, which has an absorption maximum at the wavelength of 734 nm, by reaction with most antioxidants turns into its colorless neutral form, i.e., 2,2′-azino-bis(3-ethylbenzothiazoline-6-sulfonic acid) (ABTS) [85].

The antioxidant capacity of the tested compounds was assessed as described earlier by Re et al. [91], with some modifications. The $ABTS^{\bullet+}$ is prepared by oxidation of ABTS with radical-initiator potassium persulfate. Initially, the stock solution of ABTS was obtained by dissolving it in water at a concentration of 7 μM. The ABTS radical cation ($ABTS^{\bullet+}$) was then generated by reacting the 7 μM ABTS stock solution with a 2.45 μM $K_2S_2O_8$ solution (1:1, V/V) and kept in the dark at room temperature for 12–16 h. Before analysis, this solution was diluted with ethanol and equilibrated at 30 °C to have absorbance at 734 nm, A_{734} of 0.7000 ± 0.02. Subsequently, 2 mL of $ABTS^{\bullet+}$ solution was added to 2 mL of 500 μM tested compound solution in DMSO. After 6 min, the absorbance of each sample was read at 734 nm with a UV/VIS spectrophotometer (Specord 40, Analytik Jena AG, Jena, Germany) and converted into the percentage radical scavenging activity, %*RSA*, using the formula:

$$RSA\ (\%) = (1 - \frac{A_\text{sample} - A_\text{blank sample}}{A_{ABTS^{\bullet+}\ \text{control}}}) \times 100$$

where A_sample is the absorbance of the tested compound solution with the $ABTS^{\bullet+}$ solution, $A_\text{blank sample}$ is the absorbance of the tested compound solution (without the $ABTS^{\bullet+}$ solution), and $A_{ABTS^{\bullet+}\ \text{control}}$ is the absorbance of the $ABTS^{\bullet+}$ solution (without the tested compound solution) [89,90]. All determinations were undertaken in triplicate and the results were averaged. The same antioxidant agents (AA, BHA, and BHT) were used as positive controls.

4.5.3. Ferric Reducing Power Assay

In the ferric reducing power test, is measured the reduction in the bright red color potassium ferricyanide to the intense blue color ferric ferrocyanide (Prussian blue), which occurs by means of antioxidants in acidic environments, with an increase in absorbance at 700 nm [84].

The ability of all the newly synthesized compounds to reduce iron(III) to iron(II) was determined by the modified Oyaizu method [92–94]. The sample solutions were prepared in DMSO at a concentration of 500 μM. Two milliliters of each tested compound solution were mixed with 2 mL of 0.2 M phosphate buffer (pH 6.6) and 2 mL of 1% $K_3[Fe(CN)_6]$ solution. After 20 min incubation at 50 °C, 2 mL of 10% trichloroacetic acid solution was added, and the mixture was centrifuged for 15 min at 4500 rpm. Finally, 2 mL of the upper layer was mixed with 2 mL of deionized water and 0.4 mL of 0.1% $FeCl_3$ solution, and after 5 min the absorbance was recorded at 700 nm with an Analytik Jena Specord 40 UV/VIS spectrophotometer. All analyses were performed on three replicates, and the results were averaged and compared with those of antioxidant standards used, i.e., AA, BHA, and BHT.

4.6. Daphnia magna Toxicity Bioassay

In ecotoxicology, the planktonic crustacean *Daphnia magna* Staus is specified for use in the "OECD Guidelines for the Testing of Chemicals". Test No. 202: "*Daphnia* sp., Acute Immobilization Test" is an acute toxicity study, in which young daphnids, aged less than 24 h at the beginning of the test, are exposed to different concentrations of the chemical under test for 48 h and the half-maximal effective concentration (EC_{50}) determined [95].

In this study, the young *D. magna* Straus organisms were selected according to their size from a parthenogenetic culture. All compounds were tested at six concentrations ranging from 2 to 50 µg/mL, in duplicate, with each replicate having 10 individuals. The concentration range was selected based on the compounds' solubilities and a pre-screening bioassay. L-Valine, compound **3**, and a 1% DMSO solution were used as controls. All determinations were carried out using the same conditions (25 °C, a long day photoperiod of 16 h light/8 h dark cycle) in a Sanyo MLR-351H climate test chamber (Sanyo, San Diego, CA, USA) [48,96,97]. After 24 and 48 h of exposure, the lethality was evaluated. LC_{50} was calculated for each compound based on interpolating on lethality curves which were obtained using the least square fit method. The 95% confidence intervals (95% CI) were calculated and the goodness of fit was also evaluated. The calculations were performed using GraphPad Prism v5.1 software (GraphPad Software, Inc., La Jolla, CA, USA). The prediction of LC_{50} at 48 h was made using GUSAR online application (Institute of Biomedical Chemistry, Moscow, Russia, http://www.way2drug.com/gusar/) [98].

5. Conclusions

In order to identify new bioactive compounds, we designed and synthesized novel derivatives that incorporate in the same molecule the *N*-acyl-α-amino acid, 4*H*-1,3-oxazol-5-one, 2-acylamino ketone, or 1,3-oxazole template and, a biologically active fragment derived from diphenyl sulfone. The obtained compounds' structures were confirmed on the basis of spectral studies and elemental analysis data. Taken together, the results of the qualitative and quantitative antimicrobial activity evaluation, antioxidant effect assessment, toxicity bioassay, as well as of in silico analysis revealed a promising potential of *N*-acyl-L-valine **5** and 2-substituted 4-isopropyl-4*H*-1,3-oxazol-5-one **6** for the development of novel antimicrobial agents to combat infections produced by Gram-positive bacterial strains, and in particular of *E. faecium* biofilm-associated infections.

6. Patents

Patent application no. a 2019 00668: Theodora-Venera Apostol, Ștefania-Felicia Bărbuceanu, Laura Ileana Socea, Ioana Șaramet, Constantin Drăghici, Valeria Rădulescu, Mariana-Carmen Chifiriuc, Luminița Gabriela Măruțescu, Octavian Tudorel Olaru, and George Mihai Nițulescu, 4-Isopropyl-1,3-oxazol-5(4*H*)-one Derivatives Containing a Diaryl sulfonyl Substituent in Position 2 with Antimicrobial Action, published in RO-BOPI No. 9/2021, from 30 september 2021.

Supplementary Materials: The following supporting information can be downloaded at: https://www.mdpi.com/article/10.3390/pr10091800/s1, Figure S1: The ^{1}H NMR spectrum of 2-{4-[(4-bromophenyl)sulfonyl]benzamido}-3-methylbutanoic acid **5**; Figure S2: The ^{13}C NMR spectrum of 2-{4-[(4-bromophenyl)sulfonyl]benzamido}-3-methylbutanoic acid **5**; Figure S3: The 2D HETCOR spectrum of 2-{4-[(4-bromophenyl)sulfonyl]benzamido}-3-methylbutanoic acid **5**; Figure S4: The ^{1}H NMR spectrum of 2-{4-[(4-bromophenyl)sulfonyl]phenyl}-4-isopropyl-4*H*-1,3-oxazol-5-one **6**; Figure S5: The ^{13}C NMR spectrum of 2-{4-[(4-bromophenyl)sulfonyl]phenyl}-4-isopropyl-4*H*-1,3-oxazol-5-one **6**; Figure S6: The 2D HETCOR spectrum of 2-{4-[(4-bromophenyl)sulfonyl]phenyl}-4-isopropyl-4*H*-1,3-oxazol-5-one **6**; Figure S7: The ^{1}H NMR spectrum of 4-[(4-bromophenyl)sulfonyl]-*N*-(3-methyl-1-oxo-1-phenylbutan-2-yl)benzamide **7a**; Figure S8: The ^{13}C NMR spectrum of 4-[(4-bromophenyl)sulfonyl]-*N*-(3-methyl-1-oxo-1-phenylbutan-2-yl)benzamide **7a**; Figure S9: The 2D HETCOR spectrum of 4-[(4-bromophenyl)sulfonyl]-*N*-(3-methyl-1-oxo-1-phenylbutan-2-yl)benzamide **7a**; Figure S10: The ^{1}H NMR spectrum of 4-[(4-bromophenyl)sulfonyl]-*N*-[3-methyl-1-oxo-1-(*p*-tolyl)butan-2-yl]benzamide **7b**; Figure S11: The ^{13}C NMR spectrum of 4-[(4-bromophenyl)sulfonyl]-*N*-[3-methyl-1-oxo-1-(*p*-

tolyl)butan-2-yl]benzamide **7b**; Figure S12: The 2D COSY spectrum of 4-[(4-bromophenyl)sulfonyl]-
N-[3-methyl-1-oxo-1-(p-tolyl)butan-2-yl]benzamide **7b**; Figure S13: The 2D HETCOR spectrum of
4-[(4-bromophenyl)sulfonyl]-N-[3-methyl-1-oxo-1-(p-tolyl)butan-2-yl]benzamide **7b**; Figure S14: The
^{1}H NMR spectrum of 2-{4-[(4-bromophenyl)sulfonyl]phenyl}-4-isopropyl-5-phenyl-1,3-oxazole **8a**;
Figure S15: The ^{13}C NMR spectrum of 2-{4-[(4-bromophenyl)sulfonyl]phenyl}-4-isopropyl-5-phenyl-
1,3-oxazole **8a**; Figure S16: The 2D HETCOR spectrum of 2-{4-[(4-bromophenyl)sulfonyl]phenyl}-4-
isopropyl-5-phenyl-1,3-oxazole **8a**; Figure S17: The ^{1}H NMR spectrum of 2-{4-[(4-bromophenyl)
sulfonyl]phenyl}-4-isopropyl-5-(p-tolyl)-1,3-oxazole **8b**; Figure S18: The ^{13}C NMR spectrum of
2-{4-[(4-bromophenyl)sulfonyl]phenyl}-4-isopropyl-5-(p-tolyl)-1,3-oxazole **8b**; Figure S19: The 2D
HETCOR spectrum of 2-{4-[(4-bromophenyl)sulfonyl]phenyl}-4-isopropyl-5-(p-tolyl)-1,3-oxazole **8b**;
Figure S20: The FTIR spectrum of 2-{4-[(4-bromophenyl)sulfonyl]benzamido}-3-methylbutanoic acid
5; Figure S21: The FTIR spectrum of 2-{4-[(4-bromophenyl)sulfonyl]phenyl}-4-isopropyl-4H-1,3-
oxazol-5-one **6**; Figure S22: The FTIR spectrum of 4-[(4-bromophenyl)sulfonyl]-N-(3-methyl-1-oxo-
1-phenylbutan-2-yl)benzamide **7a**; Figure S23: The FTIR spectrum of 4-[(4-bromophenyl)sulfonyl]-
N-[3-methyl-1-oxo-1-(p-tolyl)butan-2-yl]benzamide **7b**; Figure S24: The FTIR spectrum of 2-{4-[(4-
bromophenyl)sulfonyl]phenyl}-4-isopropyl-5-phenyl-1,3-oxazole **8a**; Figure S25: The FTIR spectrum
of 2-{4-[(4-bromophenyl)sulfonyl]phenyl}-4-isopropyl-5-(p-tolyl)-1,3-oxazole **8b**; Figure S26: The
UV/VIS spectrum of 2-{4-[(4-bromophenyl)sulfonyl]benzamido}-3-methylbutanoic acid **5** dissolved
in methanol at ≈0.025 mM; Figure S27: The UV/VIS spectrum of 2-{4-[(4-bromophenyl)sulfonyl]phenyl}-
4-isopropyl-4H-1,3-oxazol-5-one **6** dissolved in methanol at ≈0.025 mM; Figure S28: The UV/VIS spectrum of 4-[(4-bromophenyl)sulfonyl]-N-(3-methyl-1-oxo-1-phenylbutan-2-yl)benzamide **7a** dissolved
in methanol at ≈0.025 mM; Figure S29: The UV/VIS spectrum of 4-[(4-bromophenyl)sulfonyl]-N-[3-
methyl-1-oxo-1-(p-tolyl)butan-2-yl]benzamide **7b** dissolved in methanol at ≈0.025 mM; Figure S30:
The UV/VIS spectrum of 2-{4-[(4-bromophenyl)sulfonyl]phenyl}-4-isopropyl-5-phenyl-1,3-oxazole
8a dissolved in methanol at ≈0.025 mM; Figure S31: The UV/VIS spectrum of 2-{4-[(4-bromophenyl)
sulfonyl]phenyl}-4-isopropyl-5-(p-tolyl)-1,3-oxazole **8b** dissolved in methanol at ≈ 0.025 mM; Figure S32:
The UV/VIS spectrum of 2-{4-[(4-bromophenyl)sulfonyl]benzamido}-3-methylbutanoic acid **5** dissolved in acetonitrile at ≈ 0.015 mM; Figure S33: The UV/VIS spectrum of 2-{4-[(4-bromophenyl)
sulfonyl]phenyl}-4-isopropyl-4H-1,3-oxazol-5-one **6** dissolved in acetonitrile at ≈0.015 mM;
Figure S34: The UV/VIS spectrum of 4-[(4-bromophenyl)sulfonyl]-N-(3-methyl-1-oxo-1-phenylbutan-
2-yl)benzamide **7a** dissolved in acetonitrile at ≈0.015 mM; Figure S35: The UV/VIS spectrum of
4-[(4-bromophenyl)sulfonyl]-N-[3-methyl-1-oxo-1-(p-tolyl)butan-2-yl]benzamide **7b** dissolved in acetonitrile at ≈0.015 mM; Figure S36: The UV/VIS spectrum of 2-{4-[(4-bromophenyl)sulfonyl]phenyl}-
4-isopropyl-5-phenyl-1,3-oxazole **8a** dissolved in acetonitrile at ≈0.015 mM; Figure S37: The UV/VIS
spectrum of 2-{4-[(4-bromophenyl)sulfonyl]phenyl}-4-isopropyl-5-(p-tolyl)-1,3-oxazole **8b** dissolved
in acetonitrile at ≈0.015 mM; Figure S38: The GC/MS spectrum of 2-{4-[(4-bromophenyl)sulfonyl]phenyl}-
4-isopropyl-4H-1,3-oxazol-5-one **6**; Figure S39: The RP–HPLC chromatogram of 2-{4-[(4-bromophenyl)
sulfonyl]benzamido}-3-methylbutanoic acid **5**; Figure S40: The RP–HPLC chromatogram of 2-{4-
[(4-bromophenyl)sulfonyl]phenyl}-4-isopropyl-4H-1,3-oxazol-5-one **6**; Figure S41: The RP–HPLC
chromatogram of 4-[(4-bromophenyl)sulfonyl]-N-(3-methyl-1-oxo-1-phenylbutan-2-yl)benzamide **7a**;
Figure S42: The RP–HPLC chromatogram of 4-[(4-bromophenyl)sulfonyl]-N-[3-methyl-1-oxo-1-(p-
tolyl)butan-2-yl]benzamide **7b**; Figure S43: The RP–HPLC chromatogram of 2-{4-[(4-bromophenyl)
sulfonyl]phenyl}-4-isopropyl-5-phenyl-1,3-oxazole **8a**; Figure S44: The RP–HPLC chromatogram of
2-{4-[(4-bromophenyl)sulfonyl]phenyl}-4-isopropyl-5-(p-tolyl)-1,3-oxazole **8b**.

Author Contributions: Conceptualization, T.-V.A.; methodology, T.-V.A., M.C.C., L.-I.S., C.D., O.T.O.,
G.M.N., D.-C.V., L.G.M., E.M.P., G.S. and S.-F.B.; investigation, T.-V.A., M.C.C., L.-I.S., C.D., O.T.O.,
G.M.N., D.-C.V., L.G.M., E.M.P., G.S. and S.-F.B.; writing—original draft preparation, T.-V.A., M.C.C.,
L.-I.S., C.D., O.T.O., G.M.N., D.-C.V., L.G.M., E.M.P., G.S. and S.-F.B.; writing—review and editing,
T.-V.A., M.C.C., C.D., O.T.O., G.M.N. and S.-F.B. All authors have read and agreed to the published
version of the manuscript.

Funding: This research was funded by the Publish Not Perish Grants at the "Carol Davila" University
of Medicine and Pharmacy, Bucharest, Romania.

Institutional Review Board Statement: Not applicable.

Informed Consent Statement: Not applicable.

Data Availability Statement: The datasets used and/or analyzed during the current study are available from the corresponding author upon reasonable request.

Conflicts of Interest: The authors declare no conflict of interest.

References

1. Christaki, E.; Marcou, M.; Tofarides, A. Antimicrobial Resistance in Bacteria: Mechanisms, Evolution, and Persistence. *J. Mol. Evol.* **2020**, *88*, 26–40. [CrossRef] [PubMed]
2. Chen, J.; Lv, S.; Liu, J.; Yu, Y.; Wang, H.; Zhang, H. An Overview of Bioactive 1,3-Oxazole-Containing Alkaloids from Marine Organisms. *Pharmaceuticals* **2021**, *14*, 1274. [CrossRef]
3. Mhlongo, J.T.; Brasil, E.; de la Torre, B.G.; Albericio, F. Naturally Occurring Oxazole-Containing Peptides. *Mar. Drugs* **2020**, *18*, 203. [CrossRef]
4. Zheng, X.; Liu, W.; Zhang, D. Recent Advances in the Synthesis of Oxazole-Based Molecules via van Leusen Oxazole Synthesis. *Molecules* **2020**, *25*, 1594. [CrossRef] [PubMed]
5. Kakkar, S.; Narasimhan, B. A comprehensive review on biological activities of oxazole derivatives. *BMC Chem.* **2019**, *13*, 16. [CrossRef] [PubMed]
6. Li, Q.; Seiple, I.B. Modular Synthesis of Streptogramin Antibiotics. *Synlett* **2021**, *32*, 647–654. [CrossRef] [PubMed]
7. Velluti, F.; Mosconi, N.; Acevedo, A.; Borthagaray, G.; Castiglioni, J.; Faccio, R.; Back, D.F.; Moyna, G.; Rizzotto, M.; Torre, M.H. Synthesis, characterization, microbiological evaluation, genotoxicity and synergism tests of new nano silver complexes with sulfamoxole X-ray diffraction of [Ag$_2$(SMX)$_2$]·DMSO. *J. Inorg. Biochem.* **2014**, *141*, 58–69. [CrossRef] [PubMed]
8. Kohlmann, F.W.; Kuhne, J.; Wagener, H.H.; Weifenbach, H. Studies on microbiology and pharmacology of sulfaguanol, a new sulfonamide with effect on the intestines. *Arzneimittelforschung* **1973**, *23*, 172–178.
9. Kim, H.J.; Ryu, H.; Song, J.-Y.; Hwang, S.-G.; Jalde, S.S.; Choi, H.-K.; Ahn, J. Discovery of Oxazol-2-amine Derivatives as Potent Novel FLT3 Inhibitors. *Molecules* **2020**, *25*, 5154. [CrossRef] [PubMed]
10. Schmitt, F.; Gosch, L.C.; Dittmer, A.; Rothemund, M.; Mueller, T.; Schobert, R.; Biersack, B.; Volkamer, A.; Höpfner, M. Oxazole-Bridged Combretastatin A-4 Derivatives with Tethered Hydroxamic Acids: Structure–Activity Relations of New Inhibitors of HDAC and/or Tubulin Function. *Int. J. Mol. Sci.* **2019**, *20*, 383. [CrossRef] [PubMed]
11. Matio Kemkuignou, B.; Treiber, L.; Zeng, H.; Schrey, H.; Schobert, R.; Stadler, M. Macrooxazoles A–D, New 2,5-Disubstituted Oxazole-4-Carboxylic Acid Derivatives from the Plant Pathogenic Fungus *Phoma macrostoma*. *Molecules* **2020**, *25*, 5497. [CrossRef] [PubMed]
12. Jakeman, D.L.; Bandi, S.; Graham, C.L.; Reid, T.R.; Wentzell, J.R.; Douglas, S.E. Antimicrobial Activities of Jadomycin B and Structurally Related Analogues. *Antimicrob. Agents Chemother.* **2009**, *53*, 1245–1247. [CrossRef] [PubMed]
13. De Koning, C.B.; Ngwira, K.J.; Rousseau, A.L. Biosynthesis, synthetic studies, and biological activities of the jadomycin alkaloids and related analogues. In *The Alkaloids: Chemistry and Biology*; Knölker, H.-J., Ed.; Academic Press: Cambridge, MA, USA, 2020; Volume 84, pp. 125–199. ISBN 978-0-12-820982-0.
14. De Azeredo, C.M.O.; Ávila, E.P.; Pinheiro, D.L.J.; Amarante, G.W.; Soares, M.J. Biological activity of the azlactone derivative EPA-35 against *Trypanosoma cruzi*. *FEMS Microbiol. Lett.* **2017**, *364*, fnx020. [CrossRef]
15. Pinto, I.L.; West, A.; Debouck, C.M.; DiLella, A.G.; Gorniak, J.G.; O'Donnell, K.C.; O'Shannessy, D.J.; Patel, A.; Jarvest, R.L. Novel, selective mechanism-based inhibitors of the herpes proteases. *Bioorg. Med. Chem. Lett.* **1996**, *6*, 2467–2472. [CrossRef]
16. Bhandari, S.; Bisht, K.S.; Merkler, D.J. The Biosynthesis and Metabolism of the N-Acylated Aromatic Amino Acids: N-Acylphenylalanine, N-Acyltyrosine, N-Acyltryptophan, and N-Acylhistidine. *Front. Mol. Biosci.* **2022**, *8*, 801749. [CrossRef] [PubMed]
17. Arul Prakash, S.; Kamlekar, R.K. Function and therapeutic potential of N-acyl amino acids. *Chem. Phys. Lipids* **2021**, *239*, 105114. [CrossRef] [PubMed]
18. Battista, N.; Bari, M.; Bisogno, T. N-Acyl Amino Acids: Metabolism, Molecular Targets, and Role in Biological Processes. *Biomolecules* **2019**, *9*, 822. [CrossRef] [PubMed]
19. Burstein, S.H. N-Acyl Amino Acids (Elmiric Acids): Endogenous Signaling Molecules with Therapeutic Potential. *Mol. Pharmacol.* **2018**, *93*, 228–238. [CrossRef]
20. Wei, C.-W.; Yu, Y.-L.; Chen, Y.-H.; Hung, Y.-T.; Yiang, G.-T. Anticancer effects of methotrexate in combination with α-tocopherol and α-tocopherol succinate on triple-negative breast cancer. *Oncol. Rep.* **2019**, *41*, 2060–2066. [CrossRef]
21. Ezeriņa, D.; Takano, Y.; Hanaoka, K.; Urano, Y.; Dick, T.P. N-Acetyl Cysteine Functions as a Fast-Acting Antioxidant by Triggering Intracellular H$_2$S and Sulfane Sulfur Production. *Cell Chem. Biol.* **2018**, *25*, 447–459.e4. [CrossRef] [PubMed]
22. Chikukwa, M.T.R.; Walker, R.B.; Khamanga, S.M.M. Formulation and Characterisation of a Combination Captopril and Hydrochlorothiazide Microparticulate Dosage Form. *Pharmaceutics* **2020**, *12*, 712. [CrossRef] [PubMed]
23. Lockbaum, G.J.; Henes, M.; Lee, J.M.; Timm, J.; Nalivaika, E.A.; Thompson, P.R.; Kurt Yilmaz, N.; Schiffer, C.A. Pan-3C Protease Inhibitor Rupintrivir Binds SARS-CoV-2 Main Protease in a Unique Binding Mode. *Biochemistry* **2021**, *60*, 2925–2931. [CrossRef]
24. Vandyck, K.; Deval, J. Considerations for the discovery and development of 3-chymotrypsin-like cysteine protease inhibitors targeting SARS-CoV-2 infection. *Curr. Opin. Virol.* **2021**, *49*, 36–40. [CrossRef]

25. Van Dycke, J.; Dai, W.; Stylianidou, Z.; Li, J.; Cuvry, A.; Roux, E.; Li, B.; Rymenants, J.; Bervoets, L.; de Witte, P.; et al. A Novel Class of Norovirus Inhibitors Targeting the Viral Protease with Potent Antiviral Activity In Vitro and In Vivo. *Viruses* **2021**, *13*, 1852. [CrossRef] [PubMed]
26. Semple, G.; Ashworth, D.M.; Batt, A.R.; Baxter, A.J.; Benzies, D.W.M.; Elliot, L.H.; Evans, D.M.M.; Franklin, R.J.; Hudson, P.; Jenkins, P.D.; et al. Peptidomimetic aminomethylene ketone inhibitors of interleukin-1β-converting enzyme (ICE). *Bioorg. Med. Chem. Lett.* **1998**, *8*, 959–964. [CrossRef]
27. Deng, H.; Bannister, T.D.; Jin, L.; Babine, R.E.; Quinn, J.; Nagafuji, P.; Celatka, C.A.; Lin, J.; Lazarova, T.I.; Rynkiewicz, M.J.; et al. Synthesis, SAR exploration, and X-ray crystal structures of factor XIa inhibitors containing an α-ketothiazole arginine. *Bioorg. Med. Chem. Lett.* **2006**, *16*, 3049–3054. [CrossRef]
28. Allen, L.A.T.; Raclea, R.-C.; Natho, P.; Parsons, P.J. Recent advances in the synthesis of α-amino ketones. *Org. Biomol. Chem.* **2021**, *19*, 498–513. [CrossRef]
29. Rashdan, H.R.M.; Shehadi, I.A.; Abdelrahman, M.T.; Hemdan, B.A. Antibacterial Activities and Molecular Docking of Novel Sulfone Biscompound Containing Bioactive 1,2,3-Triazole Moiety. *Molecules* **2021**, *26*, 4817. [CrossRef]
30. Ahmad, I. Shagufta Sulfones: An Important Class of Organic Compounds with Diverse Biological Activities. *Int. J. Pharm. Pharm. Sci.* **2015**, *7*, 19–27.
31. Apostol, T.-V.; Drăghici, C.; Socea, L.-I.; Olaru, O.T.; Şaramet, G.; Enache-Preoteasa, C.; Bărbuceanu, Ş.-F. Synthesis, Characterization and Cytotoxicity Evaluation of New Diphenyl Sulfone Derivatives. *Farmacia* **2021**, *69*, 657–669. [CrossRef]
32. Mady, M.F.; Awad, G.E.A.; Jørgensen, K.B. Ultrasound-assisted synthesis of novel 1,2,3-triazoles coupled diaryl sulfone moieties by the CuAAC reaction, and biological evaluation of them as antioxidant and antimicrobial agents. *Eur. J. Med. Chem.* **2014**, *84*, 433–443. [CrossRef]
33. Guzmán-Ávila, R.; Avelar, M.; Márquez, E.A.; Rivera-Leyva, J.C.; Mora, J.R.; Flores-Morales, V.; Rivera-Islas, J. Synthesis, In Vitro, and In Silico Analysis of the Antioxidative Activity of Dapsone Imine Derivatives. *Molecules* **2021**, *26*, 5747. [CrossRef]
34. Fernández-Villa, D.; Aguilar, M.R.; Rojo, L. Folic Acid Antagonists: Antimicrobial and Immunomodulating Mechanisms and Applications. *Int. J. Mol. Sci.* **2019**, *20*, 4996. [CrossRef]
35. Kumar Verma, S.; Verma, R.; Xue, F.; Kumar Thakur, P.; Girish, Y.R.; Rakesh, K.P. Antibacterial activities of sulfonyl or sulfonamide containing heterocyclic derivatives and its structure-activity relationships (SAR) studies: A critical review. *Bioorg. Chem.* **2020**, *105*, 104400. [CrossRef] [PubMed]
36. Madduluri, V.K.; Baig, N.; Chander, S.; Murugesan, S.; Sah, A.K. Mo(VI) complex catalysed synthesis of sulfones and their modification for anti-HIV activities. *Catal. Commun.* **2020**, *137*, 105931. [CrossRef]
37. Xu, S.; Song, S.; Sun, L.; Gao, P.; Gao, S.; Ma, Y.; Kang, D.; Cheng, Y.; Zhang, X.; Cherukupalli, S.; et al. Indolylarylsulfones bearing phenylboronic acid and phenylboronate ester functionalities as potent HIV-1 non-nucleoside reverse transcriptase inhibitors. *Bioorg. Med. Chem.* **2022**, *53*, 116531. [CrossRef]
38. Regueiro-Ren, A. Cyclic sulfoxides and sulfones in drug design. In *Advances in Heterocyclic Chemistry*; Meanwell, N.A., Lolli, M.L., Eds.; Academic Press: Cambridge, MA, USA, 2021; Volume 134, pp. 1–30. ISBN 978-0-12-820181-7.
39. Feng, M.; Tang, B.; Liang, S.H.; Jiang, X. Sulfur Containing Scaffolds in Drugs: Synthesis and Application in Medicinal Chemistry. *Curr. Top. Med. Chem.* **2016**, *16*, 1200–1216. [CrossRef]
40. Kucwaj-Brysz, K.; Baltrukevich, H.; Czarnota, K.; Handzlik, J. Chemical update on the potential for serotonin 5-HT$_6$ and 5-HT$_7$ receptor agents in the treatment of Alzheimer's disease. *Bioorg. Med. Chem. Lett.* **2021**, *49*, 128375. [CrossRef] [PubMed]
41. Apostol, T.-V.; Draghici, C.; Dinu, M.; Barbuceanu, S.-F.; Socea, L.I.; Saramet, I. Synthesis, Characterization and Biological Evaluation of New 5-aryl-4-methyl-2-[*para*-(phenylsulfonyl)phenyl]oxazoles. *Rev. Chim.* **2011**, *62*, 142–148.
42. Apostol, T.-V.; Saramet, I.; Draghici, C.; Barbuceanu, S.-F.; Socea, L.I.; Almajan, G.L. Synthesis and Characterization of New 5-Aryl-2-[*para*-(4-chlorophenylsulfonyl)phenyl]-4-methyloxazoles. *Rev. Chim.* **2011**, *62*, 486–492.
43. Apostol, T.-V.; Barbuceanu, S.-F.; Olaru, O.T.; Draghici, C.; Saramet, G.; Socea, B.; Enache, C.; Socea, L.-I. Synthesis, Characterization and Cytotoxicity Evaluation of New Compounds from Oxazol-5(4*H*)-ones and Oxazoles Class Containing 4-(4-Bromophenylsulfonyl)phenyl Moiety. *Rev. Chim.* **2019**, *70*, 1099–1107. [CrossRef]
44. Apostol, T.V.; Barbuceanu, S.F.; Socea, L.I.; Draghici, C.; Saramet, G.; Iscrulescu, L.; Olaru, O.T. Synthesis, Characterization and Cytotoxicity Evaluation of New Heterocyclic Compounds with Oxazole Ring Containing 4-(Phenylsulfonyl)phenyl Moiety. *Rev. Chim.* **2019**, *70*, 3793–3801. [CrossRef]
45. Apostol, T.-V.; Socea, L.-I.; Drăghici, C.; Olaru, O.T.; Şaramet, G.; Enache-Preoteasa, C.; Bărbuceanu, Ş.-F. Design, Synthesis, Characterization, and Cytotoxicity Evaluation of New 4-Benzyl-1,3-oxazole Derivatives Bearing 4-(4-Chlorophenylsulfonyl)phenyl Moiety. *Farmacia* **2021**, *69*, 314–324. [CrossRef]
46. Apostol, T.V.; Drăghici, C.; Socea, L.I.; Şaramet, G.; Hrubaru, M.; Bărbuceanu, Ş.F. Synthesis, Characterization and Cytotoxicity Assessment of New 4-Benzyl-1,3-oxazole Derivatives Incorporating 4-[(4-Bromophenyl)sulfonyl]phenyl Fragment. *Farmacia* **2021**, *69*, 521–529. [CrossRef]
47. Guilhermino, L.; Diamantino, T.; Silva, M.C.; Soares, A.M.V.M. Acute Toxicity Test with *Daphnia magna*: An Alternative to Mammals in the Prescreening of Chemical Toxicity? *Ecotoxicol. Environ. Saf.* **2000**, *46*, 357–362. [CrossRef]
48. Seremet, O.C.; Olaru, O.T.; Gutu, C.M.; Nitulescu, G.M.; Ilie, M.; Negres, S.; Zbarcea, C.E.; Purdel, C.N.; Spandidos, D.A.; Tsatsakis, A.M.; et al. Toxicity of plant extracts containing pyrrolizidine alkaloids using alternative invertebrate models. *Mol. Med. Rep.* **2018**, *17*, 7757–7763. [CrossRef]

49. Lagunin, A.; Stepanchikova, A.; Filimonov, D.; Poroikov, V. PASS: Prediction of activity spectra for biologically active substances. *Bioinformatics* **2000**, *16*, 747–748. [CrossRef]
50. Mavrodin, A.; Oteleanu, D.; Vorel-Stoenescu, M.; Zotta, V. Studies on the sulfone group. IV. New sulfone-hydrazide derivatives. *Pharm. Zent. Dtschl.* **1956**, *95*, 353–361.
51. Schiketanz, I.; Draghici, C.; Saramet, I.; Balaban, A.T. Aminoketone, oxazole and thiazole synthesis. Part 15.[1] 2-[4-(4-Halobenzenesulphonyl)-phenyl]-5-aryloxazoles. *Arkivoc* **2002**, *2002*, 64–72. [CrossRef]
52. Cui, S.-Q.; Liao, W.-W. Recent Advances in Transition-Metal-Catalyzed C–H Addition to Nitriles. *Synthesis* **2022**, *54*, 33–48. [CrossRef]
53. da Rosa, R.; Grand, L.; Schenkel, E.P.; Campos Bernardes, L.S.; Jacolot, M.; Popowycz, F. The Use of 5-Hydroxymethylfurfural towards Fine Chemicals: Synthesis and Direct Arylation of 5-HMF-Based Oxazoles. *Synlett* **2021**, *32*, 838–844. [CrossRef]
54. Wei, L.; You, S.; Tuo, Y.; Cai, M. A Highly Efficient Heterogeneous Copper-Catalyzed Oxidative Cyclization of Benzylamines and 1,3-Dicarbonyl Compounds To Give Trisubstituted Oxazoles. *Synthesis* **2019**, *51*, 3091–3100. [CrossRef]
55. Murru, S.; Bista, R.; Nefzi, A. Synthesis of Novel Oxazolyl Amino Acids and Their Use in the Parallel Synthesis of Disubstituted Oxazole Libraries. *Synthesis* **2018**, *50*, 1546–1554. [CrossRef]
56. Katariya, K.D.; Vennapu, D.R.; Shah, S.R. Synthesis and molecular docking study of new 1,3-oxazole clubbed pyridyl-pyrazolines as anticancer and antimicrobial agents. *J. Mol. Struct.* **2021**, *1232*, 130036. [CrossRef]
57. Metelytsia, L.O.; Trush, M.M.; Kovalishyn, V.V.; Hodyna, D.M.; Kachaeva, M.V.; Brovarets, V.S.; Pilyo, S.G.; Sukhoveev, V.V.; Tsyhankov, S.A.; Blagodatnyi, V.M.; et al. 1,3-Oxazole derivatives of cytisine as potential inhibitors of glutathione reductase of *Candida spp.*: QSAR modeling, docking analysis and experimental study of new anti-Candida agents. *Comput. Biol. Chem.* **2021**, *90*, 107407. [CrossRef]
58. Ghazvini, M.; Sheikholeslami-Farahani, F.; Hamedani, N.F.; Shahvelayati, A.S.; Rostami, Z. Bio-Fe_3O_4 Magnetic Nanoparticles Promoted Green Synthesis of thioxo- 1,3-Oxazole Derivatives: Study of Antimicrobial and Antioxidant Activity. *Comb. Chem. High Throughput Screen.* **2021**, *24*, 1261–1270. [CrossRef] [PubMed]
59. Keivanloo, A.; Abbaspour, S.; Sepehri, S.; Bakherad, M. Synthesis, Antibacterial Activity and Molecular Docking Study of a Series of 1,3-Oxazole-Quinoxaline Amine Hybrids. *Polycycl. Aromat. Compd.* **2022**, *42*, 2378–2391. [CrossRef]
60. Ansari, A.; Ali, A.; Asif, M.; Rauf, M.A.; Owais, M. Shamsuzzaman Facile one-pot multicomponent synthesis and molecular docking studies of steroidal oxazole/thiazole derivatives with effective antimicrobial, antibiofilm and hemolytic properties. *Steroids* **2018**, *134*, 22–36. [CrossRef]
61. Wales, S.M.; Hammer, K.A.; Somphol, K.; Kemker, I.; Schröder, D.C.; Tague, A.J.; Brkic, Z.; King, A.M.; Lyras, D.; Riley, T.V.; et al. Synthesis and Antimicrobial Activity of Binaphthyl-Based, Functionalized Oxazole and Thiazole Peptidomimetics. *Org. Biomol. Chem.* **2015**, *13*, 10813–10824. [CrossRef]
62. Stokes, N.R.; Baker, N.; Bennett, J.M.; Chauhan, P.K.; Collins, I.; Davies, D.T.; Gavade, M.; Kumar, D.; Lancett, P.; Macdonald, R.; et al. Design, synthesis and structure–activity relationships of substituted oxazole–benzamide antibacterial inhibitors of FtsZ. *Bioorg. Med. Chem. Lett.* **2014**, *24*, 353–359. [CrossRef]
63. Marc, G.; Araniciu, C.; Oniga, S.D.; Vlase, L.; Pîrnău, A.; Duma, M.; Măruțescu, L.; Chifiriuc, M.C.; Oniga, O. New N-(oxazolylmethyl)-thiazolidinedione Active against *Candida albicans* Biofilm: Potential Als Proteins Inhibitors. *Molecules* **2018**, *23*, 2522. [CrossRef]
64. Juhás, M.; Bachtíková, A.; Nawrot, D.E.; Hatoková, P.; Pallabothula, V.S.K.; Diepoltová, A.; Jand'ourek, O.; Bárta, P.; Konečná, K.; Paterová, P.; et al. Improving Antimicrobial Activity and Physico-Chemical Properties by Isosteric Replacement of 2-Aminothiazole with 2-Aminooxazole. *Pharmaceuticals* **2022**, *15*, 580. [CrossRef]
65. Apostol, T.-V.; Chifiriuc, M.C.; Nitulescu, G.M.; Olaru, O.T.; Barbuceanu, S.-F.; Socea, L.-I.; Pahontu, E.M.; Karmezan, C.M.; Marutescu, L.G. In Silico and In Vitro Assessment of Antimicrobial and Antibiofilm Activity of Some 1,3-Oxazole-Based Compounds and Their Isosteric Analogues. *Appl. Sci.* **2022**, *12*, 5571. [CrossRef]
66. Molchanova, N.; Nielsen, J.E.; Sørensen, K.B.; Prabhala, B.K.; Hansen, P.R.; Lund, R.; Barron, A.E.; Jenssen, H. Halogenation as a tool to tune antimicrobial activity of peptoids. *Sci. Rep.* **2020**, *10*, 14805. [CrossRef]
67. Cruz, J.C.S.; Iorio, M.; Monciardini, P.; Simone, M.; Brunati, C.; Gaspari, E.; Maffioli, S.I.; Wellington, E.; Sosio, M.; Donadio, S. Brominated Variant of the Lantibiotic NAI-107 with Enhanced Antibacterial Potency. *J. Nat. Prod.* **2015**, *78*, 2642–2647. [CrossRef]
68. Gottardi, W.; Klotz, S.; Nagl, M. Superior bactericidal activity of N-bromine compounds compared to their N-chlorine analogues can be reversed under protein load. *J. Appl. Microbiol.* **2014**, *116*, 1427–1437. [CrossRef]
69. Barbuceanu, S.-F.; Almajan, G.L.; Saramet, I.; Draghici, C.; Tarcomnicu, A.I.; Bancescu, G. Synthesis, characterization and evaluation of antibacterial activity of some thiazolo[3,2-b][1,2,4]triazole incorporating diphenylsulfone moieties. *Eur. J. Med. Chem.* **2009**, *44*, 4752–4757. [CrossRef]
70. Apostol, T.-V.; Marutescu, L.G.; Draghici, C.; Socea, L.-I.; Olaru, O.T.; Nitulescu, G.M.; Pahontu, E.M.; Saramet, G.; Enache-Preoteasa, C.; Barbuceanu, S.-F. Synthesis and Biological Evaluation of New N-Acyl-α-amino Ketones and 1,3-Oxazoles Derivatives. *Molecules* **2021**, *26*, 5019. [CrossRef]
71. Apostol, T.-V.; Chifiriuc, M.C.; Draghici, C.; Socea, L.-I.; Marutescu, L.G.; Olaru, O.T.; Nitulescu, G.M.; Pahontu, E.M.; Saramet, G.; Barbuceanu, S.-F. Synthesis, In Silico and In Vitro Evaluation of Antimicrobial and Toxicity Features of New 4-[(4-Chlorophenyl)sulfonyl]benzoic Acid Derivatives. *Molecules* **2021**, *26*, 5107. [CrossRef]

72. Zhang, W.; Liu, W.; Jiang, X.; Jiang, F.; Zhuang, H.; Fu, L. Design, synthesis and antimicrobial activity of chiral 2-(substituted-hydroxyl)-3-(benzo[d]oxazol-5-yl)propanoic acid derivatives. *Eur. J. Med. Chem.* **2011**, *46*, 3639–3650. [CrossRef] [PubMed]
73. Desai, N.C.; Vaja, D.V.; Joshi, S.B.; Khedkar, V.M. Synthesis and molecular docking study of pyrazole clubbed oxazole as antibacterial agents. *Res. Chem. Intermed.* **2021**, *47*, 573–587. [CrossRef]
74. Thatha, S.; Ummadi, N.; Venkatapuram, P.; Adivireddy, P. Synthesis, Characterization, and Antioxidant Activity of a New Class of Amido linked Azolyl Thiophenes. *J. Heterocycl. Chem.* **2018**, *55*, 1410–1418. [CrossRef]
75. Basha, N.M.; Reddy, P.R.; Padmaja, A.; Padmavathi, V. Synthesis and Antioxidant Activity of Bis-oxazolyl/thiazolyl/imidazolyl 1,3,4-Oxadiazoles and 1,3,4-Thiadiazoles. *J. Heterocycl. Chem.* **2016**, *53*, 1276–1283. [CrossRef]
76. Stankova, I.; Spasova, M. Hydroxycinnamic Acid Amides with Oxazole-Containing Amino Acid: Synthesis and Antioxidant Activity. *Z. Naturforsch. C* **2009**, *64*, 176–178. [CrossRef]
77. Kumar, G.; Singh, N.P. Synthesis, anti-inflammatory and analgesic evaluation of thiazole/oxazole substituted benzothiazole derivatives. *Bioorg. Chem.* **2021**, *107*, 104608. [CrossRef]
78. Mendez, D.; Gaulton, A.; Bento, A.P.; Chambers, J.; De Veij, M.; Félix, E.; Magariños, M.P.; Mosquera, J.F.; Mutowo, P.; Nowotka, M.; et al. ChEMBL: Towards direct deposition of bioassay data. *Nucleic Acids Res.* **2019**, *47*, D930–D940. [CrossRef]
79. Sander, T.; Freyss, J.; von Korff, M.; Rufener, C. DataWarrior: An Open-Source Program For Chemistry Aware Data Visualization And Analysis. *J. Chem. Inf. Model.* **2015**, *55*, 460–473. [CrossRef]
80. Stepanović, S.; Vuković, D.; Hola, V.; Di Bonaventura, G.; Djukić, S.; Ćirković, I.; Ruzicka, F. Quantification of biofilm in microtiter plates: Overview of testing conditions and practical recommendations for assessment of biofilm production by staphylococci. *APMIS* **2007**, *115*, 891–899. [CrossRef] [PubMed]
81. Limban, C.; Balotescu Chifiriuc, M.-C.; Missir, A.-V.; Chiriță, I.C.; Bleotu, C. Antimicrobial Activity of Some New Thioureides Derived from 2-(4-Chlorophenoxymethyl)benzoic Acid. *Molecules* **2008**, *13*, 567–580. [CrossRef]
82. Patrinoiu, G.; Calderón-Moreno, J.M.; Chifiriuc, C.M.; Saviuc, C.; Birjega, R.; Carp, O. Tunable ZnO spheres with high anti-biofilm and antibacterial activity *via* a simple green hydrothermal route. *J. Colloid Interface Sci.* **2016**, *462*, 64–74. [CrossRef] [PubMed]
83. Prodan, A.M.; Iconaru, S.L.; Chifiriuc, C.M.; Bleotu, C.; Ciobanu, C.S.; Motelica-Heino, M.; Sizaret, S.; Predoi, D. Magnetic Properties and Biological Activity Evaluation of Iron Oxide Nanoparticles. *J. Nanomater.* **2013**, *2013*, 893970. [CrossRef]
84. Munteanu, I.G.; Apetrei, C. Analytical Methods Used in Determining Antioxidant Activity: A Review. *Int. J. Mol. Sci.* **2021**, *22*, 3380. [CrossRef]
85. Ilyasov, I.R.; Beloborodov, V.L.; Selivanova, I.A.; Terekhov, R.P. ABTS/PP Decolorization Assay of Antioxidant Capacity Reaction Pathways. *Int. J. Mol. Sci.* **2020**, *21*, 1131. [CrossRef]
86. Mendonça, J.D.S.; de Cássia Avellaneda Guimarães, R.; Zorgetto-Pinheiro, V.A.; Fernandes, C.D.P.; Marcelino, G.; Bogo, D.; de Cássia Freitas, K.; Hiane, P.A.; de Pádua Melo, E.S.; Vilela, M.L.B.; et al. Natural Antioxidant Evaluation: A Review of Detection Methods. *Molecules* **2022**, *27*, 3563. [CrossRef]
87. Blois, M.S. Antioxidant Determinations by the Use of a Stable Free Radical. *Nature* **1958**, *181*, 1199–1200. [CrossRef]
88. Barbuceanu, S.-F.; Ilies, D.C.; Saramet, G.; Uivarosi, V.; Draghici, C.; Radulescu, V. Synthesis and Antioxidant Activity Evaluation of New Compounds from Hydrazinecarbothioamide and 1,2,4-Triazole Class Containing Diarylsulfone and 2,4-Difluorophenyl Moieties. *Int. J. Mol. Sci.* **2014**, *15*, 10908–10925. [CrossRef]
89. Ilies, D.-C.; Shova, S.; Radulescu, V.; Pahontu, E.; Rosu, T. Synthesis, characterization, crystal structure and antioxidant activity of Ni(II) and Cu(II) complexes with 2-formylpyridine N(4)-phenylthiosemicarbazone. *Polyhedron* **2015**, *97*, 157–166. [CrossRef]
90. Duan, X.-J.; Zhang, W.-W.; Li, X.-M.; Wang, B.-G. Evaluation of antioxidant property of extract and fractions obtained from a red alga, *Polysiphonia urceolata*. *Food Chem.* **2006**, *95*, 37–43. [CrossRef]
91. Re, R.; Pellegrini, N.; Proteggente, A.; Pannala, A.; Yang, M.; Rice-Evans, C. Antioxidant activity applying an improved ABTS radical cation decolorization assay. *Free Radic. Biol. Med.* **1999**, *26*, 1231–1237. [CrossRef]
92. Oyaizu, M. Studies on Products of Browning Reaction. Antioxidative Activities of Products of Browning Reaction Prepared from Glucosamine. *Japanese J. Nutr. Diet.* **1986**, *44*, 307–315. [CrossRef]
93. Dorman, H.J.D.; Koşar, M.; Kahlos, K.; Holm, Y.; Hiltunen, R. Antioxidant Properties and Composition of Aqueous Extracts from *Mentha* Species, Hybrids, Varieties, and Cultivars. *J. Agric. Food Chem.* **2003**, *51*, 4563–4569. [CrossRef]
94. Işıl Berker, K.; Güçlü, K.; Tor, I.; Apak, R. Comparative evaluation of Fe (III) reducing power-based antioxidant capacity assays in the presence of phenanthroline, *batho*-phenanthroline, tripyridyltriazine (FRAP), and ferricyanide reagents. *Talanta* **2007**, *72*, 1157–1165. [CrossRef] [PubMed]
95. *OECD. Test No. 202: Daphnia sp. Acute Immobilisation Test*; OECD Guidelines for the Testing of Chemicals, Section 2; OECD Publishing: Paris, France, 2004 ISBN 9789264069947.
96. Stecoza, C.E.; Nitulescu, G.M.; Draghici, C.; Caproiu, M.T.; Olaru, O.T.; Bostan, M.; Mihaila, M. Synthesis and Anticancer Evaluation of New 1,3,4-Oxadiazole Derivatives. *Pharmaceuticals* **2021**, *14*, 438. [CrossRef] [PubMed]
97. Nitulescu, G.; Mihai, D.P.; Nicorescu, I.M.; Olaru, O.T.; Ungurianu, A.; Zanfirescu, A.; Nitulescu, G.M.; Margina, D. Discovery of natural naphthoquinones as sortase A inhibitors and potential anti-infective solutions against *Staphylococcus aureus*. *Drug Dev. Res.* **2019**, *80*, 1136–1145. [CrossRef] [PubMed]
98. Zakharov, A.V.; Peach, M.L.; Sitzmann, M.; Nicklaus, M.C. A New Approach to Radial Basis Function Approximation and Its Application to QSAR. *J. Chem. Inf. Model.* **2014**, *54*, 713–719. [CrossRef]

Article

The Assessment of Anticancer and VEGFR-2 Inhibitory Activities of a New 1*H*-Indole Derivative: In Silico and In Vitro Approaches

Eslam B. Elkaeed [1], Reda G. Yousef [2], Hazem Elkady [2], Ibraheem M. M. Gobaara [3], Aisha A. Alsfouk [4], Dalal Z. Husein [5], Ibrahim M. Ibrahim [6], Ahmed M. Metwaly [7,8,*] and Ibrahim H. Eissa [2,*]

[1] Department of Pharmaceutical Sciences, College of Pharmacy, AlMaarefa University, Riyadh 13713, Saudi Arabia; ikaeed@mcst.edu.sa
[2] Pharmaceutical Medicinal Chemistry & Drug Design Department, Faculty of Pharmacy (Boys), Al-Azhar University, Cairo 11884, Egypt; redayousof@azhar.edu.eg (R.G.Y.); hazemelkady@azhar.edu.eg (H.E.)
[3] Zoology Department, Faculty of Science (Boys), Al-Azhar University, Cairo 11884, Egypt; ibraheemgobaara@azhar.edu.eg
[4] Department of Pharmaceutical Sciences, College of Pharmacy, Princess Nourah Bint Abdulrahman University, P.O. Box 84428, Riyadh 11671, Saudi Arabia; aaalsfouk@pnu.edu.sa
[5] Chemistry Department, Faculty of Science, New Valley University, El-Kharja 72511, Egypt; dalal_husein@sci.nvu.edu.eg
[6] Biophysics Department, Faculty of Science, Cairo University, Cairo 12613, Egypt; iabdelmagid@sci.cu.edu.eg
[7] Pharmacognosy and Medicinal Plants Department, Faculty of Pharmacy (Boys), Al-Azhar University, Cairo 11884, Egypt
[8] Biopharmaceutical Products Research Department, Genetic Engineering and Biotechnology Research Institute, City of Scientific Research and Technological Applications (SRTA-City), Alexandria 21934, Egypt
* Correspondence: ametwaly@azhar.edu.eg (A.M.M.); ibrahimeissa@azhar.edu.eg (I.H.E.)

Abstract: Corresponding to the reported features of anti-VEGFR-2-approved compounds, a new 1*H*-indole derivative (compound **7**) was designed. The inhibitory potential of the designed compound was revealed via a molecular docking study that showed the appropriate binding. Then, MD simulation (six studies) over a period of 100 ns was performed to confirm the precise binding and optimum energy. Additionally, MM-GBSA reaffirmed the perfect binding, exhibiting a total precise energy of −40.38 Kcal/Mol. The MM-GBSA experiments named the essential amino acids in the protein–ligand interaction, employing the binding energy decomposition and revealing the diversity of interactions of compound **7** inside the VEGFR-2 enzyme. As compound **7** is new, DFT experiments were utilized for molecular structure optimization. Additionally, the DFT results validated the coherent interaction of compound **7** with the VEGFR-2 enzyme. A good value of drug-likeness of compound **7** was acknowledged via in silico ADMET studies. Interestingly, the experimental in vitro prohibitory potential of compound **7** was better than that of sorafenib, demonstrating an IC_{50} value of 25 nM. Notably, the strong inhibitory effects of compound **10** against two cancer cell lines (MCF-7 and HCT 116) were established with IC_{50} values of 12.93 and 11.52 µM, disclosing high selectivity indexes of 6.7 and 7.5, respectively.

Keywords: 1*H*-indole; VEGFR-2 inhibitors; molecular docking; MD simulations; MM-GBSA; PLIP; DFT; ADMET; in vitro antiproliferative

1. Introduction

The WHO has estimated that cancer will become the main cause of death over the upcoming years. Additionally, they have pointed to the dominance of breast cancer (according to the discovery of new cancer cases) in 2020, with more than 2.2 million new cases [1]. Correspondingly, cancer therapy, especially breast cancer, is a challenging area for medicinal chemists to develop treatments that inhibit the growth of cancer cells by interacting with specific molecular targets and subsequently killing them. Tumor growth and reproduction are associated with increased vascularity (angiogenesis) in cancer cells [2]. Accordingly, anti-angiogenesis mechanisms were considered as one of the potential ways to combat cancer [3]. The vascular endothelial growth factor (VEGF) pathway plays an important role in the incidence of angiogenesis [4]. The vascular endothelial growth factor receptors (VEGFRs) can be divided into three subtypes; VEGFR-1, VEGFR-2, and VEGFR-3. Among the most valuable targets in cancer management, VEGFR-2 is a transmembrane tyrosine kinase receptor that is correlated with cell proliferation, division, motility, adhesion, and angiogenesis [5]. Thus, blocking the VEGFR-2 signaling cascade inhibits cancer cell proliferation [6]. The fact that cancer cells express VEGFR-2 receptors more than normal cells has enabled scientists to create safe and selective drugs to combat angiogenesis in tumor cells without affecting normal cells [7].

Using computers, computational (in silico) chemistry is a scientific approach that applies theoretical and mathematical basics to address and solve chemical problems. Computational chemistry is widely used in the pharmaceutical industry for investigating interactions between potential drugs and biomolecular targets [8]. Our team have synthesized and examined various compounds that were designed as anti-VEGFR-2, belonging to different classes such as quinazoline [9–12], nicotinamides [13], benzoxazole [14], pyridine [15] dihydroquinolones [16], thiourea-azetidine [17,18], and quinoxaline-2 (1H)-one [19–23], in addition to thieno [2,3-d]pyrimidine [24]. Furthermore, our team utilized the basics of in silico chemistry as a successful tool in molecular design and docking [25,26], structural similarity [27], toxicity [28], ADMET [29], DFT [30,31], MD [32], and pharmacophore [33] investigation.

We herein employ our former experiences in both of in silico chemistry and drug discovery to introduce an effective and selective chemotherapeutic 1H-indole derivative focusing on VEGFR-2.

Rationale

Figure 1 demonstrates some reported and FDA-approved VEGFR-2 inhibitors such as sorafenib **I** [34], sunitinib **II** [35], vorolanib **III** [36], nintedanib **IV** [37], and toceranib **V** [38]. These drugs have four key pharmacophoric features that must exist in any inhibitor to fit with the VEGFR-2 active site. The four key features are aromatic ring, spacer moiety, pharmacophore moiety (comprising hydrogen bond donor and hydrogen bond acceptor atoms), and hydrophobic group [20,39,40].

Regarding sunitinib **II**, vorolanib **III**, nintedanib **IV**, and toceranib **V**, these drugs possess 1H-indole derivatives as hetero aromatic structures that can occupy the hinge region of the VEGFR-2 active site. In all compounds, it was noticed that the 1H-indole derivatives share three characteristics. (i) The NH group at the 1-position was kept free without any substitution. (ii) Except nintedanib **IV**, the 5-position was substituted with the hydrophobic (fluoro) group. (iii) The 1H-indole moiety was attached to the rest of the structures at the 3-position. Moreover, sunitinib **II**, vorolanib **III**, nintedanib **IV**, and toceranib **V** have an amide moiety as a pharmacophore that binds the DFG motif region and forms essential hydrogen bonds with Asp1044 and Glu833.

Taking these characteristics into consideration, we designed a modified analog that kept the main characteristics of the lead compounds (**II–V**). As appeared in Figure 2, the designed molecule has a 1H-indole moiety as a hetero aromatic structure that can occupy the hinge region of the VEGFR-2 active site. The 1H-indole moiety has a free NH group with a substitution at the 5-position with a hydrophobic methoxy group. In addition, the 1H-indole moiety was attached to the rest of the structures at the 3-position as in the lead compounds. Furthermore, the designed molecule has an amide moiety as a pharmacophore that binds the DFG motif region at the VEGFR-2 active site. The linker structure of the designed compound consisted of an N'-methylene benzohydrazide moiety that can form hydrophobic and hydrophilic interactions at the gatekeeper region. Finally, a plain phenyl ring was utilized as a hydrophobic tail in the designed molecule to form a hydrophobic interaction at the allosteric binding pocket.

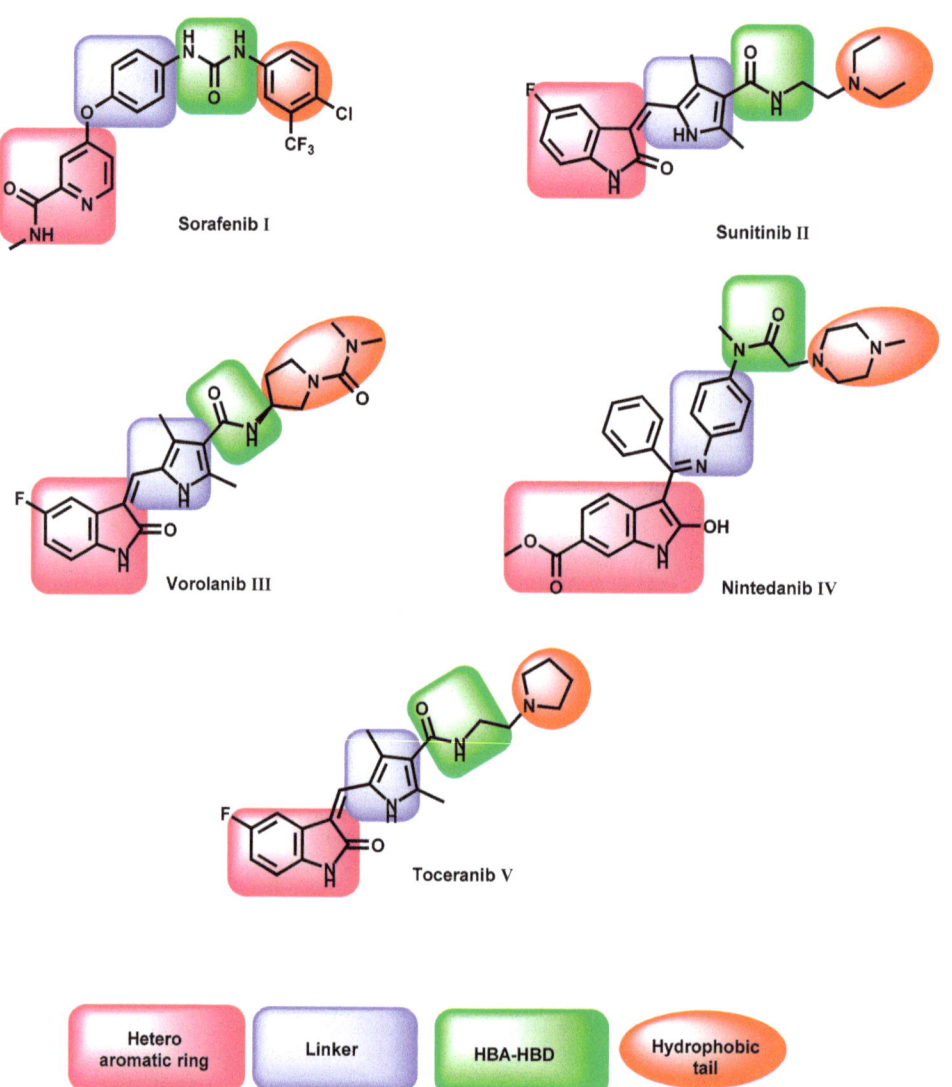

Figure 1. Sorafenib and some FDA-approved 1*H*-indole derivatives working as VEGFR-2 inhibitors show the essential pharmacophoric features.

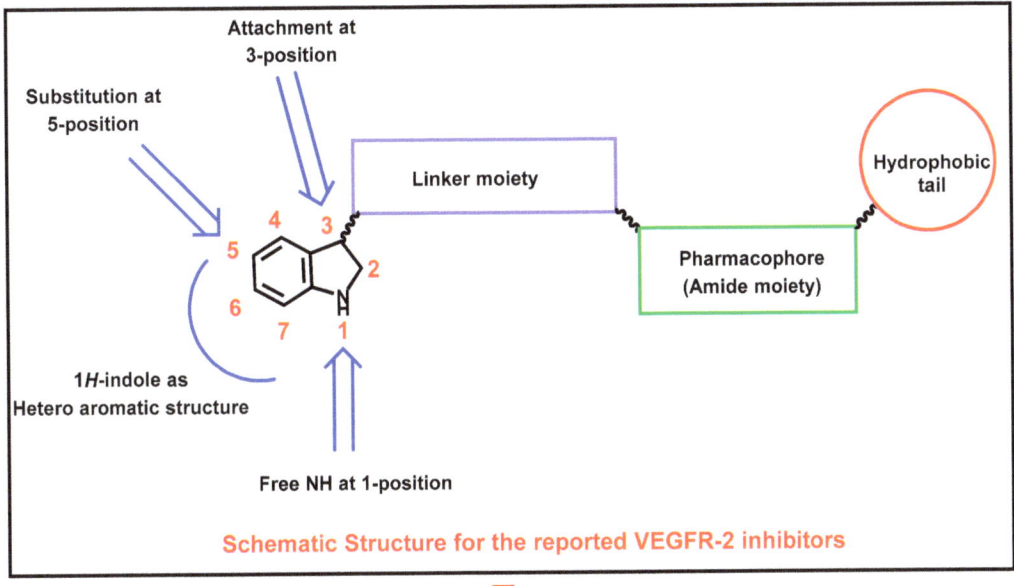

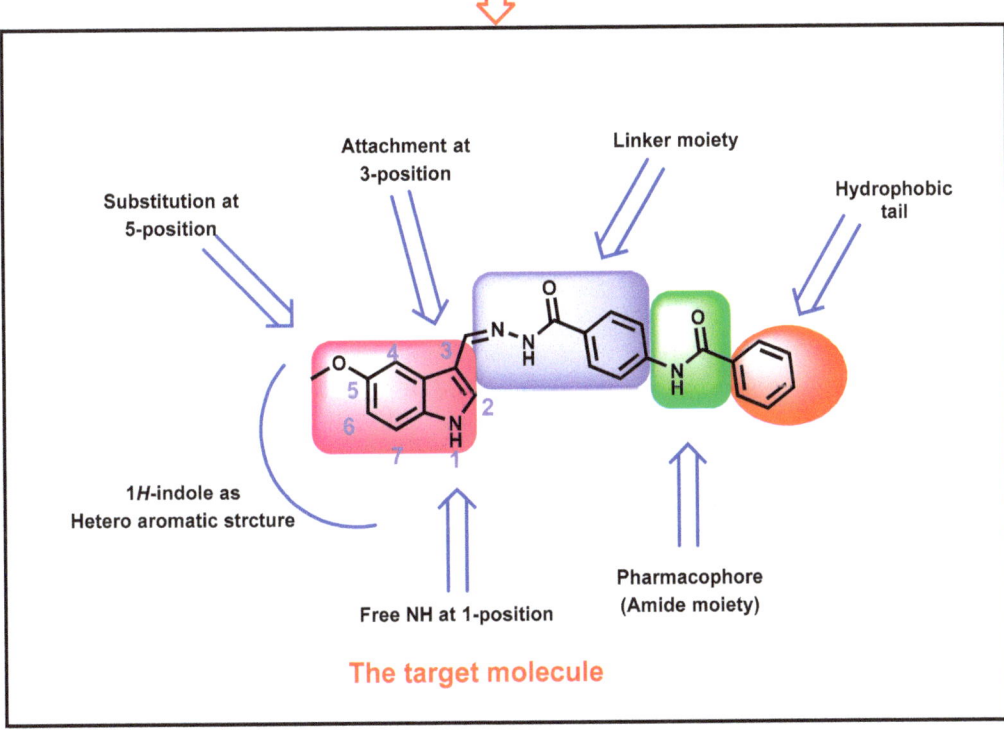

Figure 2. The design rationale of the targeted compound.

2. Results and Discussion
2.1. In Silico Studies
2.1.1. Docking Study

Molecular docking has been applied as an essential tool in the process of drug design and discovery for various bioactive compounds [39,40], including antispasmodics [41], antimicrobials [42], and antimalarials [43]. Compound **7** was docked into the VEGFR-2 ATP binding site to better understand the pattern by which it bounds to the active site [38–40]. Following the preparation of the downloaded protein (PDB ID: 2OH4, resolution: 2.05 Å), the native co-crystallized inhibitor was re-docked against the catalytic VEGFR-2 site as a validation step for the docking protocol. The validation process successfully formed an RMSD value of 0.79 that indicated the validity of the docking operation (Figure 3).

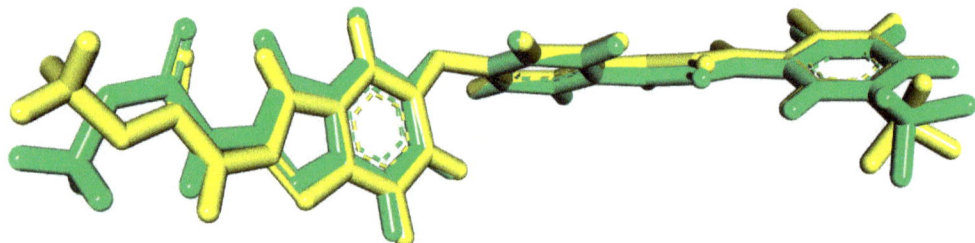

Figure 3. Validation step inside VEGFR-2 catalytic site; native ligand (green) and the obtained pose (yellow).

The surveillance of interaction types between sorafenib and the VEGFR-2 active pocket showed that it could form two main types of interactions (Figure 4). The first type was represented by three H-bonding interactions with the critical amino acids in the VEGFR-2 catalytic site. This includes Cys917 in the hinge region, and Asp1044, and Glu883 in the DFG motif. The second type of interactions comprised a huge network of π interactions between sorafenib and the hydrophobic amino acids (Val846, Val914, Val897, Phe1045, and Cys1043 in the linker region, Leu1033, Leu838, Val846, Phe1045, and Ala864 in the hinge region, and Leu1017, His1024, Ile890, Ile886, and Leu887 in the hydrophobic pocket).

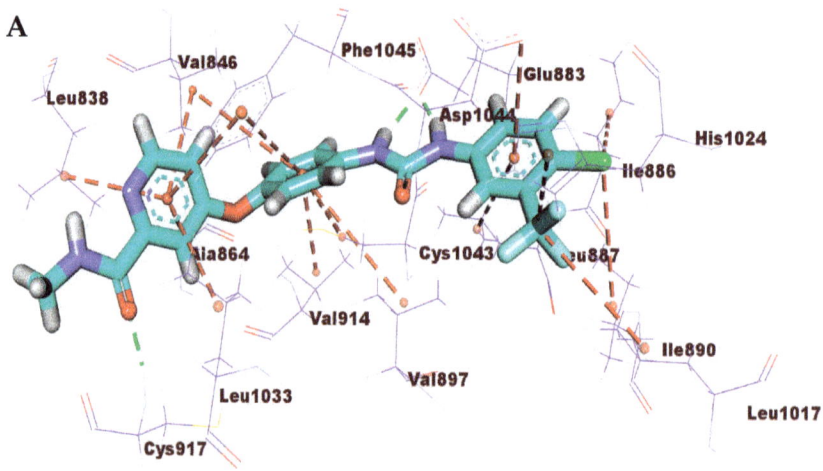

Figure 4. *Cont.*

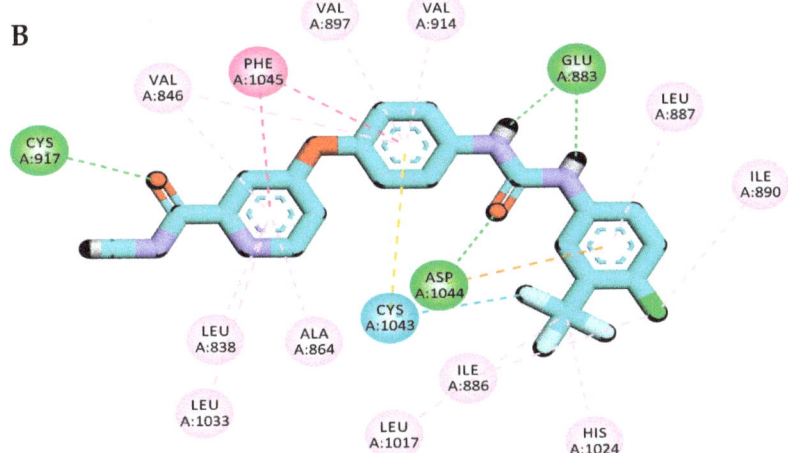

Figure 4. Sorafenib binding pattern, (**A**) 3D and (**B**) 2D.

Docking results for compound **7** revealed strong binding interactions with the VEGFR-2 enzyme, similar to sorafenib. Simply, compound **7** formed a critical hydrogen bond with Cys917 via its indole nitrogen, similar to the pyridine nitrogen of sorafenib. The methoxy group formed a hydrogen bond with Arg1049. Moreover, the amide moiety (inserted between the two phenyl groups) formed two hydrogen-bonding interactions with Glu883 and Asp1044 amino acids, similar to those of sorafenib's urea group. As well, the central phenyl moiety occupied the linker region to form hydrophobic interactions between Val897, and Val914. Similar to the 3-trifluoromethyl-4-chlorophenyl moiety of sorafenib, the hydrophobic allosteric site of the enzyme was occupied with the terminal phenyl group via interactions with Leu887 and Val897. The binding mode of compound **7** is presented in Figure 5.

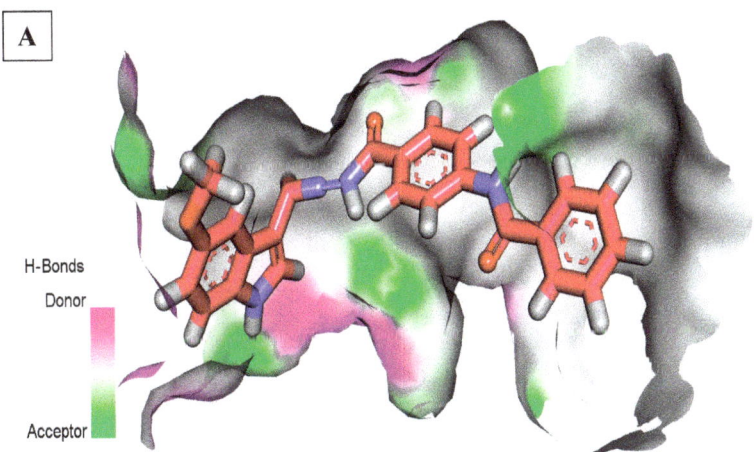

Figure 5. *Cont.*

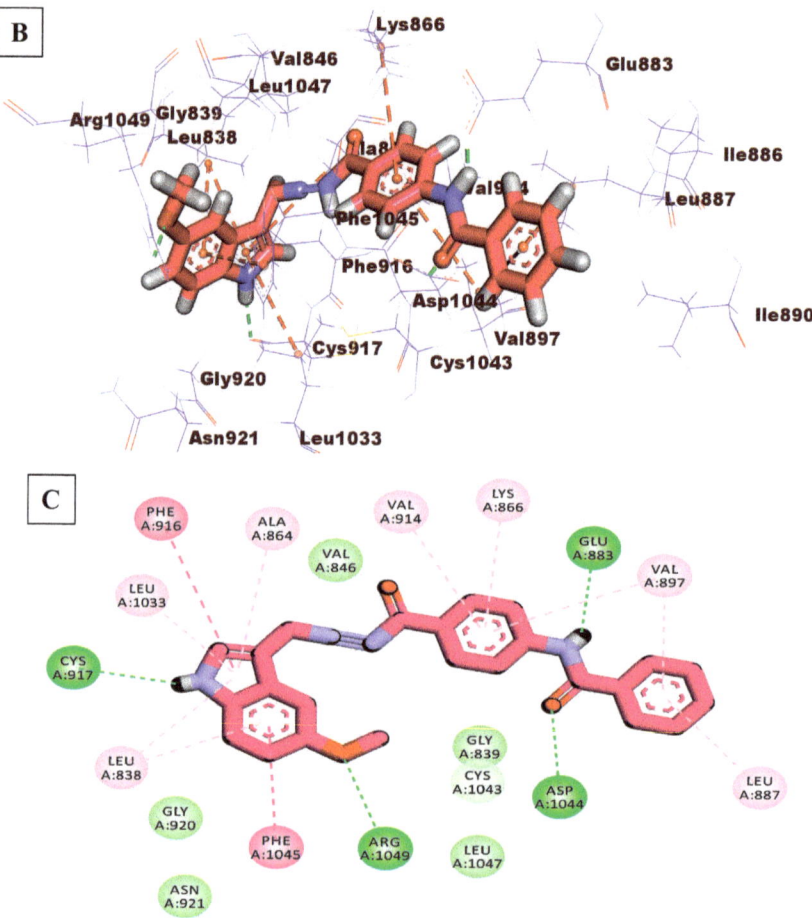

Figure 5. (**A**) Mapping surface, (**B**) 3D, and (**C**) 2D binding pattern of compound 7.

2.1.2. MD Studies

To study the stability of the compound 7-VEGFR-2 complex (7-VEGFR-2 complex), we performed an MD simulation for the complex. The trajectory was used to extract the RMSD (Figure 6A), RMSF (Figure 6B), SASA (Figure 6C), RoG (Figure 6D), the change in the number of hydrogen bonds (Figure 6E), and the distance between the center of masses between the compound **7** and VEGFR-2 (Figure 6F). The RMSD of the protein and 7-VEGFR-2 complex shows that the system was changing its conformation in the first half of the simulation before coming to a stable state, with an average of 5 Å and 5.46 Å, respectively. On the other hand, the RMSD of compound **7** shows fluctuations in approximately the first 65 ns, indicating some intrinsic movement of the ligand, before coming to stable values at around 3.2 Å. The reason for the increase in the RMSD after 50 ns is the large motion of the L1047:P1066 loop, as shown in the RMSF values. In addition, the terminals show very large fluctuations, reaching 8 Å. On the other hand, nearly most of the amino acids have fluctuations of less than 2 Å. The values of SASA (average = 17,370 Å^2), RoG (average = 20.66 Å), and the change in the number of hydrogen bonds (average = 70 bonds) show that the VEGFR-2 conformation is stable, with no unfolding or folding occurring. The distance between the center of mass of compound **7** and VEGFR-2 indicates that compound **7** is bound to VEGFR-2 during the simulation, with an average of 8.08 Å.

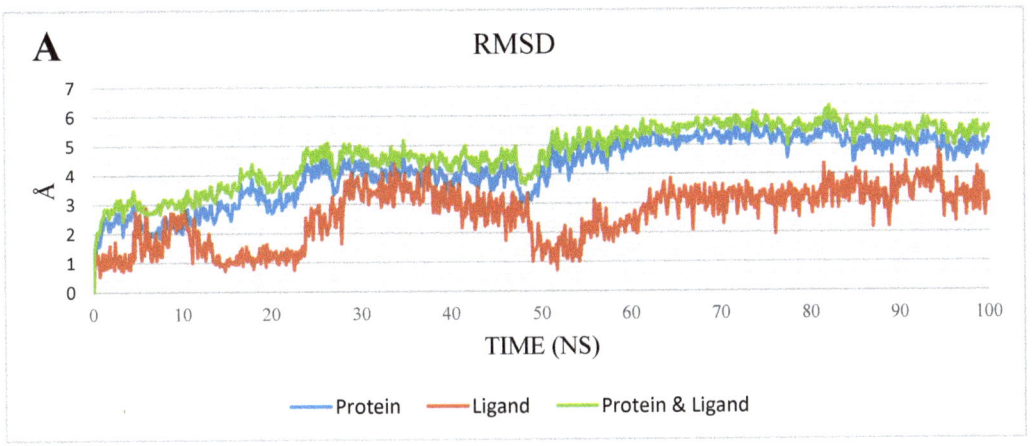

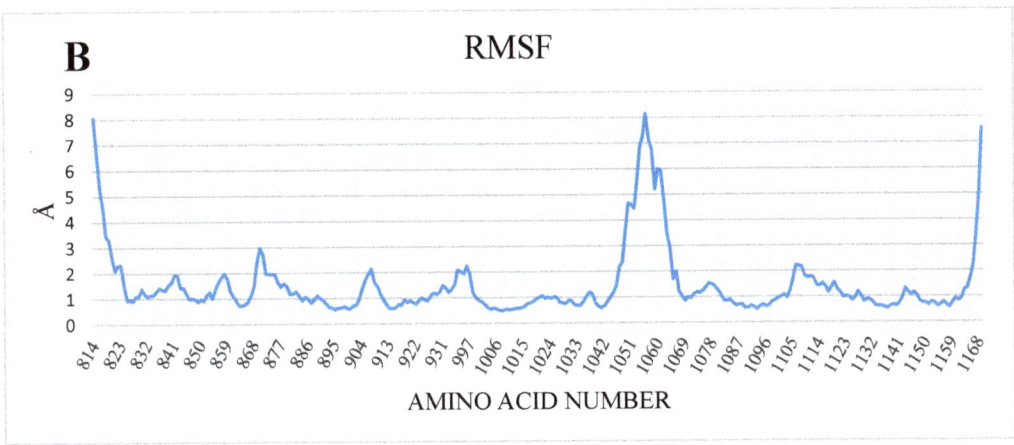

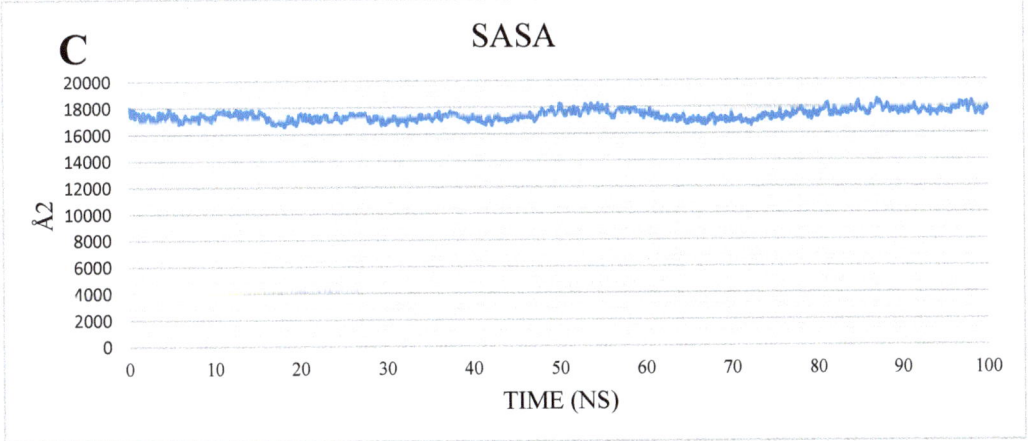

Figure 6. *Cont.*

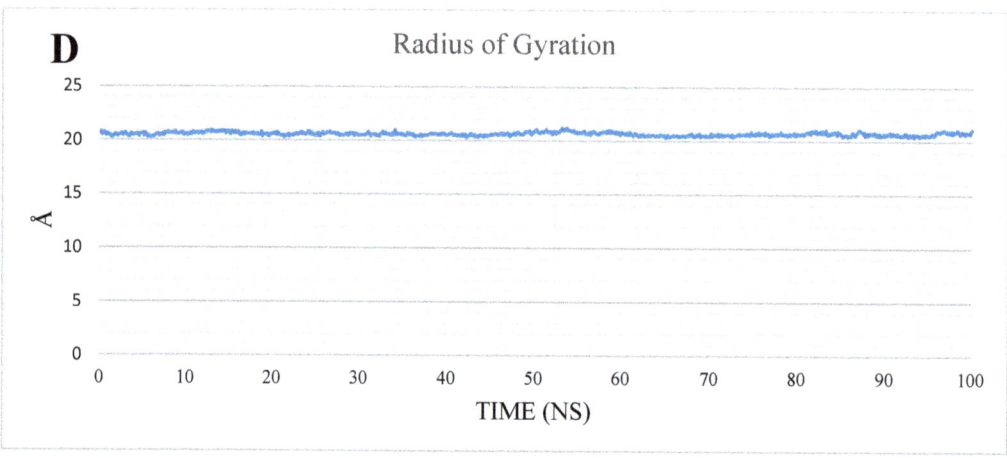
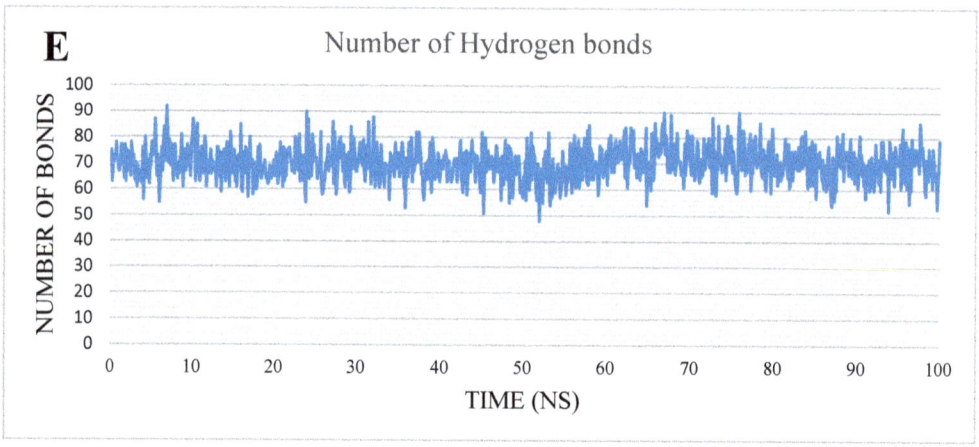
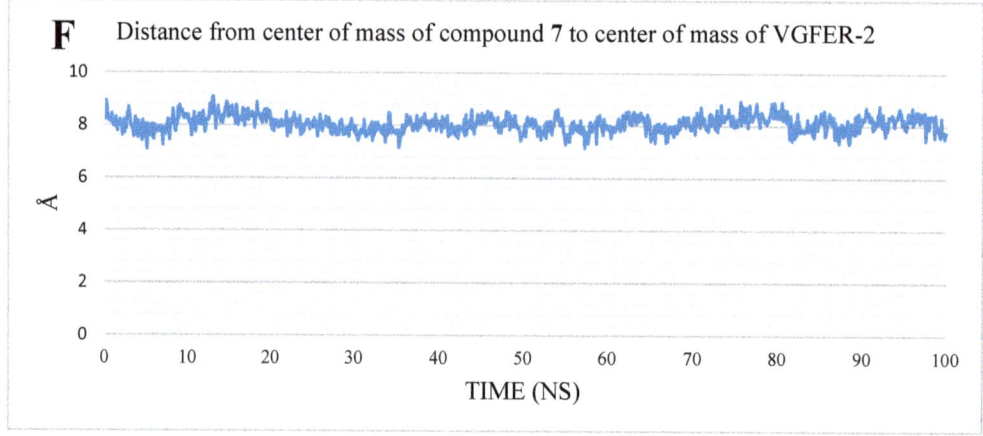

Figure 6. Analyses performed on the trajectory using VMD. (**A**) RMSD values. (**B**) RMSF. (**C**) SASA values. (**D**) RoG. (**E**) Hydrogen bond changes. (**F**) Change in the distance from the center of mass of the 7-VEGFR-2 complex.

2.1.3. MM-GBSA

To measure the strength of binding, the gmx_MMPBSA library was utilized. Figure 7 shows the values of the energy components of MM-GBSA, and their standard deviations. The binding is mostly due to the van der Waals interaction (an average of −54.8 Kcal/Mol), followed by electrostatic interactions (an average of −25.07 Kcal/Mol) and a total binding energy of −40.38 Kcal/Mol. The amino acid contribution to the binding was measured via the decomposition of MM-GBSA, to know which amino acids are contributing most to the interaction (Figure 8). Eight amino acids showed a contribution to the binding, with values of less than −1 Kcal/Mol. L838, V846, K866, L887, V897, V914, C1043, and F1045 showed binding contributions of −1.07 Kcal/Mol, −1.12 Kcal/Mol, −1.49 Kcal/Mol, −1.37 Kcal/Mol, −1.27 Kcal/Mol, −1.11 Kcal/Mol, −3.21 Kcal/Mol, and −1.80 Kcal/Mol, respectively.

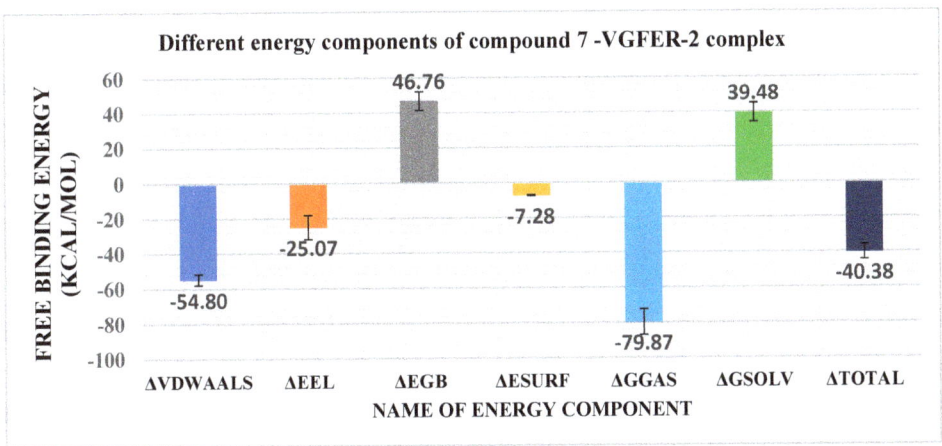

Figure 7. Different energy components obtained from the MM-GBSA analysis. Bars represent the standard deviation values.

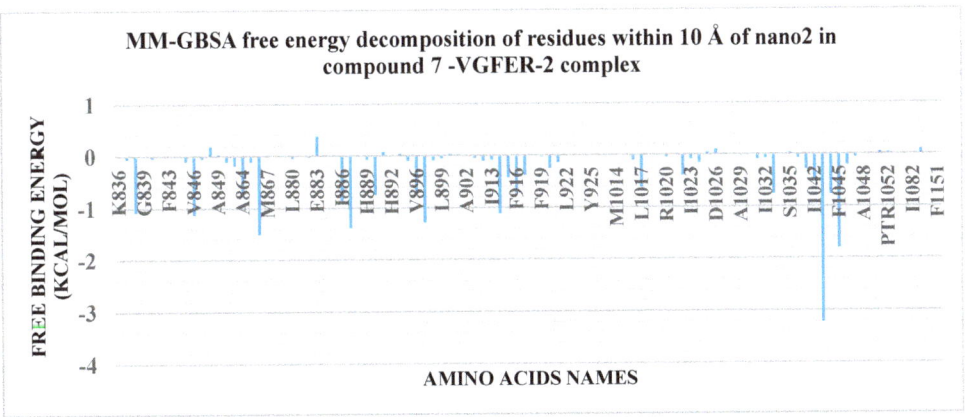

Figure 8. Free binding energy decomposition of amino acids around 10 Å of the 7-VEGFR-2 complex.

To know the numbers and types of interaction, the trajectory was clustered, and for each cluster, a representative frame was obtained that was used with the PLIP webserver. Table 1 shows the number and types of interactions for each frame. The predominant interaction is the hydrophobic interaction in all of the representative frames that support the value of the van der Waals component in MM-GBSA analysis. In addition, PLIP outputs the .pse file that shows the 3D interaction pattern for each representative frame (Figure 9).

Table 1. Variation of interactions between compound **7** and VEGFR-2, as obtained from the PLIP webserver for the representative frame of each cluster.

Cluster Number	No. of Hydrogen Bonds	Amino Acids in the Receptor	No. of Hydrophobic Interactions	Amino Acids in the Receptor
C1	3	K866–E915–D1044	6	I886–L887–I890–L1017–L1033–F1045
C2	3	K866–E915–D1044	4	I886–I890–L1017–L1033
C3	3	K866–E915–D1044	7	I886–I890 (2)–F916–L1017–D1044–F1045

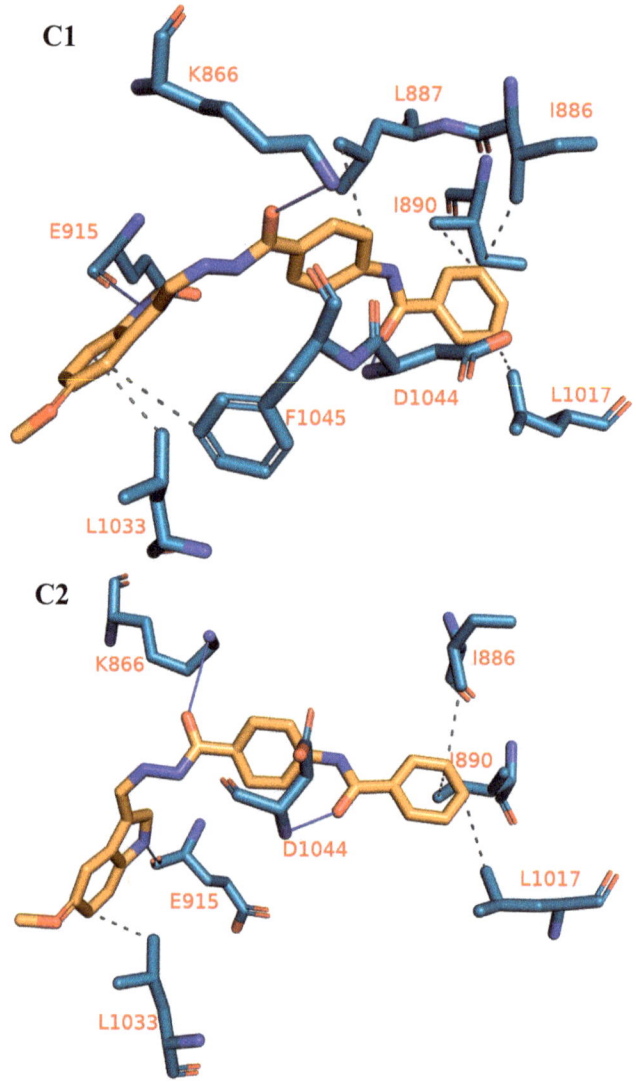

Figure 9. *Cont.*

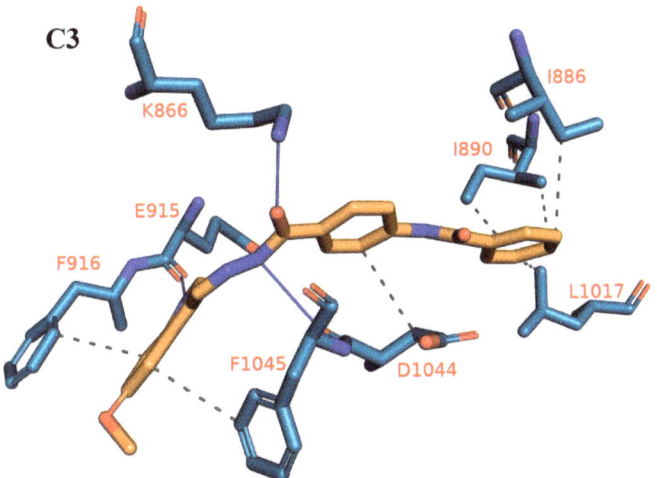

Figure 9. (**C1–C3**) Three-dimensional interaction between compound 7-VEGFR-2 complex in each of the representative frames for each cluster. Amino acids are shown as blue sticks. Compound **7** is shown as orange sticks. Grey dashed lines: hydrophobic interaction. Blue solid lines: hydrogen bonds.

2.1.4. Density Functional Theory (DFT)

Molecular Structure Optimization

The nucleophilic attack of *N*-(4-(hydrazinecarbonyl)phenyl)benzamide to 5-methoxy-1*H*-indole-3-carbaldehyde results in a Schiff base formation through the imine bond (C11-N13). The optimized structure of the formed Schiff base compound is represented in Figure 10. As shown in Figure 10, the imine bond length was found to be 1.28577 Å, while the (C9C11N13) and (C11N13C14) angles were found to be 118.21° and 129.37°, respectively.

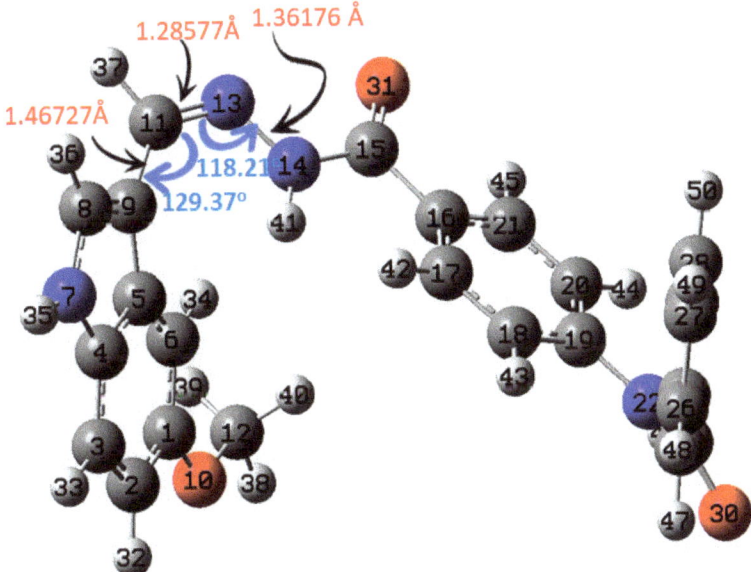

Figure 10. The optimized molecular structure of the selected compound at B3LYB/6-311++G(d,p).

Quantum Chemistry Calculations

The quantum chemistry calculations have been employed as a successful tool in the discovery of active compounds targeting various diseases such as prostate cancer [44], inflammation [45], and malignant glioblastoma [46]. The quantum chemistry calculations were conducted using the Gaussian(R) 09 program at the B3LYP level, together with the 6-311++G(d,p) basis set and the density functional theory (DFT) approach. As depicted in Figure 11, the electronic density of the highest occupied molecular orbital (HOMO) is localized on the heteroaromatic ring system, while in the lowest unoccupied molecular orbital (LUMO), the electronic density is located over the central linker and phenylbenzamide moieties. Frontier molecular orbital (FMO) theory suggested that HOMO serves as a donor, and LUMO serves as an acceptor for electrons. Both HOMO and LUMO have important roles in electronic investigations, and are essential to modern molecular biology and biochemistry when using quantum chemical calculations. A molecule is thought to be softer and more chemically reactive when its energy gap is small. A molecule is assumed to have greater chemical hardness and to be more stable when it has a large energy gap. The FMO gives very significant evidence for the stability, utilizing the difference in the energy (E_{gap}) of the frontier orbitals. Chemical quantum parameters are related to the inhibition efficiency of compound **7**, such as the chemical potential (μ), global hardness (η), maximal charge acceptance (ΔN_{max}), and energy change (ΔE); global softness (σ), electronegativity (χ), electrophilicity index (ω), ionization potential (IP), and electron affinity (EA) were calculated according to the equations of Koopmans' theory (Table 2) (the equations were detailed in Supporting Data).

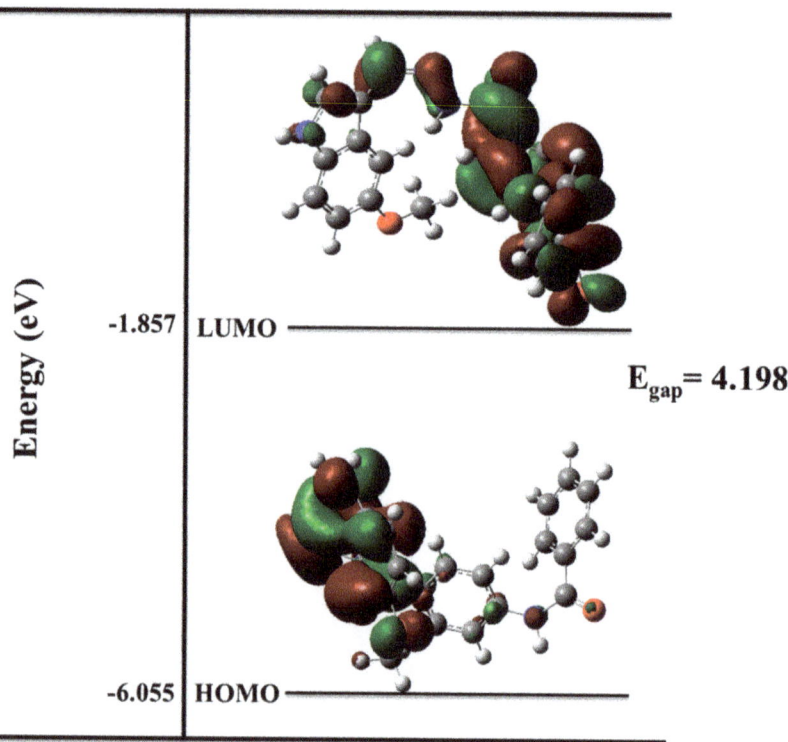

Figure 11. Energy gap (E_{gap}), frontier molecular orbitals; HOMO and LUMO at the ground state at B3LYB/6-311++G(d,p).

Table 2. The DFT calculated global reactivity parameters for comp. 7.

IP	EA	µ (eV)	χ (eV)	η (eV)	σ (eV)	ω (eV)	Dm (Debye)	TE (eV)	ΔNmax	ΔE (eV)
−6.055	−1.857	−3.956	3.956	2.099	0.476	16.421	6.472	−37,323.8	1.885	−16.421

Global quantum parameters, as well as the dipole moment (Dm) and the total ground state energy (TE), are calculated and summarized in Table 2. The results refer to the ability of comp. 7 to act as an inhibitor against VEGFR-2. For a system in equilibrium, the product of the density of states and probability distribution function gives the number of occupied states per unit volume for a given energy. This number is frequently used to study a variety of physical properties of materials. The total density of state analysis has been calculated and analyzed. The results confirmed the small energy gap of the compound under investigation, as depicted in Figure 12, which confirmed the reactivity of compound 7. When the E_{gap} of the border orbitals reduces, the inhibitor's efficiency increases [47].

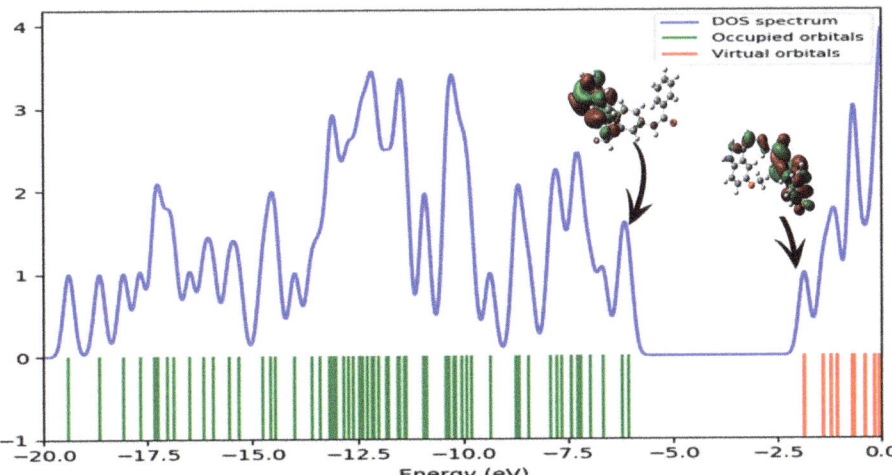

Figure 12. The total density of state analysis for the compound under investigation at B3LYB/6-311++G(d,p).

The Electron Density Maps

The reactivity strength of compound 7 can be predicted using DFT calculations based on the electron density of the donor atoms. The total electron density (TED) map, in Figure 13; represents the whole molecule's electron density. The red regions refer to the high electronegativity chemical sites, which are the O atoms of two carbonyls and methoxy groups in the investigated compound. Such active sites aid with electrophilic attack by amino acids (Cys917 and Glu833). In addition, the yellow-colored regions refer to atoms having a moderate electronegativity and that may form hydrophobic interactions, while the blue zones point to the most favorable positive regions, which accept electrons from the donor atoms of amino acids [48]. The electropositive regions are concentrated over the N-H groups. Such findings explain the nucleophilic attack of amino acids (Cys917 and Glu833) onto the NH groups of the 1H-indol and amide moieties, respectively. Furthermore, the possibility of hydrophobic interactions by the 1H-indol moiety, the central phenyl group, and the terminal hydrophobic phenyl group was supported by the yellow zones at these functional groups. The electrostatic surface potential (ESP) reveals the inhibition orientation of the molecule on the electrophilic amino acids (Figure 13), which is in the same orientation as the carbonyl and methoxy groups.

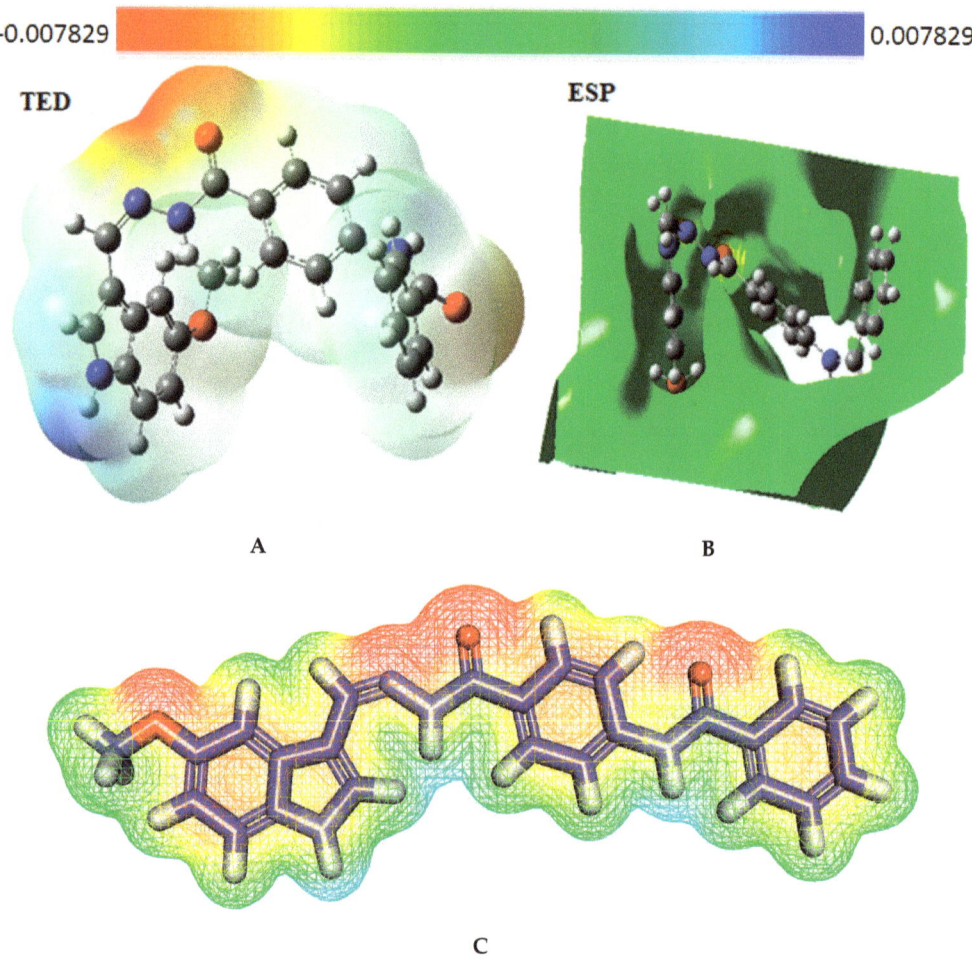

Figure 13. (**A**): Total electron density (TED) and (**B**): electrostatic potential (ESP) maps of the selected compound (Ball and line form) at the 6-311G++(d,p) basis set. (**C**): ESP maps of the compound (stick form).

2.1.5. ADMET Profiling Study

Sorafenib was used as a reference molecule, as the ADMET variables were investigated for compound **7** with Discovery studio 4.0 software. Both compound **7** and sorafenib showed remarkable similarities according to the ADMET results (Figure 14), showing a very low potential to pass the BBB, and good levels of intestinal absorption in addition to low and very low aqueous solubility levels, respectively. Similarly, both molecules presented a non-inhibitory potential against the cytochrome P-450 and CYP2D6, and revealed more than 90% binding percentage with plasma protein.

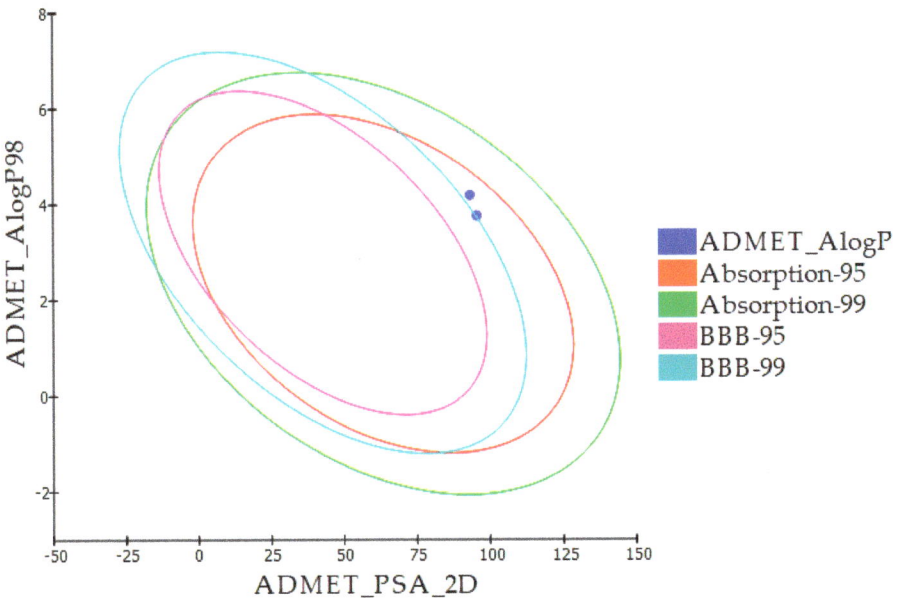

Figure 14. Computational prediction of ADMET parameters for compound **7** and sorafenib.

2.1.6. In Silico Toxicity Studies

In the presented study, five parameters of toxicity were estimated computationally, in accordance with the toxicity models built in the Discovery studio software. The employed models are; the rat-female FDA rodent carcinogenicity (RF-FDA-C) that predicts if the examined compound is carcinogenic or not, carcinogenic potency TD_{50} in a mouse model (TD_{50}-M), maximum tolerated dose in rats (MD-R), rat oral LD_{50} (R-O- LD_{50}), and rat chronic LOAEL (LOAEL-C), in addition to skin and eye irritancy. Table 3 demonstrates the expected general safety of compound **7** against sorafenib.

Table 3. In silico toxicity studies of compound **7** and reference molecule.

Comp.	RF-FDA-C	TD_{50}-M (mg/kg/day)	MD-R (g/kg)	R-O-LD_{50} (g/kg)	LOAEL-C (g/kg)	Skin Irritancy	Eye Irritancy
Compound 7	Non-Carcinogen	15.480	0.486	0.874	0.136	Non-Irritant	Mild
Sorafenib		19.236	0.089	0.823	0.005	Non-Irritant	Mild

2.1.7. Molecular Similarity

Similarity Check Using Physical Properties

A molecular similarity study using Discovery studio software was conducted to check the structural similarity between the synthesized compound and the FDA-approved 1H-inole derivatives (sunitinib **II**, vorolanib **III**, nintedanib **IV**, and toceranib **V**). This technique depends on many molecular properties to assess the structural similarity. These properties include the partition coefficient (ALog p) [49], molecular weight (M. Wt) [50], H-bond donors (HBA) [51], H-bond acceptors (HBD) [52], rotatable bond number [53], number of rings along with aromatic rings and minimum distance [54], as well as the molecular fractional polar surface area (MFPSA) [55]. Table 4 showed the different values for each tested compound, while Figure 15 showed the degree of similarity.

Table 4. Molecular properties of the tested FDA-approved drug and the synthesized compound 7.

Comp.	ALog p	M. Wt	HBA	HBD	Rotatable Bonds	Rings	Aromatic Rings	MFPSA	Minimum Distance
Toceranib V	2.761	396.458	3	3	5	4	2	0.191	1.12
Compound 7	3.745	412.441	4	3	6	4	4	0.233	0.00
Sunitinib II	2.997	398.474	3	3	7	3	2	0.179	1.35
Nintedanib IV	4.392	539.625	7	2	8	5	4	0.178	1.83
Vorolanib III	1.671	439.483	3	3	3	4	2	0.213	1.31

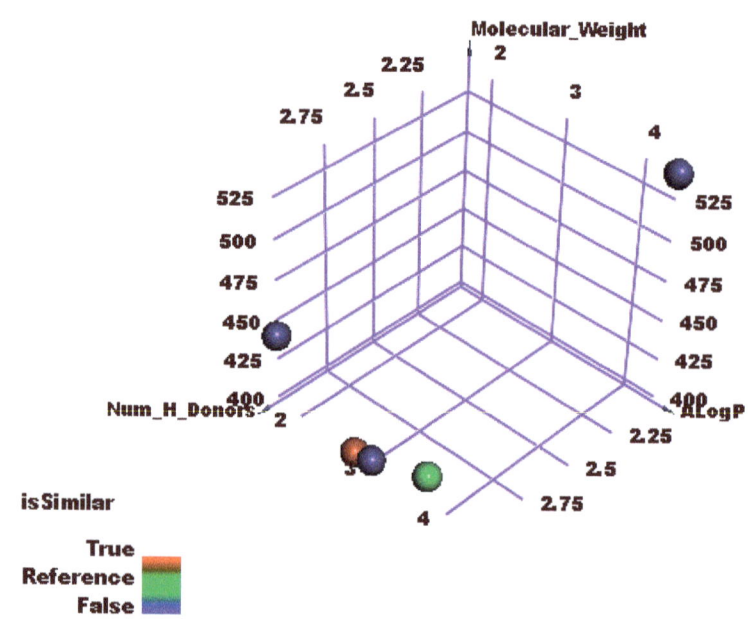

Figure 15. Molecular similarity study between the synthesized compound (green ball) and the FDA-approved 1H-inole derivatives (sunitinib, vorolanib, nintedanib, and toceranib). The red ball refers to toceranib (the most similar compound). The blue balls refer to sunitinib, vorolanib, and nintedanib (the less similar drugs).

The results revealed that toceranib **V** has the highest degree of similarity with the synthesized compound **7**.

Flexible Alignment

The 3D flexible alignment was studied for compound **7** and toceranib. The produced figure indicated a high degree of similarity between these compounds. Each feature in the tested compounds takes the same orientation of the corresponding one. In detail, the 5-fluoroindolin-2-one, central phenyl, amide, and terminal phenyl moieties of compound **7** showed the same orientations as 5-methoxyindolin-2-one, 2,4-dimethyl-1H-pyrrole, amide, and 1-ethylpyrrolidine moieties of toceranib, respectively (Figure 16).

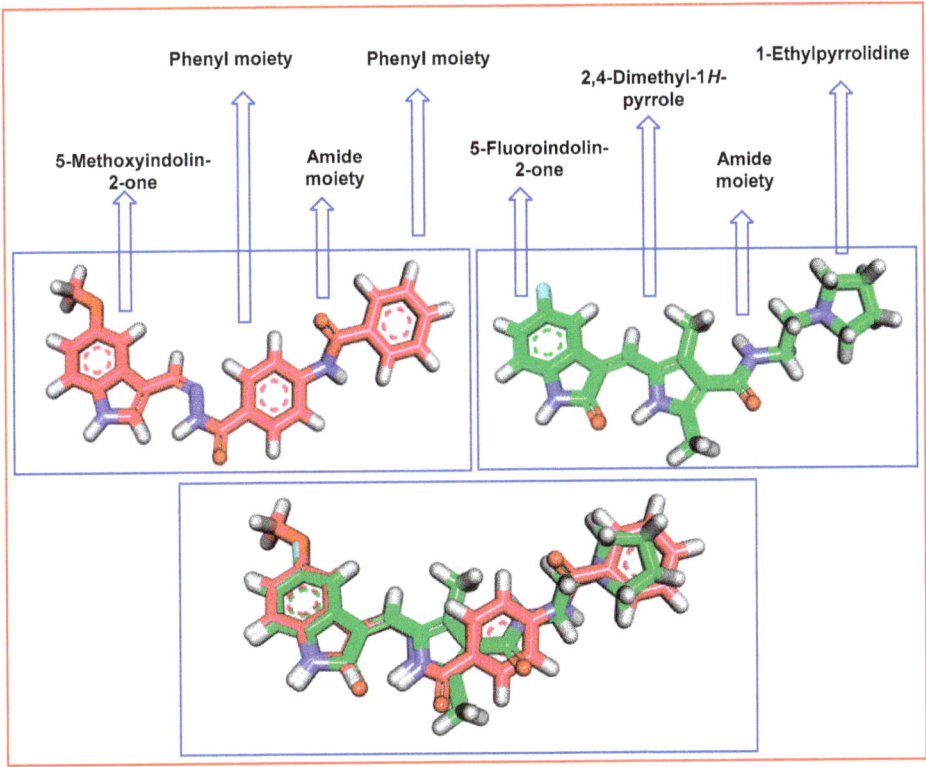

Figure 16. Flexible alignment of compound 8 (green) with sorafenib (pink), showing the same orientation.

2.2. Chemistry

To authenticate the design and the interesting results of the computational studies, the targeted lead compound was synthesized according to Scheme 1. The commercially available material (4-aminobenzoic acid **1**) was first refluxed with methanol and sulfuric acid in accordance with the reported procedure, to afford methyl 4-aminobenzoate **2**. Benzoylation of methyl 4-aminobenzoate **2** was achieved in an ice bath via dropwise addition of benzoyl chloride **3** in DCM in the presence of TEA, to afford compound **4**. Refluxing compound **4** with hydrazine hydrate in ethanol afforded the corresponding acid hydrazide derivative **5**. A mixture of the acid hydrazide derivative **5** and 5-methoxy-1H-indole-3-carbaldehyde **6** was refluxed in absolute ethanol and a few drops of glacial acetic acid, to afford the final target compound **7**.

The structure of compound **7** (Figure 17) was validated using spectral analysis. The IR spectrum showed absorption bands at 3287 cm^{-1}, and 1645 cm^{-1} attributed to NH and the amidic carbonyl group, respectively. Moreover, ^{1}H NMR spectra showed singlet signals around δ 11.48 and 10.53 ppm, corresponding to the indole NH and the two amidic NH protons. In addition, an up-field sharp singlet signal corresponding to OCH$_3$ protons appeared at δ 3.82 ppm. The three protons (6, 7, and 9) of the phenyl ring of the 1H-indol moiety resonated at δ 7.36 (d, J = 8.8 Hz), 6.88 (dd, J = 8.8, 2.4 Hz), and 7.79 (d, J = 2.4 Hz), respectively. At the same time, proton 2 was presented as a sharp singlet, peaking at δ 8.01. The characteristic proton, 10, resonated clearly as a sharp singlet signal at δ 8.62. Regarding the di para-substituted protons, protons (19, 21) and (18, 22) resonated at δ 7.58 (d, J = 8.2 Hz) and 7.99 (d, J = 8.2 Hz), respectively. Finally, the aromatic protons of the benzamide moiety resonated at the overlapped areas of 7.68 and 7.96. Matching such findings, the ^{13}C NMR spectra showed characteristic peaks corresponding to the two C=O and OCH$_3$ groups around δ 166.34, 162.36, and 55.80 ppm, respectively. The carbons (2, 3, 4, 5, 6, 7, 8, and 9) of the phenyl ring of the *1H*-indol moiety resonated at δ (128.3, 112.8, 125.4, 130.4, 112.7, 111.9, 154.8, and 104.6, respectively). The characteristic carbon, 10, resonated at δ 154.2. The di para-substituted carbons (19, 21) and (18, 22) resonated at δ 119.9 and 128.2, respectively. The aromatic carbons (26, 27, 28, 29, 30, and 31) of the benzamide moiety resonated at δ (135.1, 128.2, 128.9, 132.5, 128.9, and 128.2, respectively).

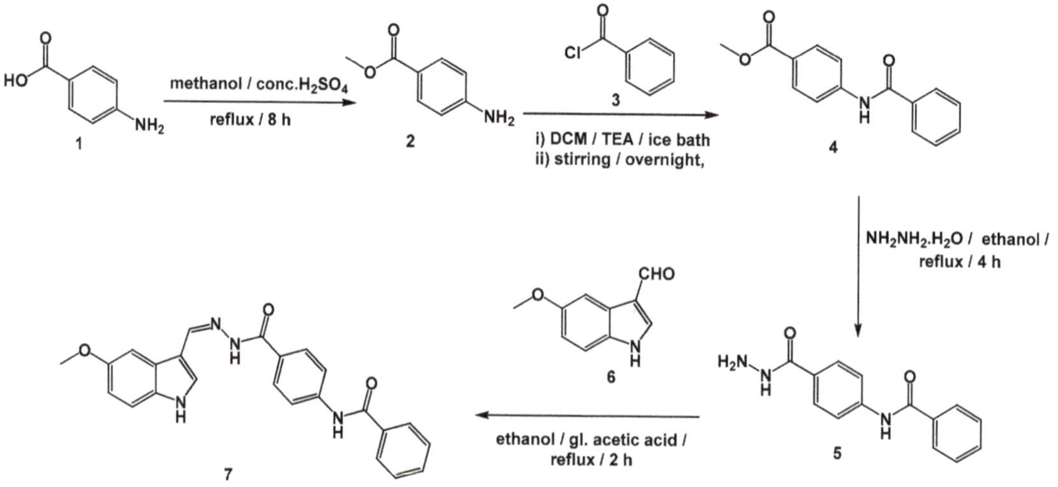

Scheme 1. Synthetic pathway of target compound 7.

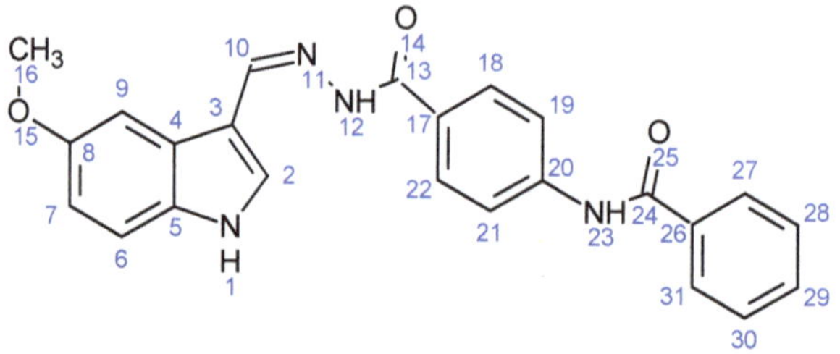

Figure 17. Chemical structure of compound 7.

2.3. Biological Examinations

2.3.1. VGFER-2 Inhibition

We assessed the in vitro inhibitory potential of compound 7 against VEGFR-2 in comparison to sorafenib, to verify our design, and the several computations conducted that indicated the high affinity of compound 7 for binding to and inhibiting the VEGFR-2 enzyme. Compound 7 showed significant inhibitory activity in vitro (an IC$_{50}$ value of 25 nM) that was lower than that of sorafenib (35 nM). The in vitro results showed that the compound had a great potential to inhibit VEGFR-2, affirming the in silico studies.

2.3.2. Cytotoxicity

To analyze the application of compound 7's VEGFR-2 inhibitory potential against cancer, in vitro cytotoxicity studies were conducted using human breast cancer MCF-7, and colon cancer HCT 116 cell lines, with sorafenib as a reference. Table 5 and Figure 18 show that compound **7** could inhibit the growth of the MCF-7 and HCT 116 cell lines, exhibiting IC$_{50}$ values of 12.93 and 11.52 µM, respectively, in comparison with sorafenib's anticancer potentialities against the same cell lines (4.32 and 7.28 µM, respectively).

Table 5. VGFER-2 inhibition and cytotoxicity of compound **7** and sorafenib.

	MCF-7 IC$_{50}$ (µM)	HCT 116 IC$_{50}$ (µM)	VEGFR IC$_{50}$ (nM)
Compound 7	12.93 ± 0.54	11.52 ± 0.70	25 ± 1.29
Sorafenib	4.32 ± 0.33	7.28 ± 0.53	35 ± 1.34

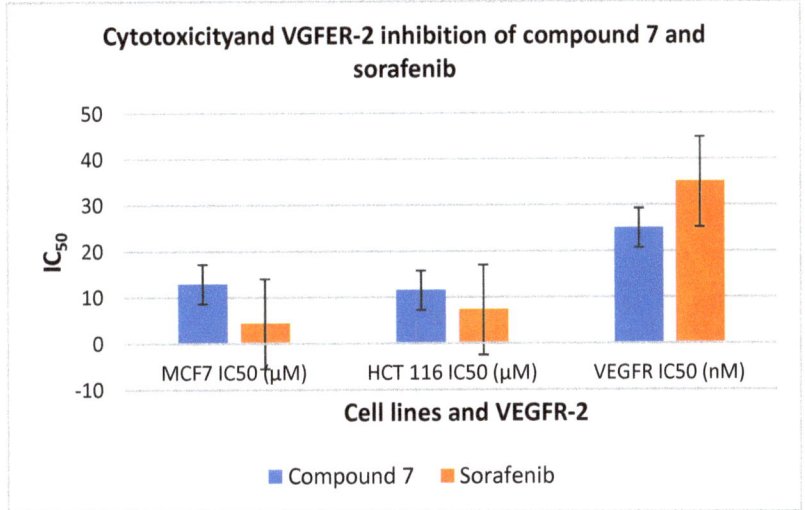

Figure 18. VGFER-2 inhibition and cytotoxicity of compound **7** and sorafenib.

2.3.3. Safety Evaluation

Against normal human cells, W138 cell lines, compound **7** was administered to test its toxicity and to determine its selectivity against cancer cell lines. Notably, compound **7** showed a very high level of safety, manifesting an IC$_{50}$ of 87.26 µM. The obtained selectivity indexes (SI) of compound **7** against MCF-7 and HCT 116 cancer cell lines were very high, at 6.7 and 7.5, respectively.

3. Conclusions

A new 1H-indole-derived lead compound has been designed, which keeps the characteristic VEGFR-2 prohibition features. The binding potential of the lead compound was suggested by docking experiments. Interestingly, accurate binding with the VEGFR-2 active pocket was verified through MD simulations (six studies), MM-GBSA (three studies), and DFT (three studies). Additionally, ADMET studies declared the lead compound's drug-likeness. Consequently, the designed lead compound was synthesized and tested for in vitro VEGFR-2 prohibition, cytotoxicity, and safety. Compound **7** exhibited a prohibitory effect against VEGFR-2, with an IC$_{50}$ value of 25 nM (lower than that of sorafenib, 35 nM), besides a promising antiproliferative potential against MCF-7 and HCT 116 cell lines, with very low IC$_{50}$ values of 12.93 and 11.52 µM, as well as high selectivity indexes of 4.32 and 7.28 µM, respectively.

4. Experimental

4.1. In Silico Studies

4.1.1. Docking Studies

Molecular docking was conducted for compound 7 against VEGFR-2 [56,57], using MOE2014 software. A thorough outline has been provided in the Supplementary Data.

4.1.2. MD Simulation

The CHARMM-GUI web server was employed, and GROMACS 2021 was used as an MD engine. A thorough outline has been provided in the Supplementary Data.

4.1.3. MM-GBSA

The Gmx_MMPBSA package was used. A thorough outline has been provided in the Supplementary Data.

4.1.4. DFT

Gaussian 09 and GaussSum3.0 programs were used. A thorough outline has been provided in the Supplementary Data.

4.1.5. ADMET Studies

An ADMET profile was determined using Discovery Studio 4.0 [58,59]. A thorough outline has been provided in the Supplementary Data.

4.1.6. Toxicity Studies

The toxicity profile was determined using Discovery Studio 4.0 [60]. A thorough outline has been provided in the Supplementary Data.

4.2. Chemistry

The solvents and fine chemicals used in the synthesis of the target molecule were purchased from Sigma-Aldrich at a purity above 99%. All chemicals and apparatuses used in this section are illustrated in the Supplementary Data. Compounds **2** and **4** were synthesized according to the previously reported methods [61].

4.2.1. General Procedure for the Synthesis of Compound **5**

To a solution of ester derivative **4** (0.002 mol) in in absolute ethanol (20 mL), hydrazine hydrate 85% (0.01 mol) was added dropwise with continuous stirring. The mixture was heated with gentle reflux for 2 h. After completion of the reaction, the mixture was cooled. The formed precipitate was filtered, dried, and recrystallized from ethanol to give compound **4**.

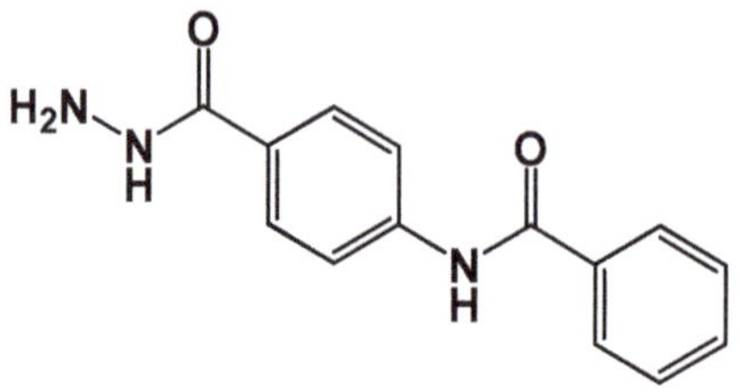

4.2.2. General Procedure for the Synthesis of Compound **7**

To a solution of compound 5 (0.001 mol) in absolute ethanol (50 mL) containing a few drops of glacial acetic acid, (0.001 mol) of 5-methoxy-1H-indole-3-carbaldehyde **6** was added. The reaction mixture was heated to reflux for 2 h. The TLC test was used to keep track of the reaction's progress. Under reduced pressure, the solvent was evaporated after the reaction was completed. The obtained powder was collected and recrystallized from ethanol to furnish the pure form of compound 7.

(Z)-N-(4-(2-((5-Methoxy-1H-indol-3-yl)methylene)hydrazine-1-carbonyl)phenyl) benzamide.

Off-white crystal (yield, 84%); m. p. = 257–259 °C; IR (KBr) ν cm^{-1}: 3287 (NH), 3093 and 3050 (CH aromatic), 2996 and 2956 (CH aliphatic), 1645 (C=O); ^{1}H NMR (400 MHz, DMSO-d6) δ 11.48 (s, 2H), 10.53 (s, 1H), 8.62 (s, 1H), 7.99 (d, J = 8.2,Hz, 2H), 7.96 (m, 4H), 7.88 (d, J = 2.5 Hz, 1H), 7.79 (d, J = 2.4 Hz, 1H), 7.68–7.61 (m, 1H), 7.58 (d, J = 8.2, 2H), 7.36 (d, J = 8.8 Hz, 1H), 6.88 (dd, J = 8.8, 2.4 Hz, 1H), 3.82 (s, 3H); ^{13}C NMR (101 MHz, DMSO-d_6) δ 166.34, 162.36, 154.88, 145.25, 142.42, 135.17, 132.49, 132.31, 131.02, 129.29, 128.95 (2C), 128.67, 128.25 (2C), 125.46, 119.99 (2C), 119.11, 112.90, 112.71, 112.01, 104.67, 55.80, for $C_{24}H_{20}N_4O_3$ (412.45).

4.3. Biological Studies

4.3.1. In Vitro VEGFR-2 Inhibition

Inhibition was performed using a Human VEGFR-2 ELISA kit. A thorough outline has been provided in the Supplementary Data.

4.3.2. In Vitro Antiproliferative Activity

The MTT procedure was employed. A thorough outline has been provided in the Supplementary Data.

4.3.3. Safety Assay

The non-cancerous cell line W138 was used. A thorough outline has been provided in the Supplementary Data.

Supplementary Materials: The detailed methods, spectral data of compound **7** and toxicity report can be downloaded at: https://www.mdpi.com/article/10.3390/pr10071391/s1.

Author Contributions: Conceptualization, A.M.M. and I.H.E.; software, E.B.E., H.E., D.Z.H., I.M.I. and I.H.E.; investigation, R.G.Y. and I.M.M.G.; writing—original draft preparation, A.M.M., H.E. and I.H.E.; writing—review and editing, A.A.A. and E.B.E.; project administration, A.M.M. and I.H.E.; funding acquisition, E.B.E. and A.A.A. All authors have read and agreed to the published version of the manuscript.

Funding: This research was funded by Princess Nourah bint Abdulrahman University Researchers Supporting Project number (PNURSP2022R116), Princess Nourah bint Abdulrahman University, Riyadh, Saudi Arabia. The authors extend their appreciation to the Research Center at AlMaarefa University for funding this work.

Institutional Review Board Statement: Not applicable.

Informed Consent Statement: Not applicable.

Data Availability Statement: Data are available with the corresponding authors upon request.

Conflicts of Interest: The authors declare no conflict of interest.

References

1. WHO Cancer, Fact Sheet. Available online: https://www.who.int/news-room/fact-sheets/detail/cancer (accessed on 13 June 2022).
2. El-Dash, Y.; Elzayat, E.; Abdou, A.M.; Hassan, R.A. Novel thienopyrimidine-aminothiazole hybrids: Design, synthesis, antimicrobial screening, anticancer activity, effects on cell cycle profile, caspase-3 mediated apoptosis and VEGFR-2 inhibition. *Bioorg. Chem.* **2021**, *114*, 105137. [CrossRef] [PubMed]
3. Quesada, A.R.; Muñoz-Chápuli, R.; Medina, M.A. Anti-angiogenic drugs: From bench to clinical trials. *Med. Res. Rev.* **2006**, *26*, 483–530. [CrossRef] [PubMed]
4. El-Zahabi, M.A.; Sakr, H.; El-Adl, K.; Zayed, M.; Abdelraheem, A.S.; Eissa, S.I.; Elkady, H.; Eissa, I.H. Design, synthesis, and biological evaluation of new challenging thalidomide analogs as potential anticancer immunomodulatory agents. *Bioorg. Chem.* **2020**, *104*, 104218. [CrossRef] [PubMed]
5. Alsaif, N.A.; Taghour, M.S.; Alanazi, M.M.; Obaidullah, A.J.; Alanazi, W.A.; Alasmari, A.; Albassam, H.; Dahab, M.A.; Mahdy, H.A. Identification of new [1, 2, 4] triazolo [4, 3-a] quinoxalines as potent VEGFR-2 tyrosine kinase inhibitors: Design, synthesis, anticancer evaluation, and in silico studies. *Bioorg. Med. Chem.* **2021**, *46*, 116384. [CrossRef]
6. Elrazaz, E.Z.; Serya, R.A.; Ismail, N.S.; Albohy, A.; Abou El Ella, D.A.; Abouzid, K.A. Discovery of Potent Thieno [2, 3-d] pyrimidine VEGFR-2 Inhibitors: Design, Synthesis and Enzyme Inhibitory Evaluation Supported by Molecular Dynamics Simulations. *Bioorg. Chem.* **2021**, *113*, 105019. [CrossRef]
7. Sana, S.; Reddy, V.G.; Bhandari, S.; Reddy, T.S.; Tokala, R.; Sakla, A.P.; Bhargava, S.K.; Shankaraiah, N. Exploration of carbamide derived pyrimidine-thioindole conjugates as potential VEGFR-2 inhibitors with anti-angiogenesis effect. *Eur. J. Med. Chem.* **2020**, *200*, 112457. [CrossRef]
8. Cavasotto, C.N.; Aucar, M.G.; Adler, N.S. Computational chemistry in drug lead discovery and design. *Int. J. Quantum Chem.* **2019**, *119*, e25678. [CrossRef]
9. Mahdy, H.A.; Ibrahim, M.K.; Metwaly, A.M.; Belal, A.; Mehany, A.B.; El-Gamal, K.M.; El-Sharkawy, A.; Elhendawy, M.A.; Radwan, M.M.; Elsohly, M.A. Design, synthesis, molecular modeling, in vivo studies and anticancer evaluation of quinazolin-4 (3H)-one derivatives as potential VEGFR-2 inhibitors and apoptosis inducers. *Bioorg. Chem.* **2020**, *94*, 103422. [CrossRef]
10. Eissa, I.H.; El-Helby, A.-G.A.; Mahdy, H.A.; Khalifa, M.M.; Elnagar, H.A.; Mehany, A.B.; Metwaly, A.M.; Elhendawy, M.A.; Radwan, M.M.; ElSohly, M.A. Discovery of new quinazolin-4 (3H)-ones as VEGFR-2 inhibitors: Design, synthesis, and anti-proliferative evaluation. *Bioorg. Chem.* **2020**, *105*, 104380. [CrossRef]
11. Eissa, I.H.; Ibrahim, M.K.; Metwaly, A.M.; Belal, A.; Mehany, A.B.; Abdelhady, A.A.; Elhendawy, M.A.; Radwan, M.M.; ElSohly, M.A.; Mahdy, H.A. Design, molecular docking, in vitro, and in vivo studies of new quinazolin-4 (3H)-ones as VEGFR-2 inhibitors with potential activity against hepatocellular carcinoma. *Bioorg. Chem.* **2021**, *107*, 104532. [CrossRef]
12. El-Adl, K.; El-Helby, A.-G.A.; Ayyad, R.R.; Mahdy, H.A.; Khalifa, M.M.; Elnagar, H.A.; Mehany, A.B.; Metwaly, A.M.; Elhendawy, M.A.; Radwan, M.M. Design, synthesis, and anti-proliferative evaluation of new quinazolin-4 (3H)-ones as potential VEGFR-2 inhibitors. *Bioorg. Med. Chem.* **2021**, *29*, 115872. [CrossRef] [PubMed]
13. Yousef, R.G.; Ibrahim, A.; Khalifa, M.M.; Eldehna, W.M.; Gobaara, I.M.; Mehany, A.B.; Elkaeed, E.B.; Alsfouk, A.A.; Metwaly, A.M.; Eissa, I.H. Discovery of new nicotinamides as apoptotic VEGFR-2 inhibitors: Virtual screening, synthesis, anti-proliferative, immunomodulatory, ADMET, toxicity, and molecular dynamic simulation studies. *J. Enzym. Inhib. Med. Chem.* **2022**, *37*, 1389–1403. [CrossRef] [PubMed]
14. El-Helby, A.G.A.; Sakr, H.; Eissa, I.H.; Abulkhair, H.; Al-Karmalawy, A.A.; El-Adl, K. Design, synthesis, molecular docking, and anticancer activity of benzoxazole derivatives as VEGFR-2 inhibitors. *Arch. Der Pharm.* **2019**, *352*, 1900113. [CrossRef] [PubMed]
15. Saleh, N.M.; Abdel-Rahman, A.A.H.; Omar, A.M.; Khalifa, M.M.; El-Adl, K. Pyridine-derived VEGFR-2 inhibitors: Rational design, synthesis, anticancer evaluations, in silico ADMET profile, and molecular docking. *Arch. Der Pharm.* **2021**, *354*, 2100085. [CrossRef]
16. Taghour, M.S.; Elkady, H.; Eldehna, W.M.; El-Deeb, N.M.; Kenawy, A.M.; Elkaeed, E.B.; Alsfouk, A.A.; Alesawy, M.S.; Metwaly, A.M.; Eissa, I.H. Design and synthesis of thiazolidine-2, 4-diones hybrids with 1, 2-dihydroquinolones and 2-oxindoles as potential VEGFR-2 inhibitors: In-vitro anticancer evaluation and in-silico studies. *J. Enzym. Inhib. Med. Chem.* **2022**, *37*, 1903–1917. [CrossRef]
17. Parmar, D.R.; Soni, J.Y.; Guduru, R.; Rayani, R.H.; Kusurkar, R.V.; Vala, A.G.; Talukdar, S.N.; Eissa, I.H.; Metwaly, A.M.; Khalil, A. Discovery of new anticancer thiourea-azetidine hybrids: Design, synthesis, in vitro antiproliferative, SAR, in silico molecular docking against VEGFR-2, ADMET, toxicity, and DFT studies. *Bioorg. Chem.* **2021**, *115*, 105206. [CrossRef]
18. Abdelgawad, M.A.; El-Adl, K.; El-Hddad, S.S.; Elhady, M.M.; Saleh, N.M.; Khalifa, M.M.; Khedr, F.; Alswah, M.; Nayl, A.A.; Ghoneim, M.M. Design, Molecular Docking, Synthesis, Anticancer and Anti-Hyperglycemic Assessments of Thiazolidine-2, 4-diones Bearing Sulfonylthiourea Moieties as Potent VEGFR-2 Inhibitors and PPARγ Agonists. *Pharmaceuticals* **2022**, *15*, 226. [CrossRef]
19. El-Adl, K.; Sakr, H.M.; Yousef, R.G.; Mehany, A.B.; Metwaly, A.M.; Elhendawy, M.A.; Radwan, M.M.; ElSohly, M.A.; Abulkhair, H.S.; Eissa, I.H. Discovery of new quinoxaline-2 (1H)-one-based anticancer agents targeting VEGFR-2 as inhibitors: Design, synthesis, and anti-proliferative evaluation. *Bioorg. Chem.* **2021**, *114*, 105105. [CrossRef]

20. Yousef, R.G.; Sakr, H.M.; Eissa, I.H.; Mehany, A.B.; Metwaly, A.M.; Elhendawy, M.A.; Radwan, M.M.; ElSohly, M.A.; Abulkhair, H.S.; El-Adl, K. New quinoxaline-2 (1 H)-ones as potential VEGFR-2 inhibitors: Design, synthesis, molecular docking, ADMET profile and anti-proliferative evaluations. *New J. Chem.* **2021**, *45*, 16949–16964. [CrossRef]
21. Alanazi, M.M.; Mahdy, H.A.; Alsaif, N.A.; Obaidullah, A.J.; Alkahtani, H.M.; Al-Mehizia, A.A.; Alsubaie, S.M.; Dahab, M.A.; Eissa, I.H. New bis ([1, 2, 4] triazolo)[4, 3-a: 3′, 4′-c] quinoxaline derivatives as VEGFR-2 inhibitors and apoptosis inducers: Design, synthesis, in silico studies, and anticancer evaluation. *Bioorg. Chem.* **2021**, *112*, 104949. [CrossRef]
22. Alanazi, M.M.; Eissa, I.H.; Alsaif, N.A.; Obaidullah, A.J.; Alanazi, W.A.; Alasmari, A.F.; Albassam, H.; Elkady, H.; Elwan, A. Design, synthesis, docking, ADMET studies, and anticancer evaluation of new 3-methylquinoxaline derivatives as VEGFR-2 inhibitors and apoptosis inducers. *J. Enzym. Inhib. Med. Chem.* **2021**, *36*, 1760–1782. [CrossRef]
23. Alanazi, M.M.; Elkady, H.; Alsaif, N.A.; Obaidullah, A.J.; Alanazi, W.A.; Al-Hossaini, A.M.; Alharbi, M.A.; Eissa, I.H.; Dahab, M.A. Discovery of new quinoxaline-based derivatives as anticancer agents and potent VEGFR-2 inhibitors: Design, synthesis, and in silico study. *J. Mol. Struct.* **2021**, *1253*, 132220. [CrossRef]
24. El-Metwally, S.A.; Abou-El-Regal, M.M.; Eissa, I.H.; Mehany, A.B.; Mahdy, H.A.; Elkady, H.; Elwan, A.; Elkaeed, E.B. Discovery of thieno [2, 3-d] pyrimidine-based derivatives as potent VEGFR-2 kinase inhibitors and anti-cancer agents. *Bioorg. Chem.* **2021**, *112*, 104947. [CrossRef] [PubMed]
25. Eissa, I.H.; Alesawy, M.S.; Saleh, A.M.; Elkaeed, E.B.; Alsfouk, B.A.; El-Attar, A.-A.M.; Metwaly, A.M. Ligand and structure-based in silico determination of the most promising SARS-CoV-2 nsp16-nsp10 2′-o-Methyltransferase complex inhibitors among 3009 FDA approved drugs. *Molecules* **2022**, *27*, 2287. [CrossRef] [PubMed]
26. Alsaif, N.A.; Dahab, M.A.; Alanazi, M.M.; Obaidullah, A.J.; Al-Mehizia, A.A.; Alanazi, M.M.; Aldawas, S.; Mahdy, H.A.; Elkady, H. New quinoxaline derivatives as VEGFR-2 inhibitors with anticancer and apoptotic activity: Design, molecular modeling, and synthesis. *Bioorg. Chem.* **2021**, *110*, 104807. [CrossRef] [PubMed]
27. Elkaeed, E.B.; Elkady, H.; Belal, A.; Alsfouk, B.A.; Ibrahim, T.H.; Abdelmoaty, M.; Arafa, R.K.; Metwaly, A.M.; Eissa, I.H. Multi-Phase In Silico Discovery of Potential SARS-CoV-2 RNA-Dependent RNA Polymerase Inhibitors among 3009 Clinical and FDA-Approved Related Drugs. *Processes* **2022**, *10*, 530. [CrossRef]
28. Mohammed, S.O.; El Ashry, E.S.H.; Khalid, A.; Amer, M.R.; Metwaly, A.M.; Eissa, I.H.; Elkaeed, E.B.; Elshobaky, A.; Hafez, E.E. Expression, Purification, and Comparative Inhibition of Helicobacter pylori Urease by Regio-Selectively Alkylated Benzimidazole 2-Thione Derivatives. *Molecules* **2022**, *27*, 865. [CrossRef]
29. Suleimen, Y.M.; Jose, R.A.; Suleimen, R.N.; Arenz, C.; Ishmuratova, M.Y.; Toppet, S.; Dehaen, W.; Alsfouk, B.A.; Elkaeed, E.B.; Eissa, I.H. Jusanin, a New Flavonoid from Artemisia commutata with an In Silico Inhibitory Potential against the SARS-CoV-2 Main Protease. *Molecules* **2022**, *27*, 1636. [CrossRef]
30. Suleimen, Y.M.; Jose, R.A.; Suleimen, R.N.; Ishmuratova, M.Y.; Toppet, S.; Dehaen, W.; Alsfouk, A.A.; Elkaeed, E.B.; Eissa, I.H.; Metwaly, A.M. Isolation and In Silico SARS-CoV-2 Main Protease Inhibition Potential of Jusan Coumarin, a New Dicoumarin from Artemisia glauca. *Molecules* **2022**, *27*, 2281. [CrossRef]
31. Eissa, I.H.; Khalifa, M.M.; Elkaeed, E.B.; Hafez, E.E.; Alsfouk, A.A.; Metwaly, A.M. In silico exploration of potential natural inhibitors against SARS-Cov-2 nsp10. *Molecules* **2021**, *26*, 6151. [CrossRef]
32. Suleimen, Y.M.; Jose, R.A.; Suleimen, R.N.; Arenz, C.; Ishmuratova, M.; Toppet, S.; Dehaen, W.; Alsfouk, A.A.; Elkaeed, E.B.; Eissa, I.H. Isolation and In Silico Anti-SARS-CoV-2 Papain-Like Protease Potentialities of Two Rare 2-Phenoxychromone Derivatives from *Artemisia* spp. *Molecules* **2022**, *27*, 1216. [CrossRef] [PubMed]
33. Alesawy, M.S.; Elkaeed, E.B.; Alsfouk, A.A.; Metwaly, A.M.; Eissa, I. In Silico Screening of Semi-Synthesized Compounds as Potential Inhibitors for SARS-CoV-2 Papain-like Protease: Pharmacophoric Features, Molecular Docking, ADMET, Toxicity and DFT Studies. *Molecules* **2021**, *26*, 6593. [CrossRef]
34. Adnane, L.; Trail, P.A.; Taylor, I.; Wilhelm, S.M. Sorafenib (BAY 43-9006, Nexavar®), a dual-action inhibitor that targets RAF/MEK/ERK pathway in tumor cells and tyrosine kinases VEGFR/PDGFR in tumor vasculature. *Methods Enzymol.* **2006**, *407*, 597–612. [PubMed]
35. Peng, F.-W.; Liu, D.-K.; Zhang, Q.-W.; Xu, Y.-G.; Shi, L. VEGFR-2 inhibitors and the therapeutic applications thereof: A patent review (2012–2016). *Expert Opin. Ther. Pat.* **2017**, *27*, 987–1004. [CrossRef] [PubMed]
36. Pedersen, K.S.; Grierson, P.M.; Picus, J.; Lockhart, A.C.; Roth, B.J.; Liu, J.; Morton, A.; Chan, E.; Huffman, J.; Liang, C. Vorolanib (X-82), an oral anti-VEGFR/PDGFR/CSF1R tyrosine kinase inhibitor, with everolimus in solid tumors: Results of a phase I study. *Investig. New Drugs* **2021**, *39*, 1298–1305. [CrossRef] [PubMed]
37. Rossi, A.; Latiano, T.P.; Parente, P.; Chiarazzo, C.; Limosani, F.; Di Maggio, G.; Maiello, E. The potential role of nintedanib in treating colorectal cancer. *Expert Opin. Pharmacother.* **2017**, *18*, 1153–1162. [CrossRef]
38. Papich, M.G. *Papich Handbook of Veterinary Drugs-E-Book*; Elsevier Health Sciences: Amsterdam, The Netherlands, 2020.
39. El-Helby, A.-G.A.; Ayyad, R.R.; El-Adl, K.; Elkady, H. Phthalazine-1, 4-dione derivatives as non-competitive AMPA receptor antagonists: Design, synthesis, anticonvulsant evaluation, ADMET profile and molecular docking. *Mol. Divers.* **2019**, *23*, 283–298. [CrossRef] [PubMed]
40. El-Helby, A.G.A.; Ayyad, R.R.; Zayed, M.F.; Abulkhair, H.S.; Elkady, H.; El-Adl, K. Design, synthesis, in silico ADMET profile and GABA-A docking of novel phthalazines as potent anticonvulsants. *Arch. Der Pharm.* **2019**, *352*, 1800387. [CrossRef]

41. Wei, K.; Louis, H.; Emori, W.; Idante, P.S.; Agwamba, E.C.; Cheng, C.-R.; Eno, E.A.; Unimuke, T.O. Antispasmodic activity of carnosic acid extracted from rosmarinus officinalis: Isolation, spectroscopic characterization, DFT studies, and in silico molecular docking investigations. *J. Mol. Struct.* **2022**, *1260*, 132795. [CrossRef]
42. Eno, E.A.; Mbonu, J.I.; Louis, H.; Patrick-Inezi, F.S.; Gber, T.E.; Unimuke, T.O.; Okon, E.E.; Benjamin, I.; Offiong, O.E. Antimicrobial activities of 1-phenyl-3-methyl-4-trichloroacetyl-pyrazolone: Experimental, DFT studies, and molecular docking investigation. *J. Indian Chem. Soc.* **2022**, *99*, 100524. [CrossRef]
43. Benjamin, I.; Udoikono, A.D.; Louis, H.; Agwamba, E.C.; Unimuke, T.O.; Owen, A.E.; Adeyinka, A.S. Antimalarial potential of naphthalene-sulfonic acid derivatives: Molecular electronic properties, vibrational assignments, and in-silico molecular docking studies. *J. Mol. Struct.* **2022**, *1264*, 133298. [CrossRef]
44. Asogwa, F.C.; Agwamba, E.C.; Louis, H.; Muozie, M.C.; Benjamin, I.; Gber, T.E.; Mathias, G.E.; Adeyinka, A.S.; Ikeuba, A.I. Structural Benchmarking, Density Functional Theory Simulation, Spectroscopic Investigation and Molecular Docking of N-(1H-pyrrol-2-yl) methylene)-4-methylaniline as Castration-Resistant Prostate Cancer Chemotherapeutic Agent. *Chem. Phys. Impact* **2022**, *5*, 100091. [CrossRef]
45. Louis, H.; Mathias, G.E.; Unimuke, T.O.; Emori, W.; Ling, L.; Owen, A.E.; Adeyinka, A.S.; Ntui, T.N.; Cheng, C.-R. Isolation, characterization, molecular electronic structure investigation, and in-silico modeling of the anti-inflammatory potency of trihydroxystilbene. *J. Mol. Struct.* **2022**, *1266*, 133418. [CrossRef]
46. Udoikono, A.D.; Louis, H.; Eno, E.A.; Agwamba, E.C.; Unimuke, T.O.; Igbalagh, A.T.; Edet, H.O.; Odey, J.O.; Adeyinka, A.S. Reactive azo compounds as a potential chemotherapy drugs in the treatment of malignant glioblastoma (GBM): Experimental and theoretical studies. *J. Photochem. Photobiol.* **2022**, *10*, 100116. [CrossRef]
47. Husein, D.Z.; Hassanien, R.; Khamis, M. Cadmium oxide nanoparticles/graphene composite: Synthesis, theoretical insights into reactivity and adsorption study. *RSC Adv.* **2021**, *11*, 27027–27041. [CrossRef]
48. Wang, T.; Husein, D.Z. Novel synthesis of multicomponent porous nano-hybrid composite, theoretical investigation using DFT and dye adsorption applications: Disposing of waste with waste. *Environ. Sci. Pollut. Res.* **2022**, 1–28. [CrossRef]
49. Turchi, M.; Cai, Q.; Lian, G. An evaluation of in-silico methods for predicting solute partition in multiphase complex fluids–A case study of octanol/water partition coefficient. *Chem. Eng. Sci.* **2019**, *197*, 150–158. [CrossRef]
50. Sullivan, K.M.; Enoch, S.J.; Ezendam, J.; Sewald, K.; Roggen, E.L.; Cochrane, S. An adverse outcome pathway for sensitization of the respiratory tract by low-molecular-weight chemicals: Building evidence to support the utility of in vitro and in silico methods in a regulatory context. *Appl. Vitr. Toxicol.* **2017**, *3*, 213–226. [CrossRef]
51. Altamash, T.; Amhamed, A.; Aparicio, S.; Atilhan, M. Effect of hydrogen bond donors and acceptors on CO_2 absorption by deep eutectic solvents. *Processes* **2020**, *8*, 1533. [CrossRef]
52. Wan, Y.; Tian, Y.; Wang, W.; Gu, S.; Ju, X.; Liu, G. In silico studies of diarylpyridine derivatives as novel HIV-1 NNRTIs using docking-based 3D-QSAR, molecular dynamics, and pharmacophore modeling approaches. *RSC Adv.* **2018**, *8*, 40529–40543. [CrossRef]
53. Escamilla-Gutiérrez, A.; Ribas-Aparicio, R.M.; Córdova-Espinoza, M.G.; Castelán-Vega, J.A. In silico strategies for modeling RNA aptamers and predicting binding sites of their molecular targets. *Nucleosides Nucleotides Nucleic Acids* **2021**, 1–10. [CrossRef] [PubMed]
54. Jain, A.N. Morphological similarity: A 3D molecular similarity method correlated with protein-ligand recognition. *J. Comput.-Aided Mol. Des.* **2000**, *14*, 199–213. [CrossRef] [PubMed]
55. Zhang, H.; Ren, J.-X.; Ma, J.-X.; Ding, L. Development of an in silico prediction model for chemical-induced urinary tract toxicity by using naïve Bayes classifier. *Mol. Divers.* **2019**, *23*, 381–392. [CrossRef] [PubMed]
56. Elkady, H.; Elwan, A.; El-Mahdy, H.A.; Doghish, A.S.; Ismail, A.; Taghour, M.S.; Elkaeed, E.B.; Eissa, I.H.; Dahab, M.A.; Mahdy, H.A. New benzoxazole derivatives as potential VEGFR-2 inhibitors and apoptosis inducers: Design, synthesis, anti-proliferative evaluation, flowcytometric analysis, and in silico studies. *J. Enzym. Inhib. Med. Chem.* **2022**, *37*, 397–410. [CrossRef]
57. Alsaif, N.A.; Taghour, M.S.; Alanazi, M.M.; Obaidullah, A.J.; Al-Mehizia, A.A.; Alanazi, M.M.; Aldawas, S.; Elwan, A.; Elkady, H. Discovery of new VEGFR-2 inhibitors based on bis ([1, 2, 4] triazolo)[4, 3-a: 3′, 4′-c] quinoxaline derivatives as anticancer agents and apoptosis inducers. *J. Enzym. Inhib. Med. Chem.* **2021**, *36*, 1093–1114. [CrossRef]
58. Elkaeed, E.B.; Youssef, F.S.; Eissa, I.H.; Elkady, H.; Alsfouk, A.A.; Ashour, M.L.; El Hassab, M.A.; Abou-Seri, S.M.; Metwaly, A.M. Multi-Step In Silico Discovery of Natural Drugs against COVID-19 Targeting Main Protease. *Int. J. Mol. Sci.* **2022**, *23*, 6912. [CrossRef]
59. Belal, A.; Abdel Gawad, N.M.; Mehany, A.B.; Abourehab, M.A.; Elkady, H.; Al-Karmalawy, A.A.; Ismael, A.S. Design, synthesis and molecular docking of new fused 1 H-pyrroles, pyrrolo [3, 2-d] pyrimidines and pyrrolo [3, 2-e][1, 4] diazepine derivatives as potent EGFR/CDK2 inhibitors. *J. Enzym. Inhib. Med. Chem.* **2022**, *37*, 1884–1902. [CrossRef]
60. NIH Methyl 4-Aminobenzoate. Available online: https://pubchem.ncbi.nlm.nih.gov/compound/12082 (accessed on 1 May 2022).
61. NIH Methyl 4-Benzamidobenzoate. Available online: https://pubchem.ncbi.nlm.nih.gov/compound/727255 (accessed on 1 May 2022).

Opinion

Hydrogen/Deuterium Exchange in Ambrox Could Improve the Long-Term Scent and Shelf Life of Perfumes

Antonio Rosales Martínez [1,*] and Ignacio Rodríguez-García [2]

1. Department of Chemical Engineering, Escuela Politécnica Superior, University of Sevilla, 41011 Sevilla, Spain
2. Organic Chemistry, CIAIMBITAL, University of Almería, 04120 Almería, Spain; irodrigu@ual.es
* Correspondence: arosales@us.es

Abstract: Ambrox is a marine natural compound with a delicious ambergris-type scent widely used in fine perfumery. The increase in the long-term scent and shelf life of perfumes has become a paramount endeavor in the fragrance industry. To the best of our knowledge, the exchange of hydrogen by deuterium to decrease the volatility of the constituents of a perfume has not yet been investigated. In this article, we propose this new use of deuteration to synthesize deuterated ambrox in order to decrease its volatility and improve the long-term scent and extend the shelf-life of perfumes.

Keywords: ambrox; deuteration; perfumes

Citation: Rosales Martínez, A.; Rodríguez-García, I. Hydrogen/Deuterium Exchange in Ambrox Could Improve the Long-Term Scent and Shelf Life of Perfumes. *Processes* **2023**, *11*, 2358. https://doi.org/10.3390/pr11082358

Academic Editors: Athanasia Varvaresou, Iliyan Ivanov and Stanimir Manolov

Received: 8 June 2023
Revised: 29 June 2023
Accepted: 3 August 2023
Published: 5 August 2023

Copyright: © 2023 by the authors. Licensee MDPI, Basel, Switzerland. This article is an open access article distributed under the terms and conditions of the Creative Commons Attribution (CC BY) license (https://creativecommons.org/licenses/by/4.0/).

1. Introduction

The deuteration of organic compounds has been viewed by synthetic organic chemists as a method to expand their horizons due to the plethora of applications attributed to deuterated organic compounds which have emerged in recent years. Deuteration has been widely used in the study of reaction mechanisms [1–7] mainly by the measurement of kinetic isotope effects, that is, the change in the speed of a chemical reaction when hydrogen is replaced by deuterium. Deuteration also plays a critical role in the analysis of organic compounds by spectroscopic techniques such as IR, NMR, and mass spectroscopy [8–10]. In mass spectroscopy, deuterated analogues are excellent internal standards as they have a chemical behavior nearly identical to that of their undeuterated counterparts. The deuterated internal standards must have a significant mass increment in order to move their signals outside of the zone where the natural mass distribution of the undeuterated analyte appears. This goal can be easily achieved by perdeuteration [11,12]. More recently, deuteration has also shown to be a powerful tool in pharmaceutical chemistry [13–15]. In this context, the most attractive technique is the incorporation of deuterium at those strategic positions of the drug where its metabolism may be affected by the kinetic isotope effect, making the deuterium–carbon bonds stronger than the hydrogen–carbon bonds. This could allow for an increase in the half-life of the drug, which can be translated into a lower dose with an identical pharmacological effect. The first deuterated drug approved by the Food and Drug Administration was deutetrabenazine for the treatment of Huntington's disease [16].

Deuterium-labeled compounds can be prepared using the base- or acid-catalyzed exchange of enolizable protons for deuteration [17], the use of transition metals and organometallic catalysis [18] such as palladium [19], ruthenium [20], or iridium complexes [21,22], and single-electron transfer systems such as Cp_2TiCl/D_2O [23,24], SmI_2/D_2O [25], and titanocene(III) complexes [26]. However, despite the multiple applications and synthetic developments carried out, to the best of our knowledge, the deuteration of organic compounds has not been used as a tool to reduce the volatility of organic compounds as a consequence of the increase in molecular weight observed when hydrogen atoms are replaced by deuterium. This new application can be very useful for the chemical and perfume industries. It is known that the three main constituents of a perfume are fixatives, fragrant oils, and solvents.

A fixative is an indispensable component in the production of perfumes, providing long-term scent, helping to mix with the other constituents, and extending the shelf-life of perfumes [27].

A commonly used fixative is ambergris, which is a waxy excretion product from sperm whales (*Physeter macrocephalus*) used since the ninth century as a valuable component of fine fragrances [28,29]. The chemical constituents within marine ambergris include a substituted homosesquiterpenoid known as ambrox (**1**) (trade name of Firmenich, the world's largest company in the flavor and fragrance business), amberlyn (trade name of Quest), and ambroxan (trade name of Henkel) [30]. The chemical structure of **1** is shown in Figure 1.

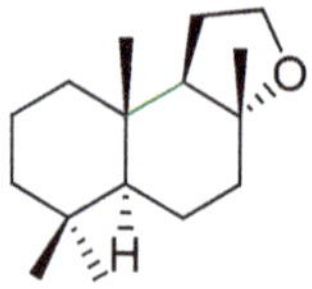

(-)-ambrox (**1**)

Figure 1. Chemical structure of natural (-)-ambrox (**1**).

The pleasantness of the smell and scents of fragrances depends on the volatility of their constituents. Volatile organic compounds are characterized by their low molecular weights, which allows for efficient evaporation [31–34]. Although volatility is a requirement to enjoy the pleasant aroma of fragrances, this property could also be an inconvenience, as too high volatilities will decrease the long-term scent and shelf life of perfumes. In this context, different research groups have dedicated significant efforts to the development of selective and effective delivery systems, which could increase the long-term scent and shelf life of perfumes. To achieve this aim, two approaches are being investigated: One is to embed the fragrance substances in polymeric matrices or microcapsules [35]. However, this approach presents as its main drawbacks low material stabilities and low perfume encapsulation capacities [36–39]. The other approach is the design of profragrances, which are nonvolatile derivatives of fragrances. These derivatives should allow for the controlled and slow release of extremely volatile compounds as a result of a selective bond cleavage initiated by an external stimulus [40,41]. However, the complexity and cost involved in the design of functional interlocked compounds programmed to release the active molecule in response to an external signal make this second proposal impractical from an industrial point of view.

Continuing our research in perfume chemistry [42], we believe that the deuteration of volatile organic compounds such as ambrox (**1**) may give rise to a new fixative of greater interest in the perfume industry. In this way, the decrease in the volatility of deuterated ambrox as its molecular weight increases should impart a greater long-term scent and improve the shelf life of perfume. To the best of our knowledge, this is the first time that the deuteration of ambrox (**1**) has been proposed to improve its fixative properties since although deuterium-labeled ambrox has been synthesized, it has only been used as an internal standard for quantification purposes in gas chromatography/mass spectrometry [43]. However, despite its simplicity, we consider that this new approach proposed in this article, aimed to improve the long-term scent and shelf life of perfumes, requires optimization of the number of hydrogen atoms exchanged for deuterium. The main reason is that a high substitution (above 50%) could modify the smell of the perfume [44] and affect the efficiency of its evaporation by increasing the molecular weight, shortening the persistence of the odor [31–34].

This manuscript is an opinion-type article which reflects the author's viewpoint on a novel application of the deuteration of volatile organic compounds. In concrete, the low deuteration of ambrox (**1**) could improve its power as a fixative in fine perfumery and is not intended to be a comprehensive review of the synthesis of ambrox (**1**). The synthesis of deuterated ambrox should proceed as shown below.

2. Future Perspectives: Synthesis Proposals for the Deuteration of Ambrox (1)

Due to its high price, various synthetic methodologies for ambrox (**1**) have been reported [45]. However, due to the presence of four chiral centers in the molecule, the most efficient synthetic routes to prepare ambrox (**1**) use enantiopure natural products as starting materials (monoterpenes, sesquiterpenoids, and diterpenes) or are based on biogenetic-type cyclizations of farnesol, farnesoic acid, monocyclofarnesoic acid, and their derivatives [45].

The main purpose of this section is to propose efficient and sustainable potential syntheses of deuterated derivatives of ambrox (**1**) with a low incorporation of deuterium. Previously described synthetic routes towards this compound could be easily modified to allow for the incorporation of deuterium into the ambrox carbon skeleton. Sclareol [46] has traditionally been the main natural chiral synthon used in the preparation of ambrox (**1**) due to its reasonable price and commercial availability as a product present in the extract of *Salvia sclarea* L. [47]. However, this compound cannot be used for the synthesis of deuterated derivatives because the described procedures are based on the oxidative degradation of the side chain present in sclareol. For that reason, we consider that four of the most viable and efficient routes to produce deuteroambrox (**2**) could be the cyclization of deuterated polyprenoids induced by chiral Brönsted acid; the manganese-catalyzed deuteration of natural sclareolide (**6**); LiAlD$_4$ reduction of sclareolide (**6**); and the reduction of a cyanide derived from albicanol (**7**) with the same reagent. It is important to mention that Chapuis and coworkers [43] have already reported the preparation of D$_4$-ambrox by the treatment of sclareolide (**6**) with MeONa in refluxing MeOD, subsequent reduction of the formed dideuterated sclareolide with LiAlD$_4$, and a final cyclization of the diol using tosyl chloride and pyridine.

3. Modified Yamamoto's Synthesis of Ambrox (1)

This retrosynthetic proposal of (D$_2$)-ambrox (**2**) by the cyclization of polyprenoids induced by chiral Brönsted acid is based on research previously developed by Yamamoto and coworkers [48] (Scheme 1). There are two key steps: (A) chiral Brönsted-acid-induced enantioselective cyclization of (D$_2$)-(*E,E*)-homofarnesyl trialkyl silylether (**3**, P = SiEt$_3$) to produce a chiral decalinic intermediate which, after diastereoselective cyclization, could form (D$_2$)-ambrox (**2**). (B) The homologation of commercially available (*E,E*)-farnesyl chloride (**4**) to generate **3**. This transformation can be accomplished by the preparation of the barium derivative of **4**; in situ treatment of this organometallic species—(*E,E*)-homofarnesylbarium(II) chloride—with excess of CO$_2$ to yield (*E,E*)-homofarnesylic acid; reduction of this acid with LiAlD$_4$; and final silylation of the primary alcohol formed to produce the desired deuterated compound **3**.

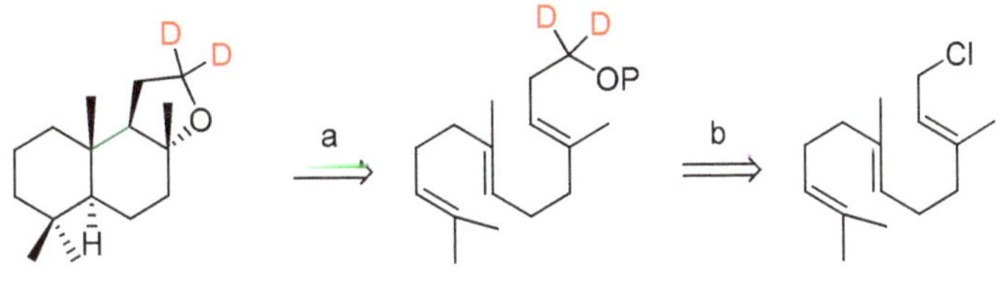

Scheme 1. Retrosynthesis of (-)-(D$_2$)-ambrox (**2**) through cyclization of a polyprenoid derivative. (a) Chiral cyclization of (D$_2$)-(*E,E*)-homofarnesyl trialkyl silylether (**3**, P = SiEt$_3$); (b) Homologation of (*E,E*)-farnesyl chloride (**4**).

This non-enzymatic enantioselective polyene cyclization of deuterated compound **3** is an attractive alternative to other multistep synthesis which use chiral natural products as starting materials. The cyclization of compound **3** should be enantioselective because it has been reported that a chiral Brönsted acid can induce the enantioselective cyclization of polyprenoids [48].

4. Modified Schaub's Synthesis of Ambrox (1)

The retrosynthetic scheme for this approach is depicted in Scheme 2. The synthesis of (D_2)-ambrox (**2**) is based on a manganese-pincer-complex-catalyzed deuteration of (+)-sclareolide (**6**) with D_2 gas to yield (D_2)-ambradiol (**5**), a procedure inspired on recent research carried out by Schaub and coworkers [49] (Scheme 2b). Subsequently, (D_2)-ambrox (**2**) could be obtained by an acid-catalyzed cyclization of (D_2)-ambradiol (**5**) (Scheme 2a).

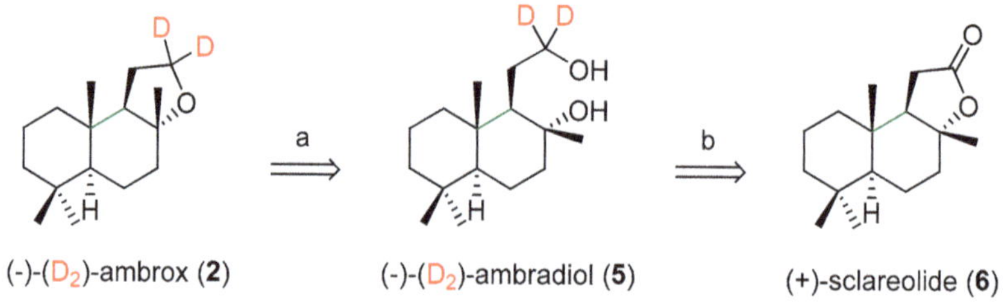

Scheme 2. Retrosynthesis of (-)-(D_2)-ambrox (**2**) through manganese-pincer-complex-catalyzed deuteration of (+)-sclareolide (**6**). (a) Acid-catalyzed cyclization of (D_2)-ambradiol (**5**); (b) Deuteration of (+)-sclareolide (**6**) with D_2 gas.

This retrosynthetic route represents a methodology in tune with the principles of green chemistry since the desired (D_2)-ambradiol (**5**) could be obtained using substoichiometric amounts of catalyst, and an almost quantitative yield is expected.

5. Modified Rosales Martínez's Synthesis of Ambrox (1)

The retrosynthetic route of (D_2)-ambrox (**2**) by the Cp_2TiCl-catalyzed radical tandem cyclization of a farnesol derivative is based on the research carried out by our research group [42] (Scheme 3). This approach comprises three retrosynthetic operations: (a) The incorporation of deuterium was achieved by reduction of nitrile **7** with $LiAlD_4$, a process that would give as intermediate (D_2)-homoalbicanol. Subsequently, this deuterated intermediate can be converted into (D_2)-ambrox (**2**) by acid-mediated cyclization. (b) The second retrosynthetic operation is the homologation of albicanol (**8**) with NaCN to form the nitrile derivative **7**. (c) Finally, albicanol (**8**) can be enantioselectively prepared by the Cp_2TiCl-catalyzed radical cyclization of enantiomerically pure epoxyfarnesyl acetate (**9**). The required epoxide **9** can be obtained from commercially available (E,E)-farnesol following the procedure reported by Spinella and coworkers [50], followed by the deoxygenation of the hydroxyl group at C3 using the Barton–McCombie deoxygenation protocol.

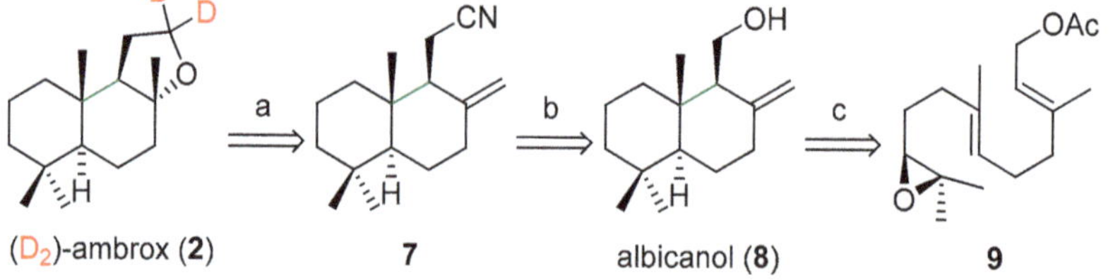

Scheme 3. Retrosynthesis of enantiomeric (D_2)-ambrox (**2**). (a) Reduction of nitrile **7** with $LiAlD_4$, and subsequently acid-mediated cyclization; (b) Homologation of albicanol (**8**); (c) Cp_2TiCl-catalyzed radical cyclization of epoxyfarnesyl acetate (**9**).

The key step of this retrosynthetic route is a highly diastereoselective Cp_2TiCl-catalyzed radical tandem cyclization of epoxide **9**.

In summary, this article intends to be a proposal for the use of deuteration as a powerful tool to decrease the volatility of deuterated derivative compounds compared to their non-deuterated analogues. This new application could be used to obtain deuterated ambrox in order to reduce its volatility and improve the long-term scent and the shelf life of perfumes. For this purpose, different retrosynthetic approaches have been proposed. We believe that this new way of preparing ambrox with a low exchange of hydrogen by deuterium may be highly attractive for the perfume industry and analytical chemistry since the deuterated ambrox derivatives can also be used as internal standards for the determination of low concentrations of ambrox (**1**) in water after biodegradability test [31].

Author Contributions: A.R.M.: design and coordination of the project, writing—original draft, writing—review and editing. I.R.-G.: review and editing. All authors have read and agreed to the published version of the manuscript.

Funding: This work was supported by the University of Seville, through the Vicerrectorado de Investigación (Projects 2020/00001014 and 2021/00000422: Ayudas a Consolidación de Grupos de la Junta de Andalucía and Project Politec-Biomat: Red de Biomateriales en la Universidad de Sevilla) and also by the University of Almería and Junta de Andalucía (Consejería de Transformación Económica, Industria, Conocimiento y Universidades) and Fondo Europeo de Desarrollo Regional (FEDER) for the Project UALFEDER 2020-FQM-B1989.

Acknowledgments: Antonio Rosales Martínez acknowledges the University of the Sevilla for his position, and for a grant for the requalification of university teaching staff at the CIQSO, University of Huelva.

Conflicts of Interest: The authors declare no conflict of interest.

References

1. Teo, W.J.; Yang, X.; Poon, Y.Y.; Ge, S. Cobalt-catalyzed deoxygenative triborylation of allylic ethers to access 1,1,3-triborylalkanes. *Nat. Commun.* **2020**, *11*, 51932. [CrossRef]
2. Matt, C.; Kern, C.; Streuff, J. Zirconium-Catalyzed Remote Defunctionalization of Alkenes. *ACS Catal.* **2020**, *10*, 6409–6413. [CrossRef]
3. Weweler, J.; Younas, S.L.; Streuff, J. Titanium(III)-Catalyzed Reductive Decyanation of Geminal Dinitriles by a Non-Free-Radical Mechanism. *Angew. Chem. Int. Ed.* **2019**, *58*, 17700–17703. [CrossRef] [PubMed]
4. Zhang, Y.-Q.; Poppel, C.; Panfilova, A.; Bohle, F.; Grimme, S.; Gansäuer, A. S_N2 Reactions at Tertiary Carbon Centers in Epoxides. *Angew. Chem. Int. Ed.* **2017**, *56*, 9719–9722. [CrossRef] [PubMed]
5. Fra, L.; Millán, A.; Souto, J.A.; Muñiz, K. Indole Synthesis Based On A Modified Koser Reagent. *Angew. Chem. Int. Ed.* **2014**, *53*, 7349–7353. [CrossRef] [PubMed]
6. Rosales, A.; Muñoz-Bascón, J.; Lopez-Sanchez, C.; Alvarez-Corral, M.; Muñoz-Dorado, M.; Rodriguez-Garcia, I.; Oltra, J.E. Ti-catalyzed homolytic opening of ozonides: A sustainable C-C bond-forming reaction. *J. Org. Chem.* **2012**, *77*, 4171–4176. [CrossRef]
7. Cuerva, J.M.; Campaña, A.G.; Justicia, J.; Rosales, A.; Oller-Lopez, J.L.; Robles, R.; Cárdenas, D.J.; Buñuel, E.; Oltra, J.E. Water: The ideal hydrogen-atom source in free-radical chemistry mediated by Ti^{III} and other single-electron-transfer metals? *Angew. Chem. Int. Ed.* **2006**, *45*, 5522–5526. [CrossRef]
8. Hartmann, B.; Müller, M.; Seyler, L.; Bäuerle, T.; Wilferth, T.; Avdievitch, N.; Ruhm, L.; Henning, A.; Lesiv, P.; Ivashkin, P.; et al. Feasibility of deuterium magnetic resonance spectroscopy of 3-O-Methylglucose at 7 Tesla. *PLoS ONE* **2021**, *16*, e0252935. [CrossRef]
9. Kostyukevich, Y.; Acter, T.; Zherebker, A.; Ahmed, A.; Kim, S.; Nikolaev, E. Hydrogen/deuterium exchange in mass spectrometry. *Mass Spec. Rev.* **2018**, *37*, 811–853. [CrossRef]
10. MacCarthy, P. Infrared Spectroscopy of Deuterated Compounds. *J. Chem. Educ.* **1985**, *62*, 633–634. [CrossRef]
11. Hewavitharana, A.K. Matrix matching in liquid chromatography-mass spectrometry with stable isotope labelled internal standards—Is it necessary? *J. Chromatogr. A* **2011**, *1218*, 359–361. [CrossRef] [PubMed]
12. Kang, Q.-K.; Shi, H. Catalytic Hydrogen Isotope Exchange Reactions in Late-Stage Functionalization. *Synlett* **2022**, *33*, 329–338.
13. Atzrodt, J.; Derdau, V.; Kerr, W.J.; Reid, M. Deuterium- and Tritium-Labelled Compounds: Applications in the Life Sciences. *Angew. Chem.* **2018**, *130*, 1774–1802. [CrossRef]
14. Prakash, G.; Paul, N.; Oliver, G.A.; Werz, D.B.; Maiti, D. C–H deuteration of organic compounds and potential drug candidates. *Chem. Soc. Rev.* **2022**, *51*, 3123–3163. [CrossRef] [PubMed]

15. Pirali, T.; Serafini, M.; Cargnin, S.; Genazzani, A. Applications of Deuterium in Medicinal Chemistry. *J. Med. Chem.* **2019**, *62*, 5276–5297. [CrossRef] [PubMed]
16. Cargnin, S.; Serafini, M.; Pirali, T. A primer of deuterium in drug design. *Future Med. Chem.* **2019**, *16*, 2039–2042. [CrossRef]
17. Mullard, A. Deuterated drugs draw heavier backing. *Nat. Rev. Drug Discov.* **2016**, *15*, 219–221. [CrossRef]
18. Murray, A., III; Williams, D.L. *Organic Syntheses with Isotopes; Part II*; Interscience Publishers: New York, NY, USA; London, UK, 1958.
19. Ma, S.; Villa, G.; Thuy-Boun, P.S.; Homs, A.; Yu, J.-Q. Palladium-Catalyzed ortho-Selective C-H Deuteration of Arenes: Evidence for Superior Reactivity of Weakly Coordinated Palladacycles. *Angew. Chem. Int. Ed.* **2014**, *53*, 734–737. [CrossRef]
20. Takakahashi, M.; Oshima, K.; Matsubara, S. Ruthenium catalyzed deuterium labelling of alpha-carbon in primary alcohol and primary/secondary amine in D2O. *Chem. Lett.* **2005**, *34*, 192–193. [CrossRef]
21. Zhou, J.; Hartwig, J.F. Iridium-catalyzed H/D exchange at vinyl roups without olefin isomerization. *Angew. Chem. Int. Ed.* **2008**, *47*, 5783–5787. [CrossRef]
22. Yung, C.M.; Skaddan, M.B.; Bergman, R.G. Stoichiometric and Catalytic H/D Incorporation by Cationic Iridium Complexes: A Common Monohydrido-Iridium Intermediate. *J. Am. Chem. Soc.* **2004**, *126*, 13033–13043. [CrossRef]
23. Rosales, A.; Rodríguez-García, I. Cp2TiCl/D2 O/Mn, a formidable reagent for the deuteration of organic compounds. *Beilstein J. Org. Chem.* **2016**, *12*, 1585–1589. [CrossRef] [PubMed]
24. Rosales Martínez, A.; Pozo Morales, L.; Díaz Ojeda, E.; Castro Rodríguez, M.; Rodríguez-García, I. The Proven Versatility of Cp_2TiCl. *J. Org. Chem.* **2021**, *86*, 1311–1329. [CrossRef] [PubMed]
25. Szostak, M.; Spain, M.; Procter, D.J. Selective Synthesys of α,α-Dideuterio Alcohols by the Reduction of Carboxylic Acids Using SmI2 and D2 O as Deuterium Source under SET Conditions. *Org. Lett.* **2014**, *16*, 5052–5055. [CrossRef]
26. Henriques, D.S.G.; Rojo-Wiechel, E.; Klare, S.; Mika, R.; Höthker, S.; Schacht, J.H.; Schmickler, N.; Gansäuer, A. Titanocene(III)-Catalyzed Precision Deuteration of Epoxides. *Angew. Chem. Int. Ed.* **2022**, *61*, e202114198. [CrossRef] [PubMed]
27. Frater, G.; Bajgrowicz, J.A.; Kraft, P. Fragrance chemistry. *Tetrahedron* **1998**, *54*, 7633–7703. [CrossRef]
28. Lederer, E.; Marx, F.; Mercier, D.; Pérot, G. Sur les constituants de l'ambre gris II. Ambréine et Coprostanone. *Helv. Chim. Acta.* **1946**, *29*, 1354–1365. [CrossRef]
29. Stoll, M.; Hinder, M. Odeur et Constitution III. Les substances bicyclohomofarnésiques. *Helv. Chim. Acta* **1950**, *33*, 1251–1260. [CrossRef]
30. Ohloff, G. In fragrance chemistry. In *The Science of the Sense of Smell*; Theimer, E.T., Ed.; Academic Press: New York, NY, USA, 1982.
31. Herrmann, A. *The Chemistry and Biology of Volatiles*; John Wiley & Sons: Chichester, UK, 2010.
32. Epstein, J.L.; Castaldi, M.; Patel, G.; Telidecki, P.; Karakkatt, K. Using Flavor Chemistry to Design and Synthesize Artificial Scents and Flavors. *J. Chem. Educ.* **2014**, *92*, 954–957. [CrossRef]
33. Francl, M. Scents and sensibility. *Nat. Chem.* **2015**, *7*, 265–266.
34. Goss, K.-U. The physical chemistry of odors—Consequences for the work with detection dogs. *Forensic Sci. Int.* **2019**, *296*, 110–114. [CrossRef] [PubMed]
35. Kaur, R.; Kukkar, D.; Bhardwaj, S.K.; Kim, K.-H.; Deep, A. Potential Use of Polymers and their Complexes as Media for Storage and Delivery of Fragrances. *J. Control. Release* **2018**, *285*, 81–95. [CrossRef]
36. Rodrigues, S.N.; Martins, I.M.; Fernandes, I.P.; Gomes, P.B.; Mata, V.G.; Barreiro, M.F.; Rodrigues, A.E. Scentfashion®: Microencapsulated Perfumes for Textile Application. *Chem. Eng. J.* **2009**, *149*, 463–472. [CrossRef]
37. Martins, I.M.; Barreiro, M.F.; Coelho, M.; Rodrigues, A.E. Microencapsulation of Essential Oils with Biodegradable Polymeric Carriers for Cosmetic Applications. *Chem. Eng. J.* **2014**, *245*, 191–200. [CrossRef]
38. Hofmeister, I.; Landfester, K.; Taden, A. Controlled Formation of Polymer Nanocapsules with High Diffusion-Barrier Properties and Prediction of Encapsulation Efficiency. *Angew. Chem. Int. Ed.* **2015**, *54*, 327–330. [CrossRef]
39. Ciriminna, R.; Pagliaro, M. Sol-gel Microencapsulation of Odorants and Flavors: Opening the Route to Sustainable Fragrances and Aromas. *Chem. Soc. Rev.* **2013**, *42*, 9243–9250. [CrossRef]
40. López-Sánchez, J.; Alajarin, M.; Pastor, A.; Berna, J. Mechanically Interlocked Profragances for the Controlled Release of Scents. *J. Org. Chem.* **2021**, *86*, 15045–15054. [CrossRef]
41. Saura-Sanmartín, A.; Andreu-Ardil, L. Recent Advances in the Preparation of Delivery Systems for the Controlled Release of Scents. *Int. J. Mol. Sci.* **2023**, *24*, 4685. [CrossRef]
42. Rosales, A.; Foley, L.A.R.; Padial, N.M.; Muñoz-Bascón, J.; Sancho-Sanz, I.; Roldán-Molina, E.; Pozo-Morales, L.; Irías-Álvarez, A.; Rodríguez-Maecker, R.; Rodríguez-García, I.; et al. Diastereoselective Synthesis of (±)-ambrox by Titanium(III)-Catalyzed Radical Tandem Cyclization. *Synlett* **2016**, *27*, 369–374. [CrossRef]
43. Chapuis, C.; Cantatore, C.; Fankhauser, P.; Challand, R.; Riedhauser, J.-J. Synthesis of Deuterium-Labeled Perfume Ingredients as Internal Standards for Their GC/MS Quantification. *Helv. Chim. Acta* **2009**, *92*, 1782–1799. [CrossRef]
44. Gane, S.; Georganakis, D.; Maniati, K.; Vamvakias, M.; Ragoussis, N.; Skoulakis, E.M.C.; Turin, L. Molecular Vibration-Sensing Component in Human Olfaction. *PLoS ONE* **2013**, *8*, e55780. [CrossRef]
45. Ncube, E.N.; Steenkamp, L.; Dubery, I.A. Ambrafuran (Ambrox^TM) Synthesis from Natural Plant Product Precursors. *Molecules* **2020**, *25*, 3851. [CrossRef] [PubMed]
46. Yang, S.; Tian, H.; Sun, B.; Liu, Y.; Hao, Y.; Lv, Y. One-pot synthesis of (-)-Ambrox. *Sci. Rep.* **2016**, *6*, 32650. [CrossRef] [PubMed]

47. Novak, J.; Blüthner, W.-D. Medicinal, Aromatic and Stimulant Plants. In *Handbook of Plant Breeding*; Chapter 12; Springer Nature: Basel, Switzerland, 2020.
48. Ishihara, K.; Ishibashi, H.; Yamamoto, H. Enantio- and Diastereoselective Stepwise Cyclization of Polyprenoids Induced by Chiral and Achiral LBAs. A New Entry to (−)-Ambrox, (+)-Podocarpa-8,11,13-triene Diterpenoids, and (−)-Tetracyclic Polyprenoid of Sedimentary Origin. *J. Am. Chem. Soc.* **2002**, *124*, 3647–3655. [CrossRef] [PubMed]
49. Zubar, V.; Lichtenberger, N.; Schelwies, M.; Oeser, T.; Hashmi, A.S.K.; Schaub, T. Manganese-Catalyzed Hydrogenation of Sclareolide to Ambradiol. *ChemCatChem* **2022**, *14*, e202101443. [CrossRef]
50. D'Acunto, M.; Monica, C.D.; Izzo, I.; De Petrocellis, L.; di Marzo, V.; Spinella, A. Enantioselective synthesis of 3(S)-hydroxy polygodial derivatives and evaluation of their vanilloid activity. *Tetrahedron* **2010**, *66*, 9785–9789. [CrossRef]

Disclaimer/Publisher's Note: The statements, opinions and data contained in all publications are solely those of the individual author(s) and contributor(s) and not of MDPI and/or the editor(s). MDPI and/or the editor(s) disclaim responsibility for any injury to people or property resulting from any ideas, methods, instructions or products referred to in the content.

MDPI
St. Alban-Anlage 66
4052 Basel
Switzerland
www.mdpi.com

Processes Editorial Office
E-mail: processes@mdpi.com
www.mdpi.com/journal/processes

Disclaimer/Publisher's Note: The statements, opinions and data contained in all publications are solely those of the individual author(s) and contributor(s) and not of MDPI and/or the editor(s). MDPI and/or the editor(s) disclaim responsibility for any injury to people or property resulting from any ideas, methods, instructions or products referred to in the content.

www.ingramcontent.com/pod-product-compliance
Lightning Source LLC
LaVergne TN
LVHW070657100526
838202LV00013B/988